A GUIDE TO THE PHENOMENOLOGY OF RELIGION

A Guide to the Phenomenology of Religion

Key Figures, Formative Influences and Subsequent Debates

JAMES L. COX

T&T CLARK INTERNATIONAL
A Continuum imprint
LONDON • NEW YORK

The Continuum International Publishing Group
The Tower Building, 11 York Road, London SE1 7NX
80 Maiden Lane, Suite 704, New York, NY 10038

www.continuumbooks.com

British Library Cataloguing-in-Publication Data
A catalogue record for this book is available from the British Library

Typeset by Data Standards Ltd, Frome, Somerset
Printed and bound in Great Britain by Biddles Ltd, King's Lynn, Norfolk

ISBN 0-8264-5290-6 (hardback)
ISBN 0-8264-5289-2 (paperback)

To Liz Leitch
A constant support for students and an invaluable research colleague

Contents

Acknowledgements

This book is born from a growing conviction, confirmed by my discussions with colleagues and supported by teaching in undergraduate and postgraduate courses, that religious studies, as a field midway between theology and the social and cultural sciences, suffers from a severe crisis of identity. In my view, this has resulted, at least in part, from the central role the phenomenology of religion has played historically in the development of the academic study of religions. This conviction was strengthened on my arrival in Edinburgh in 1993 from the University of Zimbabwe, when I was invited to deliver the phenomenology section of the Honours course on methodologies in the study of religions. It became apparent to me that theories of religion were being presented to the students on the course in sometimes contradictory ways, partly because no single book was available that brought together the various strands in phenomenology, and at the same time, which clarified critical methodological issues facing scholars of religion. Professor Alistair Kee, now retired, suggested that I consider writing such a book, and with the strong encouragement of other colleagues in Religious Studies in Edinburgh, Professors Frank Whaling and Nick Wyatt, also both now retired, I began this project. I am grateful for the inspiration and support they provided from the outset. I am also indebted to my more recent colleagues in Religious Studies in Edinburgh who have shared many of my teaching and administrative duties over the past two years and have engaged in lively discussion with me on our place within Edinburgh's School of Divinity. I am particularly grateful in this regard to Dr Jeanne Openshaw and Dr Steven Sutcliffe. In addition, Professor David Fergusson, Head of the School of Divinity has offered his unswerving support as this project has neared its completion.

I wish to express my gratitude specifically to the British Association for the Study of Religions for enabling me, as an officer in the Association for the past nine years, to participate in numerous national and international conferences, where I have encountered first hand many of the critical debates I describe in Chapter 7. In a similar vein, I owe much to my fellow officers and members of the African Association for the Study of Religions, who have persistently challenged me to consider the contexts of power out of which the phenomenology of religion has developed.

I would like to thank Continuum for agreeing to publish this book, and for the patience shown by the editors as I have prepared the manuscript. Many thanks also to Valerie Smith for her invaluable advice as I struggled to format the text I eventually submitted to the publisher, but, mostly, for her unfailing encouragement.

James L. Cox
Edinburgh

Defining the Scope: Phenomenology within the Academic Study of Religions

For many years, the religious studies programme at Edinburgh University has included in its course structure a compulsory component in which students analyse the thought of influential thinkers who have moulded the academic study of religions. A primary difficulty students have encountered in the course has resulted from the lack of a comprehensive text that serves as a guide to the principal ideas of main thinkers in religious studies when understood as a discipline in its own right. Among the available resources, students have consulted the anthology compiled by Jacques Waardenburg in his *Classical Approaches to the Study of Religion* (1973). In recent years, they have been using the selected readings in Russell McCutcheon's *The Insider/Outsider Problem in the Study of Religion* (1999) and the edited volume by Braun and McCutcheon called *Guide to the Study of Religion* (2000). For a comprehensive history of the development of religious studies, students have been referred to Eric Sharpe's *Comparative Religion: A History* (1986) and to Walter Capps' comprehensive review, *Religious Studies: The Making of a Discipline* (1995). Despite these very useful resources, no single volume exists that outlines and contextualizes the thinking of key figures in the phenomenology of religion.

Another component in the Edinburgh religious studies course has aimed at identifying some of the persistent issues that have accompanied the academic study of religions, both during the early development of the discipline and as methodological problems have come to be defined, understood and interpreted by contemporary scholars in the field. The critical concerns relevant to the subject have emerged out of the writings of key thinkers in the field and in recent years have crystallized around the legitimate place of religious studies as a distinct discipline within the humanities and social sciences. One of the most contentious issues has focused on the relationship between religious studies and theology. In the late nineteenth century, when Chantepie de la Saussaye first applied the term phenomenology to the study of religion, attempts were made to clarify this relationship. Subsequently, scholars have sought to put this matter to rest once and for all, but the issue persists today, expressed frequently in a heated debate between those who see the academic study of religions as fundamentally a theological enterprise, although often disguised as a science, and those who

1

prefer to fit any study of religious phenomena under the broad umbrella of cultural studies.

The Edinburgh course, therefore, has demonstrated clearly that the ideas of principal scholars of religion cannot be divorced from their historical contexts nor from the issues they have identified as critical for defining a distinctively religious studies methodology. This book endeavours to draw from the insights the present author has gained in constructing, revising and teaching a course on methodologies in the study of religions over many years in the University of Edinburgh and previously at the University of Zimbabwe. By so doing, I hope to fill some significant gaps students have experienced in these courses by describing, analysing and, to some extent evaluating, the ideas of key thinkers in the phenomenology of religion. At the same time, I intend to place the ideas of the key thinkers identified into historical and social contexts by examining the formative influences over their thinking and by indicating how their ideas have helped to create the debates at the core of religious studies today.

Identifying the key figures

When I proposed writing a guide that would outline in detail the principal ideas of seminal thinkers in the academic study of religions during the twentieth century, one of the first problems I encountered was which figures I should include and which I should exclude in such a review. A partial resolution to this is suggested by the title of the book itself, which limits the thinkers examined to those who can be considered 'phenomenologists' of religion. This has been done for two principal reasons. The first is suggested by the practical need to place limits on the scope of the book and the second, which is much more important, results from the central, and even overriding, role phenomenology has played in shaping religious studies as an academic field. My aim in this book, therefore, is not to provide a comprehensive list of scholars who have made significant contributions to the study of religion from within a wide variety of disciplines. In one sense, Walter Capps has attempted just this in his extremely ambitious overview of the history and development of religious studies. Of necessity, he could devote just a few pages, and in many cases just a few paragraphs, to an extremely far reaching list of scholars. Capps (1995: xii) admits that he adopts an eclectic approach: '... anthropologists, historians, sociologists, philosophers, psychologists, linguists, theologians, and others all understand themselves to have some real, legitimate place and stake in the inquiry'. In this book, I have sought to avoid such an all-embracing examination of scholars, and, in the first instance, this explains why I have restricted my investigation to one type of religious specialist, the phenomenologist of religion.

A restricted approach, similar to the one I am taking, has been adopted by Daniel Pals in his *Seven Theories of Religion* (1996), which has been extremely useful, particularly for students of anthropology. Pals explores the study of religion through the eyes of highly influential theorists: E. B. Tylor and James Frazer (taken as one), Sigmund Freud, Émile Durkheim, Karl Marx, Mircea

Eliade, E. E. Evans-Pritchard and Clifford Geertz. None of these called themselves phenomenologists of religion, although I argue in this book that Eliade was in fact a phenomenologist. The others fall into social scientific studies of religion, mostly in anthropology, with the exception of Freud, whose interests in totems and taboos resulted largely from his Oedipus theory, and Marx, whose analysis of religion was based on sociological and ideological constructs. Pals justifies his choice of figures largely by arguing that the theories he has selected represent 'seven of the more important theories of religion that have been put forward since the idea of a scientific approach to religion first caught the imagination of serious scholars in the century before our own' (1996: 10). He adds that the key thinkers he has selected 'have exercised a shaping influence not only on religion but on the whole intellectual culture of our century' (1996: 10). Pals explains his omission of important thinkers like Max Müller, Max Weber, Lucien Lévy-Bruhl and C. G. Jung on various grounds: Müller's theory that religion originated in nature worship was largely discredited in his lifetime; the theories of Weber and Lévy-Bruhl are quite complex and difficult to access and, thus for students, are more easily understood through the ideas of Durkheim, Evans-Pritchard and Geertz; because Jung was overly sympathetic to religion, a functionalist interpretation in psychology is better exemplified by Freud. Despite the limitations caused by omitting many highly important theorists of religion, Pals' main aim is largely achieved: to demonstrate how trends in the scientific study of religion developed in the nineteenth and twentieth centuries through the influence of exemplary thinkers.

This book adopts different criteria for selecting key figures than that employed by Pals. I have chosen my key figures on the basis of their identification with the phenomenology of religion. This does not resolve the problem of deciding how some are included and others excluded, a point to which I will return shortly, but it does attempt to place the scholars selected within a broad frame of reference. This avoids the difficulty of rejecting some, as Pals does, on the basis of the acceptability of their theories in their own time or with reference to complexity, or even sympathy. My choices initially have been made because each can be defined within a phenomenological framework. Of course, just as it was for Pals, the scholars I have chosen must be regarded generally as influential in religious studies and they must be able to be shown to have played significantly formative roles in the shaping of its academic approach.

I have deliberately selected phenomenology as my primary and initial criterion for including scholars on my list because, in my view, the phenomenology of religion defines the methodology that is uniquely associated with religious studies as a distinct discipline studying 'religion' itself, as opposed, for example, to studying sociology as it is applied to religion or psychology as it is applied to religion. Phenomenologists study religion in and of itself and not as an epiphenomenon of other more primary subjects. In academic institutions throughout Europe and North America, therefore, those who consider themselves primarily, or even exclusively, religious studies specialists distinguish themselves from experts in other fields. They are not social scientists, although they almost always claim to have parallel concerns with those in the social sciences. They are not philosophers, although they admit theirs entails a very precise epistemological theory; nor are they theologians, although they see themselves as highly sympathetic to the theological enterprise. They are distinguishable from historians, whose discipline is far wider, but intertwined with, the phenomenology of religion. Neither are they linguists, although the study of religions in its various forms almost always requires language and/or textual study. Only phenomenology provides for the academic study of religions a distinct methodology, justifying its claim to be a field of study in its own right, *sui generis*.

Because the scholars I consider in this book devote much of their energy to defining phenomenology and its relationship to other disciplines which study religion, I will not in this introduction define what I mean by phenomenology. This will emerge out of the study itself and, in some sense, implies what the entire book is about. I will return, therefore, to the place of the phenomenology of religion in the larger study of religions in the final section on subsequent debates and in which I offer my own evaluation of the phenomenological project in the study of religions. My point here is to claim that by analysing key figures in the phenomenology of religion, I am exposing the thought of scholars who have shaped a distinctly religious studies approach within a much broader multi-disciplinary academic framework.

Identifying key phenomenologists

Although by restricting my analysis to phenomenologists of religion I have limited the scope of this book, I have by no means solved the problem of whom to include or exclude among the key figures in this field. In this sense, the problem Pals faced in selecting his seven theories of religion remains my problem, but in a more limited sense. I must justify my choice of scholars and my methods for selecting them. That this is a formidable task can be demonstrated simply by reviewing the list of names in the index of Sharpe's *Comparative Religion: A History* (1986). Although most on his list cannot be regarded as phenomenologists of religion, in the index to his 1975 volume he includes nearly 600 figures who have contributed in greater or lesser degrees to the comparative study of religions. In his 1986 second edition, he adds another 29 names, including Edward Said, who became a central figure himself in cultural studies, and Donald Wiebe, who has sharpened the debate in North America between theological and scientific studies of religion (Sharpe, 1986: 340–41).

Clearly, it is not possible to provide an in-depth study of every important contributor to the phenomenology of religion, even if I were to attempt to review the principal thinkers mentioned in Sharpe's chapter on the phenomenology of religion. There are simply too many to achieve such an aim. A fundamentally different approach has been suggested by George James in *Interpreting Religion* (1995), in which he examines the thought of P. D. Chantepie de la Saussaye, W.

Brede Kristensen and Gerardus van der Leeuw. James calls his a 'modified case study approach' (1995: 3), which he has adopted in order to exemplify 'family resemblances' among scholars who have employed phenomenological methods in the study of religions. James characterizes the family traits shared by phenomenologists as 'a-historical, a-theological, and anti-reductive' tendencies, which are exposed clearly in the thinking of his 'foundational cases' (1995: 3). James includes van der Leeuw among his case studies because, he claims, 'there is simply no exponent of a phenomenology of religion whose influence has been more widely felt or who is more often referred to in efforts to adjudicate the nature of this approach' (1995: 3). W. Brede Kristensen has been selected because van der Leeuw was his student and the influence of Kristensen can be seen throughout van der Leeuw's writings. Chantepie de la Saussaye provides the third case study because 'he was the first to use the term phenomenology in the context of the so-called scientific study of religion' and because his influence can be seen 'clearly in the works of both Kristensen and van der Leeuw' (1995: 4–5).

Both Sharpe and James will be important for my study of key phenomenologists of religion because I have adopted a method for selecting scholars midway between Sharpe's comprehensive history of comparative religions and James' case study approach. My tactic is to identify broad 'schools' of thought, located largely in geographical regions and justified by the significance members of these schools had in the development of the phenomenology of religion. I agree with James that no discussion of key figures in the phenomenology of religion can exclude Gerardus van der Leeuw, whom I treat as part of the Dutch school of phenomenology, which originates with C. P. Tiele and P. D. Chantepie de la Saussaye, but whose principal exponents for the development of religious studies as a discipline were W. Brede Kristensen, van der Leeuw and C. J. Bleeker. Hence, one section of this book is devoted to the Dutch school as exemplified by the mutual influences and inter-relationships between Kristensen, van der Leeuw and Bleeker.

I have also identified what I am calling loosely a British school of phenomenology, which has its own history and separate development from its Dutch counterpart. I argue that the British school originates not in Britain, but in Nigeria under the leadership of Geoffrey Parrinder, who, under the influence of Edwin W. Smith, and after having established the first department of religious studies of its kind in the University of Ibadan, was appointed to a post in comparative religions at King's College, London. The African factor also extends to Scotland, where, in Aberdeen, Andrew Walls founded the first Scottish department of religious studies based largely on Parrinder's model in Ibadan and his own experience at Fourah Bay College in Sierra Leone and the University of Nigeria at Nsukka. The other major figure in British phenomenology, and undoubtedly the best known, is Ninian Smart, who founded the first religious studies department in Britain at Lancaster in 1967, three years ahead of Walls. The approaches of Edwin Smith, Parrinder and Walls were perhaps more similar than that advocated by Smart, but I shall argue that, in the cases of Parrinder, Walls and Smart, each fundamentally shared a common understanding of religious studies, separate from theology but entirely sympathetic to the perspectives of adherents. The British school, of course, was influenced from the Netherlands, but its African connections give it a particular flavour and emphasis that can be distinguished from the continental approach. I will then turn to the North American context, particularly by referring to the thought of Joachim Wach, Mircea Eliade and Jonathan Z. Smith, from the University of Chicago, and Wilfred Cantwell Smith, originally an expert on Islam, who held a professorship at McGill University in Montreal and who founded the Harvard School of Comparative Religions. It is difficult to refer to Wach, Eliade, J. Z. Smith and W. C. Smith as comprising a 'school' of thought, but emphases within the writings of each display common characteristics that reformulated the earlier phenomenology of the Dutch thinkers and which followed a quite different line from that pursued in Britain.

By taking this historical and in some senses quite limited geographical approach to the subject, I wish to outline the thought of key figures in the development of the phenomenology of religion as witnessed within their peculiar contexts and as they have interacted with one another. Of course, the limitations I am employing entail omissions, and, in some senses, it can be argued that I am following James by presenting a modified case study approach. Numerous scholars who have been extremely influential in religious studies are excluded, such as Geo Widengren from Sweden or Friedrich Heiler from Germany or Raffaele Pettazzoni from Italy. My reasons for excluding such figures are two: 1) I do not think in the history of the discipline they played so lasting a role as the ones I am considering, nor have they so fundamentally influenced the current debates; 2) I am not trying to replicate the excellent history produced by Eric Sharpe, nor restrict myself so severely as James has done, but to argue that the key figures I have identified as operating in loosely defined 'schools' have constructed the phenomenology of religion as we have come to know it in religious studies circles today and thus have created the parameters within which contemporary discussions are occurring.

Formative influences and subsequent debates

I have chosen to begin this book by examining what I am calling the 'formative influences' which preceded the development of phenomenological thinking in religious studies. I regard these as falling within three categories: philosophical, theological and social scientific. I will examine in separate chapters how each of these broad areas has established a context out of which the key figures have emerged. I am aware that other scholars have done this and that much literature exists on these background factors. Eric Sharpe, perhaps most significantly, has outlined the nineteenth-century influences so important for the comparative study of religions and, as I have noted above, George James has given detailed accounts of the relationships between history, theology and the social sciences within the thinking of Chantepie, Kristensen and van der Leeuw. Nonetheless, I do not think the 'schools' of phenomenology I have identified can be understood without a discussion of the conditions which brought them into prominence. I have attempted in my study to shed some new light on each of these areas, particularly by drawing attention to the largely unnoticed influence of the Ritschlian school of theology on the phenomenological method, and by arguing that non-Western contexts contributed significantly to moulding the perspectives of some of the key thinkers. I also take up the question of the extent to which Edmund Husserl's philosophical method influenced phenomenologists of religion, and I analyse the assertion that the anti-reductive tendency in religious phenomenology can be understood best as a theologically motivated reaction against interpretations that could offend believers.

In the concluding chapter of the book, I delineate how the key figures in the phenomenology of religion have fostered debates that in many ways have polarized contemporary scholars of religion into opposing camps. Although these debates are generated almost exclusively by those who operate in departments of theology and religious studies, the central questions posed emerge out of the philosophical, theological and social scientific contexts that produced the phenomenology of religion in the first place. The philosophical problem of the relationship between the subject and the object has been introduced forcefully into the current debate by challenges, such as those put forward recently by Gavin Flood (1999), that cast doubt on the continued philosophical credibility of the phenomenological approach to the study of religions. The religion–theology impasse occurs because the methodology that came to dominate the phenomenology of religion is depicted by so-called anti-religionists as indistinguishable from liberal Christian theology. If this conclusion is correct, the claims by phenomenologists that theirs is a genuine 'science' of religion is at best mistaken and at worst a pretence for smuggling a theological agenda into disciplines which legitimately can be called scientific. Another side of this same debate has been articulated by theologians, who see the academic study of religions as necessary for informing missiological discussions. Generally, such scholars conclude that without a transcendent reality, religion could not exist, and thus, the phenomenology of religion, with its empathetic approach towards adherents, is seen as an ally with a Christian theology of dialogue and inter-faith cooperation. This theologically inspired approach is closely akin to a third debate I identify in the final chapter, which I am calling the controversy surrounding the socially engaged scholar of religions. Here, the issue focuses less on the underlying ideological assumptions of the researcher than it does on whether or not one who studies religious communities bears any social responsibility towards them, an issue which in recent times has been exacerbated by behaviours, in many cases attributable to religious teachings, that have proved harmful both to adherents and to the wider society. Finally, I offer my own suggestions for how new ground can be broken in the study of religions in light of the development of thought advanced by the key phenomenologists of religion I have described and by the subsequent academic debates they have fostered.

References

Braun, W. and McCutcheon, R. T. (eds) (2000), *Guide to the Study of Religion* (London and New York: Cassell).

Capps, W. H. (1995), *Religious Studies: The Making of a Discipline* (Minneapolis: Fortress Press).

Flood, G. (1999), *Beyond Phenomenology: Rethinking the Study of Religion* (London and New York: Cassell).

James, G. A. (1995), *Interpreting Religion. The Phenomenological Approaches of Pierre Daniel Chantepie de la Saussaye, W. Brede Kristensen, and Gerardus van der Leeuw* (Washington, DC: The Catholic University of America Press).

McCutcheon, R. T. (ed.) (1999), *The Insider/Outsider Problem in the Study of Religion: A Reader* (London and New York: Cassell).

Pals, D. L. (1996), *Seven Theories of Religion* (New York and Oxford: Oxford University Press).

Sharpe, E. J. (1986), *Comparative Religion: A History* (London: Duckworth, 2nd edn).

Waardenburg, J. (1973), *Classical Approaches to the Study of Religion: Aims, Methods and Theories of Research. I. Introduction and Anthology* (The Hague and Paris: Mouton Publishers).

Chapter 1

Understanding Phenomena: Key Ideas in the Philosophy of Edmund Husserl

The philosophical movement associated with phenomenology was begun by the German philosopher, Edmund Husserl (1859–1938), who between 1901 and 1931 developed a methodology which he believed articulated a logical and fully scientific analysis of the way humans obtain knowledge. The term phenomenology was not invented by Husserl, but the way in which the word has been used in contemporary philosophy can be traced to him. The first use of the term has been credited to the eighteenth-century German mathematician, J. H. Lambert, who applied it to describe how 'appearances' can lead to truth. It was also employed by the great German idealist of the nineteenth century, G. W. F. Hegel, to refer to 'subjective spirit' (Kenny, 1994: 228). Following Husserl, a very loosely defined 'school of thinkers' expanded his ideas, such as Karl Jaspers in the direction of existentialism, but perhaps chiefly Husserl's former academic assistant, Martin Heidegger, whose book *Being and Time* (1962, first published as *Sein und Zeit* in 1927), which was dedicated to Husserl, has played a major influence in the development of twentieth-century existentialism and new interpretations of phenomenology. Many influential philosophers of the twentieth century studied under Heidegger and thus can be included very broadly within the movement associated with phenomenology, including Hannah Arendt, Hans-Georg Gadamer, Emmanuel Levinas and Herbert Marcuse. In France, Jean Paul Sartre read both Husserl and Heidegger to produce his own version of existentialism. Another important French philosopher, Maurice Merleau-Ponty, who can be regarded fully as a phenomenologist in Husserl's tradition, has been described by Dermot Moran in his *Introduction to Phenomenology* (2000: 391) as having 'made the most original and enduring contribution to post-Husserlian phenomenology in France'. The Algerian-born philosopher very closely associated with post-modern deconstruction, Jacques Derrida, translated one of Husserl's works and again, according to Moran, displayed throughout his writings a 'thorough familiarity with both Husserl's original texts and Husserlian scholarship' (2000: 438).

It is clear, therefore, that philosophical phenomenology as it is now understood has its roots in the works of Edmund Husserl. As the phenomenology of religion developed in the twentieth century, Husserl's philosophy must be regarded as one of its major formative influences, alongside theology and the social sciences. It has been a subject of controversy as to how far and to what extent Husserl influenced the phenomenology of religion, but almost everyone is agreed that at the very least Husserlian terminology was transposed into and utilized within phenomenological analyses of religion (see, for example, Sharpe, 1986: 224; Flood, 1999: 9–10). In this chapter, I explore the principle features in Husserl's philosophy without drawing direct connections to the thinking of key phenomenologists of religion. These relationships, whether substantial or superficial, will be discussed fully in later chapters.

Background to Husserl's phenomenology: From Descartes to Kant

In 1913 Husserl published *Ideen zu einer Phänomenologie und phänomenologischen Philosophie*, which he described in the preface to the 1931 English edition, entitled *Ideas: General Introduction to Pure Phenomenology* (1931: 11), as seeking 'to found a new science … that of "Transcendental Subjectivity"'. Dermot Moran (2000: 124) argues that Husserl 'came to see the whole of philosophy as somehow encompassed in, or founded on, this new science of phenomenology'. In order to understand the most relevant concepts within Husserl's thinking, we need first to examine briefly the background to some of the most important determining forces within the development of contemporary epistemology: the method of radical doubt, as posited by the French philosopher, René Descartes (1596–1650), the analyses of perception within the thinking of the so-called British empiricists (Locke, Berkeley and Hume), and the 'transcendental' philosophy of Immanuel Kant (1724–1804).

The branch of philosophy called epistemology, or theory of knowledge, attempts to discover the grounds on which our knowledge of the world, ourselves and our consciousness can be established, including identifying the limits to which our knowledge can extend. Epistemology thus stands at the base of all philosophical issues, including metaphysical questions (What can we know about ultimate realities?); ethics (How can we determine the right and the good?); and aesthetics (How can we determine what is beautiful?). Theories of knowledge can be regarded as the methodologies within philosophy since they define not so much *what* we know as *how* we know what we know. In their introduction to philosophy, Popkin and Stroll (1986: 204–5) define the central focus of epistemology as 'concerned to determine the basis of all knowledge-claims, and to agree upon standards for judging these claims'.

René Descartes is often associated with the beginnings of a new way of thinking in philosophy because he introduced a method for knowing which became a dominant characteristic within what is now called 'the Enlightenment'. Russell McCutcheon (2000: 129) credits Descartes with inserting into contemporary Western philosophy an emphasis on 'rationality over unthinking submission to the seemingly arbitrary rule of authority or tradition'. Walter Capps (1995: 2) argues that 'Descartes' fundamental contribution lay in formalizing the human disposition to doubt' by making 'doubt a means of access to truth'. The Cartesian method for obtaining knowledge needs to be understood in the context of phenomenology because, as Dermot Moran (2000: 129) observes, 'Husserl always fully accepts the legitimacy of Descartes's argument leading to the discovery of the *cogito* [thinking self]'.

The foundation of knowledge, for Descartes, depended on finding that which was entirely indubitable. Once this bedrock could be discovered, everything else, which we often take for granted, could be rebuilt. Through a painstaking analysis, Descartes subjected every idea, assumed fact or even postulated revelatory act to radical doubt. If it could be doubted, it could not provide the singular basis on which a theory of knowledge could be built. In this way, through a series of extended mental exercises that he called 'meditations', Descartes deconstructed what in common sense we assume to be real, such as the existence of the external world, our experience of other persons and the reality of our own bodies. Finally, after having doubted virtually everything conceivable, he reached the one indubitable fact: he was doubting, hence, he was thinking. Even though he could be completely misled into believing he had a body and that he had experienced an external world, he could not escape the fact that he was the one doing the doubting. He was, therefore, a thinking being, if nothing else. Hence, Descartes (1931, vol. 1: 220) reached his point of indubitability, expressed in the famous dictum: *cogito, ergo sum*, 'I think, therefore I am'. In his second meditation (Descartes in White, 1989: 100), he reasoned: 'But what then am I? A thing which thinks. What is a thing which thinks? It is a thing which doubts, understands, [conceives], affirms, denies, wills, refuses, which also imagines and feels'. From this mainstay of certainty, he then reasoned back into existence the realities, which he had previously naïvely taken for granted: the existence of God and the world of nature, other beings, his own body and its relations to other bodies.

Descartes' method for obtaining certainty displayed an unwavering faith in human reason as the basis on which to found knowledge. By beginning with the self, he overturned the pre-Enlightenment order whereby God provides the basis for our knowledge first of himself, then of the world and finally of ourselves in the world. Capps puts it this way:

> Previous thinkers and schools of thought – Plato, Aristotle, the medieval system-builders (such as Albertus Magnus, Bonaventura, Thomas Aquinas, to name some of the prominent ones) – approached intellectual work out of contrasting motivation. Instead of doubting everything initially, they attempted to affirm, to posit, to validate. (Capps, 1995: 2)

This fundamental reversal of method in knowing came to characterize the European Enlightenment, described by Richard King (1999: 44) as 'a particular period or movement within European intellectual history that had far-reaching sociopolitical and cultural consequences for the modern Western world'. It was a movement primarily which put the human being at the centre of knowledge, either through reason or science, and it adhered strongly to a sense that scientific rationality contributed fundamentally to the progress of humanity throughout its gradually evolving history. Such progress was seen to have been fostered by placing reason at the centre of knowledge, which in turn led to moral improvement, a tolerant society and respect for individual rights.

Descartes can also be understood as adopting a philosophical method for obtaining knowledge, called rationalism. He did not need to consult the world to arrive at his point of certitude. This was accomplished entirely by using reason. The world, which can be investigated using scientific methods, was already established by Descartes through rational processes conducted before any investigation needed to occur. Although his method of doubt was innovative and reversed prior principles of attaining knowledge, it fitted into a long philosophical tradition traceable to Plato which relied on reason to attain certain knowledge. For Plato, humans were invested with innate ideas, which could be recollected under proper philosophical training using deductive methods, until the philosopher attained a vision of the ideal forms. Popkin and Stroll explain:

> Theories of knowledge like those of Plato and Descartes are 'rationalistic' because they assert that by employing certain procedures of reason alone we can discover knowledge in the strongest sense, knowledge that can under no circumstances possibly be false. (Popkin and Stroll, 1986: 232)

This rational tradition, emphasising the use of mathematical, analytical, logical and deductive reasoning, has been contrasted in the history of philosophy to empiricism, through which knowledge of the world is attained inductively, chiefly by collecting data from the world, collating the data into comprehensible categories and eventually building up a coherent picture of reality. Empiricism thus depends largely on observation and sense perception to determine what is real. Plato's student, Aristotle, used such inductive methods and thus set up the longstanding contrast in philosophy between those who argue for the primary place of reason in obtaining knowledge of the world and those who give priority to experience.

Following Descartes, the British empiricists, writing in the seventeenth and eighteenth centuries, emphasised the importance of experience for attaining knowledge of the world, primarily through sense perception. King (1999: 44) defines empiricism as the idea that 'all human knowledge about the world is ultimately derived from empirical evidence accessible through the sense organs'. The chief exponents of this philosophical doctrine were John Locke (1632–1704), the Irish bishop George Berkeley (1685–1753) and the Scottish sceptical philosopher David Hume (1711–1776). In his *Essay Concerning Human Understanding* (1924) [1690], Locke famously argued that the mind is a blank tablet, a *tabula rasa*, on which the data of the world through sense perceptions imprint themselves. Berkeley extended this argument in his *Treatise Concerning the Princples of Human Knowledge* (1998) [1710] to indicate that nothing exists whatsoever that cannot be perceived by the senses. In response to Locke's assertion that such ideas as the extension of an object in space and time could be said to exist apart from direct and simple sense perceptions, Berkeley asked, for example, 'What would a table "be" if it were not seen, heard, touched, or even tasted or smelt?' Its existence, including its extension in space, in any meaningful way, could not be established apart from sense perceptions, which are entirely in the mind. In this way, Berkeley claimed to have demonstrated that reality is immaterial, comprising of ideas in the mind and thus we can have no knowledge apart from such ideas. This logic could have led him into extreme scepticism or even solipsism (the idea that nothing exists apart from myself, my mind and my ideas), but Berkeley's philosophical aims were theological at their root. He wanted to establish a priority for the mind of God, which was constantly perceiving everything at all times. In this way, the objectivity of the external world was preserved. Although the maxim, 'to be is to be perceived' was categorically true for Berkeley, everything is real because it exists always within the perceptions of God. Ronald Knox's famous limerick, cited by Popkin and Stroll, renders clearly Berkeley's meaning:

> There was a young man who said, "God,
> I find it exceedingly odd
> That this tree I see should continue to be
> When there's no one about in the Quad."
> Reply.
> "Dear Sir:
> *I* am always about in the Quad.
> And that's why the tree
> Will continue to be
> Since observed by
> Yours faithfully,
>   GOD." (Popkin and Stroll, 1986: 252)

Despite Berkeley's claim to have overcome scepticism by ensuring reality as situated within the mind of God, in his *Enquiries Concerning the Human Understanding* (1902) [1737], David Hume demonstrated, by reducing sense perceptions in space and time to discrete units or impressions on the mind, that no necessary connection exists between one impression and another. The mind associates them in causal relationships because of their proximity in space and time, but this does not bear the weight of a necessary connection. It remains contingent on accidental occurrences relative to temporal and spatial proximities. For example, if I drop a book on a table, a sound follows and a bit of dust scatters in the air. The sound and the scattering are associated by the mind into causal connections simply because they follow closely in time and are spatially related to my dropping the book. These are contingent relations existing not necessarily in the world of experience but in the constructions of mental processes. For Hume, if nothing can be connected causally in a necessary way, science cannot be founded on incontrovertible laws. In other words, if causality cannot be shown to subsist in the objective world, certain knowledge of the world will have been destroyed.

Immanuel Kant credits Hume with awakening him from his 'dogmatic slumber' by creating a crisis at the core of modern philosophy through his analysis of sense impressions and his seemingly inevitable conclusions resulting in scepticism. In what must be regarded as one of the most creative and influential syntheses of reason and experience, Kant sought to resolve Hume's problem by circumscribing the limits and conditions within which knowledge of the world can be obtained. Kant's resolution depends on a creative juxtaposition and reinterpretation of four key concepts: *a priori* knowledge, analytic statements, *a posteriori* knowledge and synthetic statements.

A priori knowledge refers to knowledge which is known prior to experience and which does not require empirical validation to establish its certainty. Such knowledge relates to necessary logical conclusions and to linguistic definitions, where, in the first instance, something is known to be true on the basis of an incontrovertible law of thinking or, secondly, where the predicate is already contained or implied in the subject. For example, if I say, 'It is raining or it is not raining', I do not need to look out of my window to assert that this is a true statement. I know without recourse to experience that it is either raining or it is not raining, since such a statement depends entirely for its truth on the law of non-contradiction: two opposing statements cannot logically be true at the same time. In logic, this is articulated schematically in the statement: 'A does not equal non-A'. Similarly, some statements are true simply by definition: 'Mr Jones, who is a bachelor, is unmarried'. We know by defining bachelor that Mr Jones is unmarried without having to conduct any further investigation. Another somewhat less obvious example was provided by Kant: 'All bodies are extended'. Although one might be tempted to test this statement in experience, we soon recognize that a body, by definition, to be a body, must be extended in space. Mathematical and geometric statements prefigured Kant's resolution to the problem of knowledge because, although they are necessary and *a priori*, it cannot be argued that the predicate is assumed in the subject. For example, the statement, 'the shortest distance between two points is a straight line', could not be construed as adding nothing to the subject, since the predicate (shortest distance) is not implied in the subject (straight line) (Popkin and Stroll, 1986: 163; Kant, 1977: 45).

A posteriori knowledge refers to knowledge that is gained only by recourse to experience in the world. It must be substantiated by empirical investigation and its truth or falsehood cannot be determined without such investigation. The truth of the statement, 'Mr Jones is six feet tall', can only be known by measuring the height of Mr Jones. The claim that the sun is 93 million miles away from the earth provides another example of such a type of enquiry. Historical evidence is also confirmed in this way. The assertion that the British Prime Minister in 1943 was Winston Churchill can only be verified or falsified on the basis of historical methods. Any knowledge that requires corroboration by recourse to evidence using empirical tools of investigation, such as scientific or historical processes, is knowledge gained after the fact or from experience. Such knowledge is articulated in what are called synthetic statements, which build up and coordinate evidence obtained through empirical studies. The predicate of such statements adds knowledge to the subject, and in no way can be regarded as a necessary conclusion from the subject.

In normal philosophical classifications, *a priori* knowledge is expressed in analytic statements; analytic statements are known to be true *a priori*. *A posteriori* knowledge is articulated through synthetic statements; synthetic statements are verified or falsified *a posteriori*. With perhaps the exceptions of mathematics, geometry and physics, *a priori* knowledge is not expressed in synthetic statements; *a posteriori* knowledge is not communicated through analytic formulations. These distinctions constituted the basis for the great division within epistemology, as noted above, between those who derive knowledge from first principles, deductively, generally known as rationalists and those who argue that knowledge of the world is derived from collecting scientific data and organising them into coherent systems inductively, generally known as empiricists. As we have seen, Kant inherited this problem, in new forms, from Descartes, and following him, through the British empiricists. Kant sought to overcome this central polarity in philosophy, and with it the scepticism of Hume, by postulating the possibility of knowledge that is both *a priori* and synthetic, that which is necessarily true, but which still relies on experience or empirical investigation for its validation. In his famous *Critique of Pure Reason* (1929) [1787], a condensed and densely argued short version of which can be read in Kant's *Prolegomena to any Future Metaphysics* (1977) [1783], Kant re-framed the argument by creating a transcendental world of ideas necessary for thought and existing in the mind prior to experience, called the categories of the mind, which formulate or dictate the conditions for human experience. This resulted in the remarkable and previously unthinkable (as applied to knowledge of objects in the world) 'synthetic *a priori*', defined by Popkin and Stroll as a statement in which the subject contains within it necessary knowledge, but in which, at the same time, 'the predicate of the judgment must contain some information not contained in the subject' (1986: 162).

What becomes prior to experience for Kant, therefore, cannot be the content of experience. We have already seen that such assertions must be verified empirically (such as the height of Mr Jones). The mind, however, contains within it organising principles, patterns or forms, through which experience attains coherence and order. Such forms define the grid through which all minds view the world and make sense of the data it provides. Hume's error, therefore, was to seek necessary connections in the external world, in the impressions themselves which are imprinted on the mind. He was right to conclude that such connections cannot be found in experience, or at least they cannot be demonstrated to be there. Causality, however, exists as a primary category of the human mind, through which order, connections and relations are established. This assigns the primary role to the mind in structuring reality. This is not the same as arguing that reality is in the mind; the formal structures or categories for making sense of the world reside in the mind, but the content depends on experience. This precisely defines the synthetic *a priori*. The categories exist prior to experience and are necessary for experience to make any sense at all, but the content which supplies the categories with substance are derived from experience.

This, of course, means that apart from the formative role of the mind in structuring experience, we do not know precisely what experience, or reality, actually is. We cannot know the 'thing-in-itself' apart from the imposition of organising structures supplied by the mind prior to experience. The categories of the mind or understanding, according to Kant, are the 'original pure concepts of synthesis, which belong to the understanding *a priori*, and for which alone it is called pure understanding; for it is by them alone that it can understand something in the manifold intuition, that is, think an object in it' (cited by Popkin and Stroll, 1986: 164). The mind, through its organising categories, perceives and gives content to the specific data; it 'thinks an object in it'. In this way, the foundation for universal and necessary knowledge is established and scepticism is overcome. Although we cannot know what somehow lies beyond the mind, we know the world in consistent and uniform patterns because of commonly shared human categories of thinking.

The role of philosophy, therefore, differs from that of science. The philosopher analyses necessary conditions for thought, the categories in the mind which make science possible, intelligible and coherent. Scientists utilize the categories of the mind to uncover data, to test hypotheses and to establish further areas of investigation. The categories do not define what science discovers; they establish the grounds on which such discoveries can be made and set the limits within which knowledge can be ascertained. Kant calls philosophical enquiry 'transcendental', by which he means uncovering the necessary formal conditions for empirical investigations to be undertaken. These conditions, since they are determined by *a priori* categories of the mind, transcend empirical studies and constitute the nature of a 'transcendental philosophy'. Such a term should not be confused with metaphysics, or with questions of religion, theology or ethics. The realm of 'pure reason', which is constituted and established by the categories of the mind, can yield no knowledge whatsoever of that which extends beyond experience. Questions of the existence of God, the nature of ultimate reality, even human freedom, cannot be resolved using the categories of the mind. The transcendence spoken of in pure reason or philosophical knowledge defines and determines the conditions for knowledge of what Kant calls the phenomenal world (as opposed to the noumenal world, the postulated world of metaphysics and theology). God, freedom and immortality cannot be proved using the formal categories; they exist in another realm of thinking, that which Kant calls 'practical' reason – necessary postulates for moral behaviour.

Husserl's transcendental turn away from realism

In 1901, Edmund Husserl was appointed Professor in the University of Göttingen, where he spent the next 15 years and during which he published just one book, *Ideas*. In his analysis of Husserl's philosophical development, David Bell (1990) argues that at Göttingen, Husserl underwent a fundamental change in orientation away from naturalism, the position 'that everything real belongs to physical nature or is reducible to it' (Moran, 2000: 142), towards a 'transcendental' philosophy, in part at least, because he had begun an intensive study of Kant's writings. Bell comments:

> During the winter semester of 1905–6 ... he taught courses on Kant's philosophy every day of the week except Sunday – but in any case in 1907 he delivered a series of five lectures, which for the first time, made public that his philosophy had taken a 'transcendental turn' away from naturalism. (Bell, 1990: 153)

Husserl's student, the Polish philosopher Roman Ingarden (1975: 4–5) argues that in Husserl's earliest works, particularly *The Philosophy of Arithmetic* (1891) and *Logical Investigations* (1900), he was a realist, a position he abandoned after what Bell calls his 'transcendental turn'. Ingarden, who became well known as a twentieth-century phenomenologist and realist in his own right, explains that his book, *On the Motives which led Husserl to Transcendental Idealism* (1975), was written because he had often asked himself 'why Husserl, really, headed in the direction of transcendental idealism from the time of his *Ideas* whereas at the time of the *Logical Investigations* he clearly occupied a realist position' (1975: 2). 'Realists' maintain that the objects perceived by the mind exist apart from the mind, its mental acts and states, whereas idealists, according to Popkin and Stroll (1986: 145), hold 'that the most important element in the nature of reality is mind or spirit'. Ingarden notes that 'the controversy between realists and idealists concerning the existence of the real world is not about whether the real world, the material world in particular, exists in general ... but about the mode of the world's existence and what its existential relation is to acts of consciousness in which objects belonging to this world are cognised' (1975: 5). In other words, both realists and idealists affirm in some sense the objective existence of the external world; they differ on how and to what extent we can know the world and in what forms it can be apprehended through perception. Although Ingarden's assertion that Husserl was fully a realist prior to his move to Göttingen is disputed by David Bell (1990: 154), it is important to trace the stages Husserl followed towards transcendental idealism in order to understand the context and meaning of the key concepts which eventually were adopted into most phenomenologies of religion.

Husserl began his academic career as a mathematician, writing his Ph.D thesis in 1882 for the University of Vienna on 'Contributions to the Theory of the Calculus of Variations' (Bell, 1990: 3). After a short period in Berlin as a mathematician, he returned to Vienna to study philosophy under Franz Brentano, whose specialization today might be called philosophy of mind or consciousness. Under Brentano's influence, Husserl directed his intellectual interests towards philosophy rather than mathematics, particularly philosophy as it was understood at the end of the nineteenth century as including psychology. Ingarden argues that it was precisely Brentano who influenced Husserl in *Logical Investigations* to first employ the term phenomenology as 'descriptive psychology'. David Bell (1990: 4) notes that although 'during the second half of the nineteenth century the discipline of psychology ... had not yet clearly distinguished itself from philosophy ... it was nevertheless ... a discipline capable of generating a great deal of intellectual excitement', partly because it defined its subject as a 'science of the mind', which, again in Bell's words, 'could be just as empirical – perhaps even just as experimental – as the more familiar sciences of nature'. One of Husserl's overriding philosophical concerns, to found knowledge on a clearly defined and rigorously applied methodology, emerged from this interpretation of epistemology as an empirical science. This meant, according to Ingarden (1975: 5–6), that Husserl sought to address satisfactorily the fundamental and prior question: What gives unity to thought and thus makes science possible?

Husserl's concept of a rigorous science can be traced directly to the influence of Descartes, since Husserl wished to secure a firm or unshakeable foundation for knowledge. In *Logical Investigations*, this was found in the term 'ideations', which Ingarden (1975: 10) defines as 'a mode of knowing what is ideal, the so-called species', for example, mathematics. Mathematical statements as we noted in our discussion of Kant's categories of the mind, are not analytical in the sense of necessary formulations known prior to experience in which the predicate adds nothing substantial to the subject, but they can be described as ideal, that is providing pure forms or logical patterns for knowing or apprehending the world. They are, in this sense, formal categories of the mind denoting relation in time and space, both *a priori* and synthetic. This was Husserl's early understanding of mathematics: a species of thought, ideations, ways through which we perceive what is real. During the first decade of the twentieth century, Ingarden argues (1975: 10–11) that Husserl moved away from analysing ideations towards a detailed study of what he termed 'outer perception', that is what the perceiving subject observes as objects in the real world. This placed Kant's central question back at the centre of the discussion: Can we know what the objects of outer perception are in themselves? Husserl concluded that outer perception cannot provide indubitable, foundational knowledge, in a Cartesian sense. He then proceeded to examine inner perception, an experience so close, so immanent, to the subject doing the perceiving that it could not be doubted. It is just at this point that Ingarden identifies Husserl's transcendental turn in the direction of 'final subjectivity'. He explains:

By invoking the help of 'eidetic' cognition it was relatively easy to reach the idea of pure transcendental phenomenology which through its appeal to the final subjectivity of pure consciousness was to discover not only the final source of all knowledge of the real world but also by a simple transposition of the problem ... was to make possible a deduction of the real world from the ultimate source of pure consciousness. (Ingarden, 1975: 11)

The natural attitude

In order to achieve 'immanent perception', Husserl employed a method he called the phenomenological reduction, which begins with an analysis of the 'natural thesis', otherwise referred to as the natural attitude or the natural standpoint, by which he means accepting the world 'out there' as given. This entails a pre-theoretical acceptance of the world as it appears or seems to appear. The natural standpoint is assumed in common sense thinking, but it also underpins all naturalistic interpretations of the world. In *Ideas*, Husserl defines the natural standpoint in the following way:

> I find continually present and standing over against me the one spatio-temporal fact-world to which I myself belong, as do all other men found in it and related in the same way to it. This 'fact-world', as the world already tells us, I find to *be out there*, and also *take it just as it gives itself to me as something that exists out there.* (emphasis his) (Husserl, 1931: 106)

The basis of the natural attitude, therefore, is just the assumption that consciousness is something inside or immanent to the observer and objects are outside or transcendent to the one who observes them (Moran, 2000: 145). This defines the fundamental presupposition underlying what George James calls 'all projects directed toward certainty in the world', from everyday activities to scientific investigation. Husserl (1931: 106) explains: 'To know it more comprehensively, more trustworthily, more perfectly than the naïve lore of experience is able to do, and to solve all the problems of scientific knowledge which offer themselves upon its ground, that is the goal of the *sciences of the natural standpoint*' (emphasis his). In his 1974 Ernst Cassirer Lectures delivered in Yale University, the Polish philosopher Leszek Kolakowski argues that by drawing attention to the natural standpoint, Husserl was actually criticizing the assumptions that had come to dominate the empirical sciences at the beginning of the twentieth century. As technology increases, Kolakowski (1975: 34) asserted, the ways we gain understanding of the world actually decrease. 'The sciences measure things without realizing what they measure; in carrying out cognitive acts, they are incapable of grasping these very acts'.

Kolakowski (1975: 34–5) identifies 'naturalism' as one of the chief culprits in fostering technology without understanding. This is precisely because naturalism regards consciousness as an object in the world, to be examined just like any other object, and thus possesses no independence from the natural world. As we have seen, naturalism represents a philosophy that holds that the universe contains nothing but phenomena that are known purely by methods operating within the natural sciences. In the context of Husserl's discussion of the natural standpoint, as David Bell (1990: 155) notes, 'the most important ingredient in naturalism is the thesis that the human mind is just a common-or-garden part of the natural order of things'. This produces the logical conclusion, according to Bell, that the mind cannot be regarded in any way as 'foundational' or 'constitutive' with respect to the natural world; the mind can be explained, like everything else, by the natural sciences.

Husserl believed that naturalism cannot provide an adequate understanding of the world precisely because the assumptions for a naturalistic position must be founded on a thorough analysis of how we obtain knowledge in the first place. Dermot Moran (2000: 143–4) argues that for Husserl, 'consciousness should not be viewed naturalistically as part of the world at all, since consciousness is precisely the reason why there was a world there for us in the first place'. Moran notes that this position must not be interpreted to mean that consciousness *creates* the world because this entails a naturalistic assumption that mind causes nature. Rather, Moran explains, 'the world is opened up, made meaningful, or disclosed through consciousness'. The natural standpoint thus must be subjected to doubt, just as Descartes did, but as Husserl explains in *Ideas*, 'with an entirely different end in view' (1931: 107). Husserl claimed that he employed doubt, not in a Cartesian sense to set up 'an absolutely indubitable sphere of Being', but '*only as a device of method*', that is, to provide a means for attaining an adequate understanding of the world (emphasis his) (1931: 107).

Epoché: Going beyond Cartesian doubt

Husserl's method of doubt was not aimed at establishing whether or not an object is actually there, that it actually possesses being. He was concerned to examine the process of doubting itself, which he argued entails a suspension of judgement about what is taken for granted in the natural attitude, that the objects of perception are actually out there independent from perception. The suspension of judgement does not abandon the natural standpoint; no change in commitment to the existence of an external world is required. Husserl explains: 'We set it as it were "out of action", we "disconnect it", "bracket it"' (*Ideas*, 1931: 108). The natural standpoint in this method remains experience as it is lived, but in its bracketed condition, 'we make no use of it' (1931: 108). What remains, Husserl says, is a 'unique form of consciousness' that 'transvalues' lived experience 'in a quite peculiar way' (1931: 109). This method differs from that employed by Descartes who understood doubt in terms of establishing the antithesis to what is normally assumed. In other words, Descartes set Being against Non-Being so strongly that, in Husserl's view, 'his universal attempt at doubt is just an attempt at universal denial' (1931: 109). Husserl argues that his method does not attempt to exhaust analytically every component contained within the antithesis to Being: 'We extract only the phenomenon of "bracketing" or "disconnecting"'. This process is best referred to as *epoché*, defined by Husserl as 'a certain refraining from judgment', through which the natural thesis is 'put out of action' and thereby 'passes off into the modified status of a "bracketed thesis"' (1931: 110).

In this way, the phenomenological *epoché* steps 'into the place of the Cartesian attempt at universal doubt', but with a difference (1931: 110). The *epoché* operates in a limited sense. It is not intended to put every conceivable thesis into brackets, but precisely to 'put out of action the general thesis which belongs to the natural standpoint' (1931: 110). The method of *epoché*, therefore, places in brackets 'this entire natural world ... which is continually "there for us", "present to our hand"'. This does not eliminate the natural standpoint at all; it remains 'a "fact-world" of which we continue to be conscious, even though it pleases us to put it in brackets' (1931: 110). The phenomenological *epoché* establishes a new mode of consciousness because it 'bars me from using any judgment that concerns spatio-temporal existence' (1931: 111).

Up to this point, Husserl's use of the *epoché* must be seen in some sense as negative by telling us primarily that he wishes to make no use in the phenomenological reduction of the natural sciences in the way that the natural sciences understand themselves, that is from the viewpoint of the natural attitude.

> I make absolutely no use of their standards, I do not appropriate a single one of the propositions that enter into their systems, even though their evidential value is perfect, I take none of them, no one of them serve me for a foundation – so long, that is, as it is understood, in the way these sciences themselves understand it, as a truth concerning the realities of this world. I may accept it only after I have placed it in the bracket. (Husserl, 1931: 111)

It is what follows the phenomenological reduction, the bracketing out of the natural attitude, that leads to Husserl's positive contribution to a new science of the mind. Ingarden (1975: 12) observes: 'After having carried out this reduction, we find ourselves ... in the area of pure transcendental consciousness inside which we carry out all epistemological investigations'.

Husserl's transcendental idealism

Husserl proceeds beyond the phenomenological reduction, which thus far has told us only what is put aside or made no use of, by opening up a new field of study, that into consciousness itself. He asks: 'What can remain over when the whole world is bracketed, including ourselves and all our thinking?' (1931: 112). The answer is found in terms like 'pure experiences', 'pure consciousness' and 'pure Ego', through which seeing into a new realm of experience is made possible when the natural standpoint is placed in brackets. 'Thus we fix our eyes steadily upon the sphere of Consciousness and study what it is that we find immanent in it' (1931: 113). In contrast to the assumptions of the naturalists, by employing the method of *epoché*, Husserl thus established an independent field of enquiry, a new domain for investigation: 'Consciousness in itself has a being of its own which in its absolute uniqueness of nature remains unaffected by the phenomenological disconnexion' (1931: 113). That which is left over after bracketing out the natural standpoint, Husserl calls 'phenomenological residuum', the content of which, he argues, can become 'the field of a new science – the science of Phenomenology' (1931: 113). In this way, the phenomenological *epoché* ceases to operate simply as a negation, placing aside the natural standpoint, and becomes 'the necessary operation which renders "pure" consciousness accessible to us' (1931: 114).

When the phenomenological reduction has taken place, it becomes possible to analyse consciousness in a new and fresh way. Husserl begins with an example to illustrate how this occurs (1931: 116–17). He describes a white paper lying in a dim light in front of him. The perception of the paper, in its relative obscurity and particular angle with respect to his perception, is what Husserl calls a *cogitatio* (a conscious experience). The paper itself, which is 'objective' due to its qualities of extension in space and because of its particular relation to his own body is a *cogitatum* (something perceived). The background features which surround the white paper, such as pencils, an ink well and books, are also 'perceived'; they are, Husserl explains, 'perceptually there, in the "field of intuition"'. Yet, by singling out the white paper for concentration, the background features are in a sense not apprehended by the consciousness, 'not even in a secondary sense', but operate as 'a zone of background intuitions'. These background intuitions can also be referred to as a *cogitatio*, because they comprise a conscious experience of what consists objectively in the background. The conscious experience is different from that which is perceived, since involved in consciousness are variations, such as that of perspective (from which angle I see the paper when I move my head in one direction or another or the changing light which might be shed on the paper) or other considerations such as memory (how I recall viewing a white paper in the past), anticipation (how I might view it in future) or even fanciful perceptions (I may imagine images on the paper in the dim light).

In his discussion of this well-known example drawn from Husserl's *Ideas*, Dermot Moran (2000: 153) emphasizes that in each of the variations of perception suggested by Husserl a white paper is referred to, even if certain images are imagined on it or even if the paper does not exist at all. 'It is still referred to; something of the very essence of a mental process is being grasped.' We know the difference intuitively between that which is purely imagined, like a nymph or a fairy, and the white paper before us. This is demonstrated further by the fact that the observer is conscious of the difference between the paper and its quality of being white. Clearly, something is given to the consciousness, in this case the white paper; it is grasped through a conscious experience which includes a singling out, background features and the variations in its perception. This consciousness of something is referred to by Husserl as 'intentional experience', because, he explains, 'in so far as they [conscious acts] are a consciousness of something they are said to be "intentionally related" to this something' (1931: 119). It is important to note in this example that Husserl is examining consciousness as he regards it in a pure state, and thus is not exploring how the mind and objects of perception interact as an empirical psychologist would. Husserl is concerned with conscious experiences that are *a priori*, necessary for experience and not dependent on experience for validation. Otherwise, he would be reverting back to analysing consciousness as if it were just another part of natural phenomena to be studied and classified by the natural sciences. The natural attitude remains bracketed so that the essence of experiences can be grasped. Leszek Kolakowski explains:

> What remains after the reduction are the contents of phenomena *and* the place where they appear, or the transcendental, not nonempirical Ego, the pure subject of cognition, the recipient of phenomena, something having none of the properties usually attributed to psychological subjects, and retaining, nevertheless, the intentional relation to its object. (Kolakowski, 1975: 40)

The concept of intentionality, in its simplest form, refers to the subject actively perceiving or apprehending an object, for example as Husserl explains, 'every presenting refers to a presented, every judgment to something judged' (1931: 256). The situation, however, when seen from the viewpoint of the phenomenological reduction, appears much more complex than this. 'It is evident that intuition and the intuited, perception and the thing perceived, though essentially related to each other, are in principle and of necessity not really and essentially one and united' (1931: 130). To illustrate this point, Husserl describes perceiving a table. As he moves around the table, changing his own perspective on it, he keeps in his mind the one, same unified table. Each changing perception does not create a new table; it remains unchanged throughout the movements of the observer. It is the perception of the table that is in flux. Or, says Husserl, if he closes his eyes, he has no perception of the table, but when he opens them, the table returns. The table remains unchanged to the consciousness which connects the new perception with the memory of the perception before the eyes were closed. Hence, Husserl concludes, 'the perception itself is what it is within the steady flow of consciousness, and is itself constantly in flux; the perceptual now is ever passing over into the adjacent consciousness of the just-past, a new now simultaneously gleams forth' (1931: 130).

In order to distance his analysis of consciousness sharply from empirical psychology, as Moran notes (2000: 155), between 1907 and 1913, Husserl introduced a new terminology derived from the Greek terms *noesis* and *noema*, to help clarify his use of the term 'intentionality'. *Noesis* refers to the act of thinking (earlier referred to as a *cogitatio*) whereas *noema* indicates what is thought (as above, a *cogitatum*). This distinction forms a part of Husserl's analysis of the process of consciously perceiving objects in the world in order that we may direct our attention not to what is perceived but, in Husserl's words, to 'the perceiving itself' (1931: 256). Moran argues that 'for Husserl, the most important thing to emphasize is that *noesis* and *noema* are correlative parts of the structure of the mental process' (2000: 155). Husserl explains that the distinction is based on an analysis of consciousness which reveals a difference between 'the proper components of the intentional experiences, and their intentional correlates, or the components of them' (1931: 257). In other words, a phenomenological analysis of consciousness examines the experience of thinking with full recognition that the experience is 'consciousness of something'. The noetic phase of thinking entails the ascription of meaning or meanings to what is thought. It also involves holding the apprehension of what is thought in the memory while other objects are perceived. Gradually, in the act of thinking, a complex process with ever deepening levels or constitutive parts, emerges, including noting relationships between objects, making presumptions about them and introducing value judgements. Clearly, these acts form a part of the mental process and can be distinguished from that which consciousness thinks about. Perceiving, recollecting, correlating data and making judgements form a part of the noetic act of consciousness; each also has its *noema*, in which something is 'perceived as such', 'recollected as such' or 'judged as such'.

It is important to remind ourselves continually that Husserl is conducting this analysis of consciousness while employing the *epoché*. Hence, he is not concerned with the existence of objects as such, but with the way in which consciousness, in its transcendental form, operates. In his introduction to Husserl's essay, 'Philosophy as Rigorous Science', published in 1911, Quentin Lauer (1965: 10) explains that 'experience alone cannot answer the most important questions about experience but must seek the answers in a theory of cognition from which every positing of existence has been eliminated'. This eliminating the question of existence is done by the phenomenological bracketing or *epoché*, only after which, as Lauer (1965: 12) explains, does it become clear that 'the physical identity of an object is intelligible only through the acts in which the object is present to consciousness, since it is precisely the relation to consciousness (in these acts) that makes it an object'. Roger Scruton (2000: 61) explains that 'in the phenomenological attitude the whole transcendent world is placed into brackets', which means that 'the only thing that is left over is the experience (*cogito*) [above, *cogitatio*], which has an intention to the corresponding intentional object (the *cogitatum*) and the ego'.

Although he accepted the Kantian view that consciousness itself is transcendent and forms the core for philosophical analysis, by performing the phenomenological reduction, Husserl sought to avoid the dualism of Kant between that which is thought and the 'thing-in-itself'. By positing the formal categories of the mind, Kant constituted the way the world is perceived prior to experience, but the actual objects of perception, although understood as causing perception, could never be known in themselves. For Husserl, although acts of thinking are correlated to what is thought about, this is quite different from positing that what is thought about actually represents an object in itself. Kolakowski explains this quite fundamental distinction between Husserl and Kant:

> The difference from the Kantian concept of phenomenon is unmistakable: to Kant, the phenomenon is an appearance *of* something; that phenomena revealed things was to him obvious, direct; we do not know how the thing is in itself, but we do know immediately that it is revealed in the phenomenon; as if – though Kant does not say so – the existence of things was an analytical truth, included in the very sense of the word 'phenomenon'. This is not implied in Husserl's concept, existence being excluded from acceptable immediacy. (Kolakowski, 1975: 38–9)

Husserl's analysis of noetic and noematic acts occurs in the realm of pure consciousness made possible by the phenomenological reduction. Again, according to Kolakowski: 'Within such a purified field, I know neither the world nor consciousness as belonging to it; I know only phenomena as intentional correlates of my conscious acts' (1975: 39).

In his discussion of noetic and noematic acts of consciousness in *Ideas*, Husserl provides a lucid example of his sharp distinction between phenomenology as the analysis of pure consciousness and the natural attitude, which concerns itself with questions of what the objects of consciousness are in themselves. Husserl asks his reader to contemplate a scene where one is looking with appreciation at an apple tree in full blossom. He notes that the observer is perceiving the apple tree and experiencing pleasure at the same time in doing so. From the natural standpoint, the analysis would proceed to posit the reality of the apple tree and how it produces a psychic state of appreciation and well being. It is conceivable under this standpoint to suggest that the apple tree has been imagined by the perceiver, that it forms part of an hallucinatory experience. The psychic feeling and its 'objective' cause are separated. 'Nothing remains but the perception; there is nothing *real* out there to which it relates' (1931: 259). The phenomenological standpoint contrasts sharply from the natural attitude precisely because the objective reality of the tree and its relation to a psychological feeling of appreciation have been bracketed. 'Yet a relation between perception and perceived (as likewise between the pleasure and that which pleases) is obviously left over, a relation which in its essential nature comes before us in "pure immanence"' (1931: 259). In the state of bracketed consciousness, the observer is now ready to analyse the experience of perceiving the apple tree, but from the standpoint of its essential character. 'From our phenomenological standpoint we can and must put the question of essence: What is the "perceived as such"? What essential phases does it harbour in itself in its capacity as noema?' (1931: 260). The answer lies in describing faithfully, within the bracketed consciousness, what appears, what is given. This description, Husserl admits, sounds very similar to what might be described from the natural standpoint, but a radical difference must be underscored. 'The tree plain and simple, the thing in nature, is as different as it can be from this perceived tree as such, which as perceptual meaning belongs to the perception, and that inseparably' (1931: 260). This becomes evident if we consider that the tree could wither and die, burn away or change its chemical elements, but the meaning of the particular perception – one of full blossom with appreciation – cannot wither and die, burn away or change its chemical composition. 'It has no chemical elements, no forces, no real properties' (1931: 261).

We discover from an analysis of conscious acts, such as Husserl's apple tree, their essence or 'Eidos', which cannot be derived from an analysis of the tree from the natural standpoint, but only seen or intuited within conscious acts. As Quentin Lauer notes (1965: 23), 'the essence of whatever is is ultimately contained ideally and objectively in the very subjective acts whereby what is is present to consciousness'. Discovering the Eidos, therefore, results from the pure immanence of the experience within consciousness. This ensures that within phenomenological reflection only that which actually appears within conscious acts is subject to analysis. Roman Ingarden (1975: 70) claims that by this fundamental distinction between the natural and phenomenological standpoints, Husserl obtained his full and pure transcendental idealism:

> What is real is nothing but a constituted noematic unity (individual) of a special kind of sense which in its being and quality (*Sosein*) results from a set of experiences of a special kind and is quite impossible without them. Entities of this kind exist only for the pure transcendental ego which experiences such a set of perceptions. The existence of what is perceived (of the perceived as such) is nothing 'in itself' (*an sich*) but only something 'for somebody', for the experiencing ego. (Ingarden, 1974: 70)

The Eidos

As we have seen, Husserl's aim was to establish a fully scientific method in the sense of discovering necessary, as opposed to contingent, laws on which to base knowledge. Thus far in his argument, we have seen that this can occur only within the phenomenological reduction and under an analysis of immanent consciousness which enables the subjective observer to intuit the essence of conscious acts. Phenomenology as a science of consciousness, however, must go beyond individual conscious acts in order to provide a foundation for consciousness in general. In other words, the eidos or essence of particular conscious acts must in some sense be capable of universalization. To achieve this, Husserl employed the eidetic intuition or eidetic seeing (*Wesenerschauung*) whereby, as Dermot Moran (2000: 134) notes, 'a singular experience, appropriately regarded, could yield absolutely evident insight and *universal* truth' (emphasis his). The process involves seeing in an individual perception, its essence or universal quality. In his *Cartesian Meditations*, the original text of which was prepared by Husserl in 1928, and subsequently first translated into English in 1950, Husserl dealt at length with 'the method of eidetic description', which he portrayed as pervading 'the whole phenomenological method' (1977: 69). Eidetic description, otherwise referred to as eidetic seeing or intuition, applies for Husserl to all intentional acts: 'perception, retention, recollection, declaration, liking something, striving for it, and so forth' (1977: 70). For example, Husserl says, he is looking at the table in front of him. As he perceives the table, he is free to change it imaginatively by assigning it a different colour or shape, as he says, 'quite arbitrarily, keeping identical only its perceptual appearing'. The table thus perceived becomes pure possibility, one among many possibilities, irrespective of the actual existing table. 'We, so to speak, shift the actual perception into the realm of non-actualities, the realm of the as-if, which supplies us with "pure" possibilities, pure of everything that restricts to this fact or to any fact whatever' (1977: 70). By removing perceptions from facts, it becomes the 'pure "*eidos*" perception, whose "*ideal*" extension is made up of all ideally possible perceptions' (emphasis his) (1977: 70). The essence thus is not derived from actual perceptions as such (of the table, in Husserl's example) but just as plausibly from an imagined or remembered table. Dermot Moran argues that for Husserl, therefore, 'the science of essences has nothing to do with actual existence, but moves in the sphere of pure possibilities' (2000: 135). David Bell summarizes Husserl's meaning as follows:

> We take either a perception or an imaginative presentation of one instance of the universal we wish to see, and we vary that instance in imagination. We vary it quite freely – though always making sure that it remains, precisely an instance of the universal in question. (Bell, 1990: 194)

Husserl is aware that this method could be confused with a Platonic concept of forms where the essence is abstracted from its particular manifestation. All objects in the world, for Plato, imitate their essential and pure form. Any particular chair, for example, participates in the ideal form of 'chairness'. For Plato, the form is the real; the particular expressions are mere appearances, not real in the same sense as the ideal forms. Husserl rejected this kind of dualism by eliminating altogether the question of existence or reality of the forms. Rather, by analysing cognitions within the transcendentally reduced consciousness, he sought to establish how essences are derived from particular cognitions, quite distinct from both Platonic idealism and naturalistic empiricism.

The method of seeing essences can be clarified further by Husserl's investigation of adumbrations, which he first introduced in his *Logical Investigations*, and which can be defined as representing an outline, faintly indicating an object, foreshadowing what will or could appear or indicating an aspect of an object perceived. In any perception, as Roman Ingarden (1975: 21) notes, 'many different constitutive "layers" have to be distinguished'. Adumbrations illustrate how the layers are construed by the mind and how they lead in the direction of eidetic or essential seeing. An example of an adumbration is provided by Ingarden: 'A ball which is uniformly red is given to us when we perceive a certain set of adjacent shades of red and other colours with varying degrees of illumination' (1975: 22). The red ball is distinguished from other shades of red because our minds see into the essence of its redness, which is foreshadowed or adumbrated by the different shades of red surrounding it. Ingarden adds that the function of an adumbration 'increases in strength if the subject cognizes a certain series of adumbrations but not an isolated separated adumbration'. For example, an observer recognizes a rectangle by anticipating its shape on a trapeze or a rhomboid. Adumbrations provide examples of how the mind actively foreshadows the essence by distinguishing it or seeing it in the act of cognition. A rectangle does not 'exist' as such on a trapeze, but the mind intuits its essence on it. Ingarden explains: 'Changes taking place in the deep constitutive layer can be for e.g. differences of "sharpness", "expressiveness", saturation, a different degree of blurring going along with different degrees of concentration of attention' (1975: 23).

David Bell (1990: 194) calls Husserl's example (noted above) of intuiting the essence of a table by drawing attention to its unlimited or possible adumbrations, 'fatuous and utterly naïve', and indeed circular. This is because apprehending the essence in its adumbrations 'clearly depends upon our already having a very firm grasp of what is and what is not essential to perception'. Bell suggests, however, that Husserl may have intended by an eidetic science something much more credible, 'a specifically *philosophical* technique, aimed at making clearer or more distinct the universal "ideas" which we already possess and use in everyday life, albeit in a more or less indistinct and unclear way' (emphasis his) (1990: 195). If this is the case, Husserl's position does not entail a circular argument because the mind already possesses 'a tacit, pre-theoretical grasp of universal ideas'. This view appears consistent with Husserl's position, which, after all, aims at analysing conscious acts and their relationships to that which is cognized. The Kantian sense of universal categories or forms already present in the mind before experience would in this sense be affirmed without falling into a kind of dualism (either Platonic or Kantian) which either separates reality from appearances or divides acts of perception from the real thing-in-itself which in some sense 'causes' the perception.

By overcoming such dualisms and uniting subject with object in consciousness, Husserl thus achieves a universal science of cognition, or in Moran's (2000: 135) words, 'we rise from the temporally flowing, unified whole of a concrete, monadic, conscious life to grasping the meaning-essences that constitute valid scientific knowledge'. This latter point is fundamental for understanding Husserl's final claim to objectivity. Although eidetic seeing is not contingent on factual or empirical objects, all empirical or factual science depends on an eidetic intuition because eidetic sciences unveil essential or universal laws of cognition without which the natural sciences could not proceed.

Intersubjectivity

One final question remains: since the existence of other selves, like all other objects of cognition, has been placed in brackets or made subject to the *epoché*, how can we be certain that the eidetic seeing of the one who is performing particular acts of cognition conforms to the same intuition of essences for other selves? In other words, if the analysis of consciousness can occur only immanently in subjective perception, can we use the phenomenological reduction to ensure that knowledge is both objective and universal? This problem is summarized clearly by Quentin Lauer (1965: 55): 'An objectively valid cognition is one that is recognized as compelling not only for the subject who has the cognition but also for any possible subject who thinks properly (anything less would be meaningless in a rationalistic framework)'. Yet, under Husserl's structure of consciousness, 'a knowing subject must be subjectively constituted, and a known subject must be objectively constituted, and a known world must be commonly constituted' (1965: 56). This poses the problem of establishing a philosophical framework for analysing 'intersubjective' consciousness, which Lauer notes, Husserl had to address 'to rise above the level of solipsistic phenomenology' (1965: 56).

Husserl addressed this problem directly in his Fifth Meditation in *Cartesian Mediations* (1977: 89), where he asked: 'When I, the meditating I, reduce myself to my absolute transcendental ego by phenomenological epoché do I not become *solus ipse*; and do I not remain that, as long as I carry on a consistent self-explication under the name phenomenology?' Resolving the problem entailed in this question is critical for the entire phenomenological programme to succeed, since phenomenology ultimately aims, in Husserl's words, to 'solve the transcendental problems pertaining to the *Objective world*' (emphasis his) (1977: 89). Somehow, if such objectivity is to be achieved, the reality of other selves must be established. 'A path from the immanency of the ego to the transcendency of the Other' must be found.

This problem must be addressed by employing the very same phenomenological techniques which have established the possibility of transcendental consciousness and the seeing into essences. The 'Other', or alter ego, is experienced in the cognitions of the ego, or the one who perceives and experiences the Other. This Other is experienced in consciousness differently from other objects in nature as what Husserl calls ' "psychophysical" Objects' (1977: 91). They are experienced as objects, but also as subjects. My cognitions recognize them also as performing acts of cognition, including cognitions of me as an object and as a subject capable of cognitions. This means that a phenomenological analysis of others leads me to experience the world as intersubjective and not as locked within or behind my subjectivity alone. In the fullest sense, I cannot enter into the consciousness of the other and experience the other's cognitions. I am always the experiencing agent and I am the one finally who experiences someone else. However, through a process of what Husserl calls 'transcendental "Empathy" ', I can enter into the other's cognitions, in a way of speaking, and establish an intersubjective transcendental consciousness (1977: 92). Moreover, whenever he speaks of an objective world, Husserl means that the same world is there for every ego. It is, he says, a 'thereness for everyone'. Other selves fall within the category of objects with 'spiritual predicates', usually referring to other subjects who themselves intentionally constitute objects in the world. It is even possible in this context to introduce the idea of cultures, which represent objects with spiritual predicates through, for example, 'books, tools, works of any kind'. Husserl acknowledges, moreover, that these have different expressions within different cultural communities (1977: 92).

How does this differ from the natural attitude, where the distinction between self and others is assumed? What makes this distinctly phenomenological? Husserl suggests that in the area of intersubjectivity, we must perform '*a peculiar kind of epoché*'. Other selves are indeed put in brackets: 'we disregard all constitutional effects of intentionality relating immediately or mediately to other subjectivity'. In the transcendentally reduced consciousness, in what Husserl calls my 'peculiar owness, I experience myself as Other'. This occurs not on the plane of the natural standpoint, where self and others are differentiated and where I can abstract myself to the place of being ultimately alone, as if the plague had wiped out everyone but me. Husserl explains: 'What concerns us is, on the contrary, an essential structure, which is part of the all-embracing constitution in which the transcendental ego, as constituting an Objective world, lives his life'. We find through an analysis of the phenomenologically reduced transcendental ego, just the alter ego, a kind of mirror image of my ego. 'The "Other", according to his own constituted sense, points to me myself; the other is a "mirroring" of my own self' (1977: 93–4).

This intuition of the other in the self involves a special type of empathy, that which could be called interpolation, although Husserl does not employ the term. One definition of the verb 'to interpolate' is to insert or introduce between other things as in interleaving or interjecting. I intuit the reality of the Other from my own subjective experience and am able to establish the possibility of intersubjectivity thereby. Husserl suggests that this can be likened to a meditation on my own body which can perceive its essence using different sensory organs. For example, 'Touching kinesthetically, I perceive "with" my hands; seeing kinesthetically, I perceive also "with" my eyes' (1977: 97). Although such perceptions are immediate, to refer to them both as acts of perceiving involves an interpolation, inserting one type of perception for another, and thereby gaining understanding. Clearly, this does not revert to the natural attitude, because the analysis is occurring within the transcendentally reduced consciousness of the subject: 'It is all exclusively what is mine in my world-experience, pervading my world-experience through and through and likewise cohering unitarily in my intuition' (1977: 98). Nevertheless, through empathetic self-reflection, by interpolating the other into the self, by under-standing the alter ego in the light of the ego, the transcendental reduction produces a genuine intersubjectivity that establishes, in so far as is possible, the basis for positing universal essences.

Summary

From this analysis of Husserl's phenomenology, certain key points have emerged. These are: rejecting the natural attitude, employing the technique of *epoché*, allowing the objects of perception to present themselves in new and fresh ways in the state of the transcendentally reduced consciousness, seeing or intuiting essences and establishing objectivity on the basis of an empathetic rendering of intersubjectivity. Each of these key components in Husserl's philosophy will emerge again and again in various ways in the writings of the phenomenologists of religion we examine in later chapters. It is important at this stage to keep these main points firmly in mind, since they became, however indirectly, formative influences in the phenomenology of religion.

We have seen in this chapter that the natural standpoint or the natural attitude for Husserl referred not primarily to a naïve tendency to accept the world as it seems to be given in common sense, but more fundamentally to a tendency within the natural sciences to explain everything, including consciousness, in terms of physical manifestations. A chief culprit in this regard for Husserl was psychology because it reduced the mind to just another species within the natural world that could be studied empirically. Husserl wanted to give priority to a philosophical analysis of consciousness as the basis on which any natural science could be conducted. The reaction against naturalism, therefore, needs to be underscored as one of the primary motivating factors beneath Husserl's transcendental phenomenology.

His method for conducting an enquiry into consciousness began with the technique of bracketing out the natural world, not considering it, suspending all judgements about it. He called this technique, the *epoché*, derived from the Greek term to hold back, a term he used synonymously with the transcendental reduction. By putting the natural standpoint into brackets, Husserl believed a new and direct access to consciousness could be gained. Although parallels between what he called 'factuality', the empirical analysis of objects in the world, and the phenomenological analysis of consciousness could be drawn, by employing the *epoché*, he maintained always a sharp distinction between the two. Phenomenological analysis enquired into conscious acts and was not concerned with the question of existence. After the transcendental reduction had been performed, the subject was free to examine the interaction between the act of knowing and that which is known, the *noesis* and *noema*. This analysis uncovered the way consciousness intentionally apprehends not only objects in the world, but recollections, longings, perspectives, in short the pure phenomena of the experiencing subject. The world becomes as it is because the knowing subject comprehends the variations and constitutive factors of perception. What results ultimately is a seeing into the essential nature of objects apprehended in conscious acts, the eidetic vision or intuition. Essences are not ideals in a Platonic sense, nor are they mere fantasies. They occur through the analysis of consciousness in which the subject and object interact immanently. That essences are seen in the mind is supported by the way the mind foreshadows phenomena, how it sees the outline and fills in the parts into a spatio-temporal whole. The ever deepening layers of perception include not only direct observations and their adumbrations, but 'spiritual' elements pointing towards human feeling, cultural constructs and intersubjective consciousness. The idea of intersubjectivity is necessary to establish that the eidetic intuition provides an objective description of essential characteristics of the objects that present themselves in consciousness. In a final sense, consciousness always remains individual, the experience of the subjective ego. No subject can see literally with the vision of another subject. By employing empathy, however, a sense can be built up of a common or intersubjective understanding. When empathy is accompanied by an interpolation, similar to how my various organic senses experience common perceptions, an immanent, shared consciousness becomes possible.

In all of this, we must remind ourselves that Husserl was developing a method for knowing and thus that phenomenology remains through and through an approach to the way knowledge is gained. Husserl did not attempt to supply the content of knowledge. His contribution remains largely in the area of epistemology, how we know, rather than in other branches of philosophy, such as metaphysics, ethics or aesthetics. Nevertheless, he was intent on avoiding the naturalistic reduction, which makes consciousness nothing other than a part of nature. By elevating thought to the primordial place, he insisted that no type of knowledge, empirical or logical, synthetic or analytic, could proceed without a clearly defined and rigorous analysis of consciousness. In these ways, Husserl claimed to have built up the foundations for knowledge itself, or as Quentin Lauer (1965: 23) observes, Husserl made phenomenology into the ' "science of science", since it alone investigates that which all other sciences simply take for granted (or ignore) – the very essence of their own objects'.

References

Bell, D. (1990), *Husserl* (London and New York: Routledge).

Berkeley, G. (1998) [1710], *A Treatise Concerning the Principles of Human Knowledge* (Oxford: Oxford University Press).

Capps, W. H. (1995), *Religious Studies: The Making of a Discipline* (Minneapolis: Fortress Press).

Descartes, R. (1912) [1637], *A Discourse on Method* (translated by J. Veitch and with an introduction by A. D. Lindsay) (London and New York: Everyman's Library).

Descartes, R. (1931) [1911], *The Philosophical Works of Descartes, vols. I and II* (rendered into English by E. S. Haldane and G. R. T. Ross) (Cambridge: Cambridge Univ. Press).

Descartes, R. (1989), *Meditations I and II*, in J. E. White, *Introduction to Philosophy* (St Paul, Minnesota: West Publishing Company).

Flood, G. (1999), *Beyond Phenomenology. Rethinking the Study of Religion* (London and New York: Cassell).

Heidegger, M. (1962), *Being and Time* (translated by J. Macquarrie and E. Robinson) (Oxford: Basil Blackwell).

Husserl, E. (1931), *Ideas. General Introduction to Pure Phenomenology* (translated by W. R. B. Gibson) (London: George Allen and Unwin Ltd).

Husserl, E. (1965), *Phenomenology and the Crisis of Philosophy* (translated with an introduction by Q. Lauer) (New York: Harper and Row).

Husserl, E. (1970) [1913–1922], *Logical Investigations. Volumes I and II* (translated by J. N. Findlay) (London: Routledge and Kegan Paul).

Husserl, E. (1977), *Cartesian Meditations. An Introduction to Phenomenology* (translated by D. Cairns) (The Hague: Martinus Nijhoff).

Ingarden, R. (1975), *On the Motives which Led Husserl to Transcendental Idealism* (translated by A. Hannibalsson) (The Hague: Martinus Nijhoff).

James, G. A. (1995), *Interpreting Religion. The Phenomenological Approaches of Pierre Daniel Chantepie de la Saussaye, W. Brede Kristensen, and Gerardus van der Leeuw* (Washington, DC: The Catholic University of America Press).

Kant, I. (1929) [1787], *Immanual Kant's Critique of Pure Reason* (translated by N. K. Smith) (London: Macmillan and Company, 2nd edn).

Kant, I. (1977) [1783], *Prolegomena to Any Future Metaphysics That Will Be Able*

to *Come Forward as Science* (the Paul Carus translation extensively revisec
J. W. Ellington) (Indianapolis, Indiana: Hackett Publishing Company, Inc

Kenny, A. (ed.) (1994), *The Oxford Illustrated History of Western Philosop*, (Oxford: Oxford University Press).

King, R. (1999), *Orientalism and Religion. Postcolonial Theory, India and 'The Mystic East'* (London and New York: Routledge).

Kolakowski, L. (1975), *Husserl and the Search for Certitude* (New Haven and London: Yale University Press).

Lauer, Q. (1965), 'Introduction', in E. Husserl, *Phenomenology and the Crisis of Science* (translated with an introduction by Q. Lauer) (New York: Harper and Row), pp. 1–68.

Locke, J. (1924) [1690], *An Essay Concerning Human Understanding* (abridged and edited by A. S. Pringle-Pattison) (Oxford: Clarendon Press).

McCutcheon, R. T. (ed.) (1999), *The Insider/Outsider Problem in the Study of Religion. A Reader* (London and New York: Cassell).

Moran, D. (2000), *Introduction to Phenomenology* (London and New York: Routledge).

Popkin, R. H. and Stroll, A. (1986), *Philosophy: Made Simple* (London: Heinemann, 2nd edn).

Scruton, R. (2000), *On Husserl* (Belmont, California: Wadsworth).

Sharpe, E. J. (1986), *Comparative Religion: A History* (London: Duckworth, 2nd edn).

Chapter 2

The Universal Experience of Religion in Ritschlian Theology

We saw in the last chapter that René Descartes altered radically the direction of Western philosophy by establishing all knowledge on the one foundational and indubitable fact of the self-subsisting ego. From the existence of his own self, Descartes was able to reconstruct the existence of other selves, the external world and God. In a similar way, theologians beginning in the eighteenth century sought to isolate the irreducibly religious factor in human experience, that without which religion could not be said to exist. Walter Capps (1995: 5) calls this the attempt to 'uncover and isolate a *sine qua non* ... about the nature of religion'. Immanuel Kant's 'practical reason' played an important role for nineteenth-century theologians who identified the singular religious instinct with the human sense of duty or, more broadly, morality. This interpretation can be found particularly in the writings of the prominent nineteenth-century German theologian, Albrecht Ritschl (1822–1889). Another explanation of the irreducible element in religion can be traced to a 'feeling' for the transcendent, often displayed in a quest for a satisfying experience of an infinite reality. The most important exponent of this view was the German theologian Friederich Schleiermacher (1768–1834), whose influence was felt widely throughout nineteenth and twentieth-century liberal Protestant theology, but whose power was particularly strong among members of the so-called 'Ritschlian school', notably Wilhelm Herrmann and Julius Kaftan. Rudolf Otto's influential interpretation of the core of religion as the transcendental 'numinous', which produces a universal 'creature-feeling' among humans, must also be understood as falling within the broad line of thinking associated with the Ritschlians. A related interpretation of the essence of religion was developed by A. G. Hogg (1876–1954), who applied the thinking of Ritschl, Herrmann and Kaftan within a context of Hindu–Christian encounter during the early part of the twentieth century. Hogg posited a fundamental distinction between the universal core of religion, which he called faith, and faiths (or beliefs) through which the universal experience is apprehended and expressed.

In this chapter, the thought of three Ritschlians, Albrecht Ritschl himself, Wilhelm Herrmann and A. G. Hogg, will be examined as formative influences necessary for grasping in later chapters the ideas of key phenomenologists of religion. I will also argue that the German theologian Rudolf Otto, whose direct influence on phenomenological interpretations of the essence of religion is undoubted, should not be discussed in isolation from Ritschlian theology. It should be noted at this point, that the influence of Ritschl and his followers has not been acknowledged widely in prior literature. Capps (1995: 18–20) credits Ritschl with 'up-dating' the Kantian position for nineteenth-century thinking, and Sharpe (1986: 126) calls the influence of Ritschl 'considerable', but Ritschl receives no mention at all in some of the most important recent discussions of religious studies (McCutcheon, 1999; James, 1995; Braun and McCutcheon, 2000; Flood, 1999; King 1999, Fitzgerald, 2000). By attempting to establish the school of Ritschl as a fundamental formative theological influence for later phenomenologies of religion, therefore, I am seeking to rectify an important omission in histories of religious studies.

Before analysing the core of religion as understood by Ritschl, Herrmann, Otto and Hogg, it will be important to outline the intellectual background against which they constructed their theologies. In particular, the moral philosophy of Immanent Kant, the romantic idealism of Friederich Schleiermacher and the grand idealistic system of the great German philosopher G. W. F. Hegel need to be examined.

Kant and the *Critique of Practical Reason*

In his *Critique of Pure Reason*, Kant believed that he had demonstrated the limits of human knowledge. The categories of the mind establish the basis of knowledge within what Kant called the phenomenal world, by which he meant the empirical world perceived through the senses. The mind structures the phenomenal world by making causal connections, relating events in time and space and inserting order. As we have noted, what lies beyond the categories of the mind, that which causes the sensations, the so-called 'thing-in-itself', cannot be known. All questions of metaphysics, such as the substance or essence of what the mind apprehends, the nature and purpose of the cosmos, the existence of God and life after death, belong to another realm of knowing, the noumenal world. Knowledge of the noumenal world cannot be attained by logic, reason or experience, or what Kant calls pure reason. Neither analytic nor synthetic statements offer any insight into their truth or falsehood.

To demonstrate this, Kant (1977: 80) analysed what he called four fundamental and universal cosmological ideas to which equally true theses and antitheses can be developed.

1. Thesis: The world has, as to time and space, a beginning.
   Antithesis: The world is, as to time and space, infinite.
2. Thesis: Everything in the world is constituted out of the simple.
   Antithesis: There is nothing simple, but everything is composite.
3. Thesis: There are in the world causes through freedom.
   Antithesis: There is no freedom, but all is nature.

4. Thesis: In a series of world-causes, there is some necessary being.
   Antithesis: There is nothing necessary in the world, but in this series all is
   contingent.

The first idea contrasts the two theories that either the world is limited in time
and space or that it has no limits. The second posits that our perceptions of the
world are either composed of simple and discrete impressions striking the mind
or they are complex, amalgamations of parts. The third idea poses the issue of
freedom: do things happen by chance or is everything determined? The fourth
idea sets the belief in the existence of God as the one necessary being in the
universe on which everything else depends against the assertion that there is
nothing at all on which everything else depends. Kant refers to the first two ideas
as mathematical and the latter two as theological and hence transcendental. In a
quite detailed argument, he then proceeds to show how by the use of reason each
thesis and antithesis can be shown to be true and false at the same time: 'The
thesis, as well as the antithesis, can be shown by equally clear, evident, and
irresistible proofs', thereby exposing all metaphysical propositions as violating
the most fundamental principles of reason (1977: 81). 'Contradictory propositions cannot both be false, unless the concept lying at the ground of both of
them is self-contradictory' (1977: 82). For this reason, Kant refers to the
cosmological propositions as 'antinomies', defined by Popkin and Stroll (1986:
166) as 'conclusions which can be both proven and disproven', and concludes
that if we think such ideas actually enlarge our knowledge of experience, 'we
suffer from a mere misunderstanding in our estimate of the proper application of
our reason and of its principles' (Kant, 1977: 90; see also Scruton, 1982: 48–51).

If pure reason cannot provide answers to these most basic questions of human
meaning and destiny, another type of reason must be found that can address
them satisfactorily. Kant calls this alternative way of knowing 'practical' reason,
which, unlike the analytic and synthetic statements of pure reason, provides a
moral justification for asserting metaphysical truths. In Kant's view, for humans
to behave morally, they must believe and hence act as if they are free; they must
believe and act as if they will be able to achieve moral perfection in an infinite
progression; and they must believe and act as if God exists as the ultimate moral
agent who in himself both originates and manifests perfection. None of these
beliefs can be demonstrated by pure reason; they provide nonetheless the
necessary prerequisites for morality.

Despite its 'as if' character, practical reason is fully rational and operates
according to its own laws. It is founded in the first instance on what Kant (1909:
101) refers to as 'the grounds of determination of will'. Just as the critique of
pure reason demonstrated the need for categories of the mind that are both *a
priori* and synthetic, so does the critique of practical reason discover the
constitutive rules of thinking necessary for determining moral behaviour. In both
pure and practical reason, therefore, certain formal categories are required: in the
case of pure reason, these are needed to assure the universality of knowledge; in
the case of practical reason, these are needed to assure the universality of moral

judgements. The will thus performs the same role for moral judgements as do cognitions for consciousness. The will is subject to universally binding categories, just as apprehensions of the world in the thinking process are controlled by the universal categories of causality, relation and order. The moral will is determined by the maxims called by Kant the 'categorical imperatives', which are neither hypothetical nor contingent on any natural cause, but which depend entirely on the postulates of practical reason: immortality, freedom and God.

For Kant, moreover, morality is not expressed in specific laws or regulations, such as a prohibition against murder, but on universal principles that determine what constitutes a moral action within particular situations. The categorical imperative thus is purely formal; it defines the universal principle on which the moral subject must make decisions. In his *Fundamental Principles of the Metaphysics of Morals*, first published in 1785, Kant summarizes the categorical imperative under the principle that for an action to be judged moral it must be capable of being willed universally, or as Popkin and Stroll (1986: 45) put it: 'Every action must be judged in the light of how it would appear if it were to be a universal code of behaviour'. Kant (1909: 38) explains: 'There is ... but one categorical imperative, namely this: act only on that maxim whereby thou canst at the same time will that it should become a universal law'. Another formulation of the same principle is expressed by Kant (1909: 47) from a different perspective: 'So act as to treat humanity, whether in thine own person or in that of any other, in every case as an end withal, never as a means only'.

That the categorical imperative represents a formal pattern for behaviour without content can be demonstrated by a simple example. If a person under certain circumstances thinks that telling a lie could be deemed appropriate, the person would need to subject the action to the universalizing principle as dictated by the categorical imperative. Could lying be legislated as morally binding for every human being as a duty? The answer quite obviously is that it could not. Truth telling, on the other hand, could be willed as binding on every moral subject as a duty to be followed. By acting on the universalizing principle, lying clearly is shown to be immoral and truth telling moral. The secondary formulation of the categorical imperative confirms this. To lie to another person is to use that person as a means to an end, whereas truth telling possesses its own intrinsic value and is not undertaken for personal gain. This confirms that telling the truth can be regarded as a universally binding duty, whereas lying could never fall into such a category. In this way, a formal maxim quite clearly determines moral behaviour in specific instances.

Once the categorical imperative is understood as binding on the will, it is clear that one is duty bound to follow its directives and, moreover, it must be capable of being legislated for every moral subject. In this sense, Kant regarded the principle of universality as compelling every moral act to be undertaken as if the actor were also a lawmaker and could make the same act compulsory for every other moral agent. This is why Kant's ethical theory is called deontological, rooted in duty for duty's sake. The moral agent, once aware of the dutiful action, must follow it irrespective of subjective feelings about the action, including personal desire, disposition for or against the action, or whether or not it is likely to produce individual happiness. Subjective feelings about the act do not affect the duty to perform it nor are consequences entailed in determining its morally binding character. In his *Critique of Practical Reason* Kant explains:

> For the will of any finite rational being, [the moral law] ... is a law of duty, of moral constraint, and of the determination of its actions by respect for the law and reverence for its duty. No other subjective principle must be assumed as a motive, else while the action might chance to be such as the law prescribes, yet it does not proceed from duty, the intention, which is the thing properly in question in this legislation, is not moral. (Kant, 1909: 175)

To exclude consequences in the determination of a moral act does not eliminate the idea that the moral will attains to a goal, what Kant calls the 'highest good' or *summum bonum*, which he defined as 'the perfect accordance of the mind with the moral law' (1909: 218). The moral agent, in other words, seeks perfection, a total conformity between duty and actions. Since perfection is not possible for finite beings living in this world, practical reason must postulate 'an endless progress to that perfect fitness' which in turn entails the necessity of an 'endless duration of the existence and personality of the ... rational being' (1909: 218). For quite practical reasons, therefore, Kant argues that the highest good 'is only possible on the supposition of the immortality of the soul' (1909: 219). Note, this is not regarded as some reward for behaving in a moral way, but as the necessary condition for finite beings in this world to progress towards the perfect moral will. Without such a condition, morality in the purest sense would not be possible or conceivable. The first postulate of practical reason on which the categorical imperative depends thus is the immortality of the human soul.

For moral agents to act in genuinely moral ways, their wills must be free to choose to follow duty or to reject their moral obligation. A world in which the will was determined to act in various ways would remove individual responsibility for actions. Without the assumption of freedom, no agent would be capable of causing anything, since moral causes would be seen to exist outside the control of the individual. Although the actual freedom of the individual cannot be demonstrated using pure theoretical reason, it is clear that the freedom of the will must be postulated as a necessary condition for the moral law. The second postulate of practical reason thus is the presupposition that humans are free to act morally or immorally.

Although performing moral actions is not done to obtain personal happiness in the finite world, in an ultimate sense, there ought to occur a perfect correspondence between morality and happiness. This can be achieved only if we postulate the existence of God, in Kant's words, 'as the necessary condition of the possibility of the *summum bonum*' (1909: 221), since God himself is the perfect moral legislator. In this world, moral action taken to achieve personal happiness is opportunistic and violates the disinterested commitment to duty, but by postulating the existence of one on whom all things depend and who is the cause of the natural world, 'the exact harmony of happiness with morality' is assured

(1909: 221). Since it is the duty of the moral agent to will always to act morally and thereby to seek in immortality to achieve the highest good, this must in fact be possible. This can be guaranteed only if there exists a highest original good, God, whose existence unites duty with perfection resulting in the absolute coincidence between morality and happiness. The existence of God thus becomes the third necessary postulate of practical reason.

By the end of Kant's *Critique of Practical Reason*, what he has taken away from us in pure reason, metaphysical and theological certainty, he has reinstated through practical reason. We should keep in mind that by restoring immortality, freedom and God, Kant has not made these actual. The 'thing-in-itself' always remains unknown and unknowable. By giving us a different kind of knowledge, an alternative pathway to truth, Kant makes it clear that humans cannot live practically without positing the basis for the moral life. Analytical proofs for immortality, freedom and God have been destroyed. Morality, however, ensures their absolute necessity, or as Kant famously put it in his preface to the second edition of *The Critique of Pure Reason* (1929 [1787]: 29 [BXXX]): 'I have found it necessary to deny knowledge in order to make room for faith'.

Friederich Schleiermacher: Religion as feeling

In 1799 Friederich Schleiermacher published *On Religion: Speeches to Its Cultured Despisers*, in which, as Richard Crouter (Schleiermacher, 1988: 1) notes, Schleiermacher reflected 'the tensions between the religious thought of the Enlightenment and romanticism'. Like many of the thinkers we have reviewed thus far, Schleiermacher spent many years studying Kant's philosophy, particularly his moral thought. In an unpublished essay entitled 'On the Highest Good', written in 1789, Schleiermacher (Schleiermacher, 1988: 20–2) rejected what he saw as Kant's dualism between the phenomenal and noumenal worlds, between cognitions and moral judgements. Human desire cannot be subordinated to reason; the two are conjoined in the human personality. In *On Religion*, Schleiermacher takes this argument further, in Crouter's words (Schleiermacher, 1988: 30), by attacking vehemently 'Kant's teaching that an autonomous moral law is the locus of the supersensible world within human experience'. By rejecting the basis of Kant's moral philosophy, Schleiermacher opposed the fundamental conviction of Enlightenment thinkers, beginning with Descartes, that reason alone prescribes the limits to and the possibility for all human knowledge, both theoretical and practical.

By the time Schleiermacher wrote *On Religion*, the influence of the German romantic school had become widespread in the arts, literature and philosophy, through such figures as J. H. Herder (1744–1803), a philosopher and historian concerned with the 'folk spirit', G. E. Lessing (1729–1781), a literary critic who advocated the 'dramatist's vision of life', and perhaps most important for understanding Schleiermacher, Friedrich Schlegel (1772–1829), a poet and philosopher who defined romantic literature as a 'progressive universal poetry' (Benét, 1977). The romantics emphasized feeling, inner vision and the human

40

spirit in contrast to the 'cold rationalism' of the Enlightenment thinkers. H. R. Mackintosh (1937: 33) defines romanticism as 'an impassioned return to natural instincts, to life, to freedom, to individual predilection to the spontaneity of the creative fancy', and contends that among the eighteenth and early nineteenth-century German romantics, Schleiermacher 'made himself the champion in the religious field'.

Crouter suggests that *On Religion* must be understood primarily as a polemical text, one which attacks the Enlightenment views of religion in a way that employs irony leading often to 'one-sided claims' (Schleiermacher, 1988: 42). It is divided into five speeches covering a wide range of subjects, including the essence of religion, the social aspects of religion and the plurality of religions, all written in a style that is both argumentative and persuasive. Nevertheless, the interpretation of religion which Schleiermacher expounds in *On Religion* performed a positive, even groundbreaking, role not only within the later development of liberal Protestant theology, but in a foundational way also for the phenomenology of religion. In Capps' view, if Kant understood the *sine qua non* of religion as the moral understanding, Schleiermacher defined it 'as a kind or quality of feeling ... something akin to deep sensitivity' (1995: 13). In his second speech against the so-called 'despisers' of religion, Schleiermacher contends that the essence of religion is found neither in metaphysics nor morality, what he calls 'thinking' and 'acting', but in 'intuition and feeling' (1988: 102). Intuition is not defined in a precise way, but represents the human capacity to grasp the universe and to be filled by its 'immediate influences in childlike passivity'. As such, religion stands alongside speculation and practice, metaphysics and morals, as a third way of knowing and experiencing the world, which in no way is 'slighter in worth and splendour' than the other two. Intuition defines the principal religious method for gaining knowledge. Schleiermacher admits that 'it is the hinge of my whole speech' (1988: 104). Intuition depends on an originating force, that which is intuited affecting the one doing the intuiting. In religion, the originating force is the universe, which 'exists in uninterrupted activity and reveals itself to us every moment' (1988: 105). The history of religions is comprised of humans putting into words, beliefs and rituals their core intuition of the universe. Thus, when what Schleiermacher calls 'the ancients' faced human limitations in space and time, they intuited the universe as unlimited and regarded it as the 'work and reign of an omnipresent being'. Similarly, when they felt their dependence on something greater than themselves, they named it their god 'and built its own temple to it', thereby expressing through religion a 'connection to an infinite totality' (1988: 105). Intuition begins with individual feelings of finitude and dependency, feelings which Schleiermacher says 'must accompany everyone who really has religion' (1988: 107). When an individual experiences finitude and senses a total dependency on the universe, that individual becomes 'a new priest, a new mediator' and in this sense bears testimony to the common human need for religion, not as a system of beliefs or dogmas, but as that which resides at the core of the human spirit. Because intuition occurs within an individual's apprehension of the infinite, everything in the universe potentially becomes a source for religious experience, or as Schleiermacher puts it, 'Everything that can be is for it [religion] a true indispensable image of the infinite' (1988: 109).

In a famous passage, called the 'love scene' (1988: 112–13), Schleiermacher likens the religious sentiment to physical love. Religious feeling is both 'indescribable' and 'fleeting', 'as modest and delicate as a maiden's kiss, as holy and fruitful as a nuptial embrace'. The religious experience transcends human love. 'I lie on the bosom of the infinite world. At this moment I am its soul, for I feel all its powers and its infinite life as my separate form'. When this 'holy embrace' is released, like a lover, the one who has experienced such profound religious feelings, surveys the beloved and 'it mirrors itself in my open soul like the image of the vanishing beloved in the awakened eye of a youth'. This intensity of religious feeling can be known only to those who themselves have experienced it. It cannot be fabricated, nor can those who have not known such experiences claim to understand the essence of religion. If a person claims to understand religion and even to believe devoutly in it, but has not such an experience, Schleiermacher proclaims, 'then prevail upon yourself and me no further, for surely it is not so; your soul has never conceived'.

In these passages, we find the main tenets central to Schleiermacher's core understanding of religion. The source of religion is found in the human feeling of dependency, the vivid sense of experiencing oneself as a finite being face to face with an infinite universe. As these feelings express themselves in religious responses, a sense of deep satisfaction results, deeper than that experienced by lovers, between the individual and the universe where finitude is swallowed up in infinity. The dependence which is felt and experienced in religion thus achieves its complete satisfaction when the one who is dependent unites with that which is unconditioned, or as Schleiermacher puts it at the conclusion of his second speech: 'To be one with the infinite in the midst of the finite and to be eternal in a moment, that is the immortality of religion' (1988: 140).

By the early nineteenth century, the Kantian interpretation of religion as a necessary postulate for morality and Schleiermacher's romantic intuition of religion as a sense and feeling for the infinite stood side by side as parallel influences on the development of later nineteenth-century theology. Although Kant and Schleiermacher may appear to provide opposing and even antagonistic perspectives on religion, they in fact share a common starting point. Both find the core of religion in humanity, not in revelation. For Kant, religion originates in the rationality necessary for moral judgements; for Schleiermacher, its source lies in the human intuition of the infinite. Walter Capps (1995: 10) notes that after Descartes the controlling assumption in Western philosophy was 'that religion is rooted not in divine revelation or some form of ecclesiastical authority, but in something eminently natural and human'. It was the task of both Enlightenment and Romantic thinkers to determine 'that particular human power, faculty, or capacity to which religion belongs intrinsically'.

Crouter (Schleiermacher, 1988: 104) observes that both Kant and Schleiermacher understood intuition as referring to the mind's grasp of objects which are given in perception. In Schleiermacher's case, the universe prompts the religious response; for Kant, the 'thing-in-itself' causes the intuition, although we can never know what it actually is in itself. Schleiermacher accepted the limits imposed by Kant's pure reason, but, as we have seen, posited a third way of knowing, that of feeling, as a truer guide to understanding the essence of religion than Kant's postulates of practical reason. Nevertheless, the differences between their positions should not be exaggerated. Kant's conclusion in the *Critique of Practical Reason* could even be said to resonate with Schleiermacher's vision of the infinite, where Kant (1909: 260) wrote: 'Two things fill the mind with ever new and increasing admiration and awe, the oftener and more steadily we reflect on them: the starry heavens above and the moral law within'. We will see in the remainder of this chapter that the theological school of Albrecht Ritschl was influenced by both Kant and Schleiermacher. Although distinct emphases will become evident as we consider the Ritschlians, they demonstrate nonetheless that a greater consonance than difference existed between Kant and Schleiermacher, who when considered together, clearly situated the origin of religion, its sustaining power and its confirmation squarely between the twin pillars of moral reason and religious sentiment.

Hegel, history and the Absolute Spirit

We have seen that following Descartes Western philosophy gradually moved towards a profound scepticism regarding metaphysical statements. By the early part of the nineteenth century, questions concerning the absolute nature of the world, its ultimate purpose and the goal of existence were addressed either through moral understanding or in a romantic sense of feeling for the infinite. Metaphysics, however, had an almost immediate revival following Kant through the thought of the great German philosopher of the nineteenth century, G. W. F. Hegel (1770–1831), whom the French phenomenologist, Marcel Merleau-Ponty, writing in 1964, credited as having originated 'all the great philosophical ideas of the past century – the philosophies of Marx and Nietzsche, phenomenology, German existentialism and psychoanalysis' (cited by Morris, 1987: 5). H. R. Mackintosh (1937: 101) notes that Hegel and Schleiermacher frequently are contrasted, since Hegel's philosophy can be regarded as entirely speculative and derived from pure thought, whereas Schleiermacher 'went near to interpreting piety as no more than an emotional state of mind'. Nevertheless, Hegel and Schleiermacher shared a similar interpretation of religion, which, also like Kant, reduced religion to one singular core reality, that without which religion could not be said to exist. For Kant, this was morality; for Schleiermacher, feeling; for Hegel, in Mackintosh's words, it was 'the Infinite Spirit rising to consciousness in the finite' (1937: 101).

Hegel's philosophy has been called notoriously complex, and hence difficult to summarize. Popkin and Stroll (1986: 147) go so far as to claim that Hegel became the chief 'villain' for many twentieth-century philosophers due to what they call 'the abstruse complexities in his writings'. This same point has been made by Brian Morris, who in his *Anthropological Studies in Religion: An*

Introduction, devotes an entire chapter to the importance of Hegelian metaphysics for the development of the Western social sciences. Morris (1987: 6) observes: 'Hegel's writings have had an enormous impact on the modern world, yet almost everyone has acknowledged that they are obscure and difficult to understand'. This may have been because Hegel himself clearly was a romantic, who had been influenced by many of the same figures in the German romantic movement that were so important for Schleiermacher, but unlike Schleiermacher, Hegel attempted to translate the romantic vision into what Morris calls a 'rational understanding of a cosmic process' (1987: 6). For our purposes, it is important to outline the main tenets central to the Hegelian system, since the Ritschlian school of theology must be understood in the light not only of Kant and Schleiermacher, but also from the perspective of Hegel's interpretation of history.

Roger Scruton (1984: 164) refers to the philosophical tradition embodied by Hegel as 'the spiritual phenomenon which we know as the romantic movement', a tradition Scruton closely associates with the thinking of the late eighteenth-century philosopher, J. G. Fichte (1762–1814). Of particular importance for understanding Hegel is Fichte's division of philosophy into two types: idealism and dogmatism. The idealist seeks the explanation of the world in thought or intelligence, whereas the dogmatist endeavours to find it in Kant's 'thing-in-itself', the futility of which Kant had already demonstrated. Since intelligence points in the true direction for interpreting the world, it follows, according to Fichte, that the proper starting point for philosophy resides in the conscious subject or ego, a fact demonstrated convincingly earlier by Descartes. As such, Fichte defined self-knowledge, or what he called 'intuiting' the self, in Scruton's (1984: 162) words, as a ' "positing", or creating for itself, a world of objects through which it understands its own activity'. Once the self has been conceived, logic demands that there also be posited a not-self, and thus self-consciousness establishes immediately a dichotomy between the self and the not-self, the subject and the object, which could be advanced as oppositions – a thesis and an antithesis. Popkin and Stroll (1986: 146–7) put it this way: 'By the time one develops any awareness of consciousness of the meaning of [the] nature of one's experience, it is already in the form of object (the outside world) and subject (oneself)'. This occurs because once the self has been made the object of self-consciousness, it is transformed into an object and becomes in that sense the non-self of which the self as subject is aware. A synthesis is achieved out of this dichotomy for Fichte through a third causal agent or mental source, which created thought in the first place and makes it comprehensible. The process of consciousness whereby the self becomes aware of itself and at the same time of the non-self has been effected by a transpersonal mind or consciousness. This means that reality is not simply the thought processes of an individual mind and hence solipsistic, but that consciousness itself demands the necessity of a transcendent mind, what Popkin and Stroll call 'the source of all aspects of the universe' (1986: 147).

Hegel's philosophical system, as outlined in his famous book, *The*

Phenomenology of Mind (1910) [1807], sometimes translated in English as *The Phenomenology of Spirit* (1977), builds on two aspects of Fichte's idealism: the analytical processes of identifying a thesis and its corresponding antithesis which then become synthesized, and on Fichte's assertion of a transcendental mind beyond all individual minds. Fichte thus inspired the celebrated Hegelian dialectic through which Hegel interpreted the movement or progression within history of all events, ideas and experiences towards their ultimate culmination in the Absolute or Infinite Spirit. Popkin and Stroll summarize this key element or central assertion within Hegel's metaphysics as follows: 'Everything in the universe can be understood only in terms of an objective or absolute mind which has been evolving throughout the world's history into a transcendent self-contained being' (1986: 147). Roger Scruton (1984: 178) offers a similar summary but he underscores how the process proceeds from the individual mind or spirit towards the Infinite: 'The real subject-matter of [Hegel's *The Phenomenology of Spirit*] is not the concrete, sceptical, solipsistic self, but the universal, affirmative spirit (*Geist*), whose progress towards realization in an objective world is something in which you and I may participate, but which we do not otherwise affect'.

The Hegelian dialectical method is outlined in detail in his *Science of Logic* (1969), published as three 'books' in 1812, 1813 and 1816, in which thesis and antithesis are identified as originating in the very act of thinking. Thought processes require abstraction, the capacity to conceive in general, non-specific ways, or, at the very least, by removing the thought from that which is thought about. Hence, by abstracting, immediately an opposition is established between the general and what Hegel calls the 'determinate'. At the very core of consciousness, the mind becomes aware of a thesis (the abstraction) and an antithesis (the particular or determinate). Out of these a synthesis is created. The mind becomes conscious of both the abstract and determinate nature of reality and thereby in the synthesis constructs a more complete or 'truer' apprehension than existed before the antithetical relationship was recognized. For example, if I consider any object of my perception, such as the desk on which my computer is situated in space and which at this time I am facing, I obtain immediately in the perception two contrary ideas: the particular configuration of this desk, and the concept desk, in general. Desk, in other words, is an abstract concept, a generalization, but it is always applied in a specific case as referring immediately to this or that desk. In this way, generalizations are pitted against particular instances; the desk is a desk and not a desk at the same time. It is the abstract, non-specific 'deskness' and, at the same time, not the general but the concrete desk I am observing in this place at this time.

The central philosophical concept 'being' follows this same conceptual pattern. Being is an abstract, immediately apparent term, which, however, communicates nothing without reference to concrete or particular beings. When we use the concept abstractly to speak about being in general, even though we understand its meaning, it says nothing about any concrete being in space or time. This creates the dialectic at the core of being itself. My cat, which is curled at my feet as I type these words, possesses being in space and time; it is a determinate object and thereby possesses determinate being, although it participates in the general meaning of being in the abstract. Being and non-being, thus, construct a fundamental opposition at the core of conscious thought; they form a central thesis and antithesis. Like Fichte, Hegel argued that these oppositions can be synthesized only by understanding how one can be and not be at the same time; just as one could be the self and the non-self simultaneously, or be abstract and particular in the same moment. In a similar way, the individual consciousness, or what Hegel calls spirit, is constructed dialectically as it becomes aware of its relationship to the world and to other spirits. The individual responds to internal sensations like desire, longing and will, which inevitably conflict with desires, longings and wills belonging to other selves. Social conflicts result from such antithetical individual interests and can be overcome only through a synthesis of the opposing forces. In turn, the synthesis creates another thesis, which entails an antithesis, requiring a further synthesis, creating a pattern replicated ever onward in historical progression. It is important to note that the thesis and antithesis are depicted by Hegel not so much as forces in absolute opposition, but each as incomplete in itself, each capable of being fulfilled only through the synthesis. The logic of the Hegelian account of consciousness, which possesses an inherent antithesis, in this way, is translated into concrete historical processes.

The end of history occurs in the final synthesis through which the self-consciousness of the Absolute Spirit overcomes all prior antitheses. Popkin and Stroll call this 'complete self realization' by the Absolute Spirit characterized by an 'all-encompassing synthesis' in which all 'partial truths' become enveloped by 'one vast truth' (1986: 149). H. R. Mackintosh claims that the Hegelian system demonstrated that 'opposites are really moments or constituents in a living process of thought, that negation, followed by deeper affirmation, sounds the keynote of all development' (1937: 104). Every abstraction and particularity, each social conflict and historical process finds its culmination and ultimate fulfilment in the Absolute Spirit, leading Mackintosh to conclude: 'In this sense the famous aphorism holds true: "The rational is the real, and the real is the rational"'. Scruton (1984: 178–9) depicts Hegel's overriding theory of history as operating according to 'the dialectic of reason', which 'advances from pure, immediate being through all the determinations of being which in sum constitute reality, so to consummate itself in the absolute idea'. According to Scruton, Hegel posited 'this absolute idea as the whole of reality, the truth of the world, and God Himself' (1984: 150). Hegel must be seen, therefore, as a final climax to a way of thinking begun by Descartes and carried through Berkeley and Hume to Kant, in which the theory is finally established that only the mind is real. Popkin and Stroll summarize the significance of the Hegelian system by referring to this strategic location in the history of modern Western philosophy:

Hegel completely objectified thought and mind into the basic independent entity ... The Hegelian absolute mind becomes the real universe, manifesting itself outwardly as world history, and inwardly as the rational dialectical process, marching forward toward full self-realization. (Popkin and Stroll, 1986: 149)

Albrecht Ritschl and the theory of religious cognition

It was against the backdrop of late eighteenth-century and early nineteenth-century philosophy that Albrecht Ritschl constructed one of the most influential theological movements of the nineteenth century, which by the early twentieth century had become synonymous with liberal Protestantism. After completing his theological studies at Tübingen, Ritschl was appointed in 1846 as a lecturer in New Testament studies at Bonn, but by the end of the 1840s his interests had shifted to history, and particularly to reformation studies. In 1853 he was appointed as a Professor of Dogmatics in Bonn, a post he carried with him when he moved to Göttingen in 1864, where he stayed until his death in 1889. Throughout these years, in the words of Philip Hefner (Ritschl, 1972: 30), Ritschl brought together 'more creatively than any of his contemporaries the theological concerns of his age and its recent past' and was able 'to synthesize those concerns with his own categories in the formulation of a message persuasive to a great variety of his contemporaries'. After the First World War, Ritschl's influence waned substantially, and during the 1930s and 40s was completely overtaken by Barthian theology. In the last 30 years, a sort of Ritschlian revival has occurred, in which the value of Ritschl for contemporary theological discussions is being reappraised. This can be witnessed, for example, in a book edited by Darrell Jodock (1995), in which Ritschl's thought is applied to contemporary understandings of history, community and even to the relationship between science and religion. More than a decade earlier, James Richmond (1978: 13) published *Ritschl: A Reappraisal* in which he referred to 'the Ritschl renaissance'.

In his discussion of Ritschl's importance for the current science and theology debate, Richard Busse (1995: 166) asks if Ritschl can be regarded as an enduringly distinctive thinker. He claims that in one sense he cannot, since he was 'a historian concerned with interpreting Christianity for his church'. Philip Hefner (1972: 30) adds that 'deep-seated piety and belief in what we would call traditional, supernatural Lutheran Christianity permeated his life'. Yet, as Darrell Jodock (1995: 3–4) observes, Ritschl 'challenged the principal ecclesiastical parties of his own day' and his theological arguments 'were distinctive enough to be academically enticing, theologically invigorating, and highly controversial'. Jodock contends that Ritschl's lasting legacy was not so much in the theological system that he produced but resulted more from the way he 'inspired his followers to go beyond his own findings' and thus from the manner by which he 'reconfigured the task of theology'. Busse (1995: 166) concludes that despite the apparent traditional Lutheran piety found throughout his writings, Ritschl's 'epistemology may be more germane to the late twentieth century than to his own era'. For our purposes in this chapter, the significance of Ritschl and some of his key followers for the development of the phenomenology of religion needs to be examined. Central to this theme is Ritschl's theory of 'religious cognition' or 'religious knowledge', which he developed in his major theological work, *The Christian Doctrine of Justification and Reconciliation*, written between 1870 and 1874. H. R. Mackintosh and A. B. Macaulay (Ritschl, 1900: v), who edited the 1900 English translation, call *Justification and Reconciliation* 'monumental' and the most important 'dogmatic treatise' since 'Schleiermacher published his *Christliche Glaube* in 1821' (see Schleiermacher, 1922).

In the section on 'The Doctrine of God' in *Justification and Reconciliation*, Ritschl (1900: 195) identifies the goal of all religion as the attainment of blessedness or the highest good through superhuman powers. He explains: 'Feeling, as pleasure or pain, as blessedness or suffering, is the personal gain or the personal presupposition which impels individuals to participate in religious fellowship' (1900: 198). The benefits sought vary according to how each religion interprets the highest good, and, again, whether or not it is achieved in this world, outside of this world or both within and beyond earthly experience (1900: 195). Examples cited by Ritschl include what he calls 'orgiastic faiths', in which worship consists almost entirely of immediate feeling or sense gratification; ancient Rome, where ceremonial actions correctly performed were required to produce personal satisfaction; and ancient Greece, where a sense of peace or serenity was achieved by worship of the gods (1900: 198). Doctrines, acts of worship, feelings and moods combine to produce the essential characteristics found in every religion. By doctrines, Ritschl refers to 'a particular view of the world' (1900: 199). Worship is understood as a communal experience of those sharing this particular view of the world. Feelings express the level of satisfaction (or dissatisfaction) achieved through doctrines. Moods are affected by removing the worshipper from ordinary experience and in various ways transforming life with extraordinary meaning. Taken together, these four elements – doctrine, worship, feeling and mood – comprise religion; none is more important than the other (1900: 198–9).

Religion originates from a common human predicament, what Ritschl calls 'the contradiction in which man finds himself, as both a part of the world of nature and a spiritual personality claiming to dominate nature' (1900: 199). Simply put, humans find themselves powerless to control or influence natural events, from the weather on which crops depend in agricultural societies, to the universal experience of death. These common experiences produce a sense of alienation from natural forces, giving rise to the universal need to control them for human well being. This creates the felt need for superhuman help, in Ritschl's words, through which 'the power which man possesses of himself is in some way supplemented, and elevated into a unity of its own kind which is a match for the pressure of the natural world' (1900: 199). Such alienation exists between humans and nature even in societies where the superhuman is associated with natural forces, because such forces are always personified proving 'that it is in the spiritual personality of the gods that man finds the foothold which he seeks for in every religion' (1900: 199).

It is precisely at this point that the Christian doctrine of God demonstrates its superiority over beliefs in other religions, which struggle against nature despite a universal urge to achieve wholeness or unity with it. Only by asserting the oneness of God in which the whole created order will find its ultimate consummation, according to Christian doctrine, can humans achieve a sense of what Ritschl calls 'totality' with respect to nature (1900: 200). Examples derived from various religions demonstrate this clearly. In the Old Testament, a world-end is conceived, but this is just 'the perfecting of the one chosen people in moral, political, and economical independence' (1900: 200). The individual Israelite is not considered, and thus a total sense of satisfaction cannot be achieved. In 'heathen and polytheistic religions' efforts are made to find unity between the natural and spiritual worlds, but always this is intended to supplement human efforts and never achieves a completely satisfying result. In 'Brahminism' the return of original Being to 'distinctionless unity' has in its own way a holistic aim, but this is conceived as other worldly or remains against nature, since ultimately the individual self is absorbed into the universal self. Brahminism thus expresses religion through asceticism or world denial (1900: 200). Christianity, alone among the religions of the world, 'guarantees to believers that they shall be preserved unto eternal life in the Kingdom of God' (1900: 200).

For the Christian, the Kingdom of God defines the highest and greatest good. As God's revealed end of the world, the Kingdom assures that the contradiction between humans and nature will be overcome 'in the full sense that man is thus in the Kingdom of God set over the world as a whole in his own order' (1900: 200). This assurance produces in humans the most complete satisfaction, the fullest sense of purpose and gratitude and assures blessedness. 'Not only the Christian's tone of feeling, but also his estimate of self is determined by this highest and all-inclusive good' (1900: 200). In other religions, the highest good is sought, but it is subject to 'passionate impulses', 'vacillation between changing tones of feeling arising from confused ideas' and 'voluptuous alternations of aesthetic pleasure and pain' (1900: 200). The Kingdom of God can thus be described as the primary Christian doctrine or view of the world, which results in the unique form of Christian worship: 'thanksgiving for God's grace, prayer for its continuance, and service of God in His Kingdom' (1900: 201). The feeling produced is one of complete satisfaction resulting from an 'ideal bond between a definite view of the world and the idea of man as constituted for the attainment of goods or the highest good' (1900: 201). The mood elevated by Christianity sets the worshipper into an extraordinary experience, characterized by a 'bondage to what is good, and liberty to give God thanks and to act aright' (1900: 201).

In Ritschl's view, despite its qualitative superiority, Christianity remains fully a religion, consistent with all other religions, since the religious impulse in Christianity is no different from the impulse in other religions. The aim is also identical in principle, to achieve the highest good. The source of religion in the estrangement between humans and nature is also at the core of Christianity. What distinguishes Christianity from other religions is that its particular view of the world produces a feeling of complete satisfaction, expressed uniquely in acts of thanksgiving and gratitude which are then embodied in moral action in the world.

From this analysis, we can determine how for Ritschl religious knowledge can be distinguished from theoretical or scientific knowledge. In his analysis of perception, the mind or spirit (*Geist*) receives sensations from the world, which the mind in all cases must evaluate, although from different perspectives and for different purposes. On the one hand, the mind places a value on the sensations which are aroused in it, according to the pleasure or the pain such sensations produce for the individual Ego. On the other hand, the mind seeks to determine the causes of such sensations and its connections to other causes, and thereby constructs scientific knowledge. These two functions, whereby the mind apprehends the world, operate simultaneously and in this sense both result from feeling. Scientific knowledge, although it appears to operate in a disinterested manner, accrues from judgements made by the observer as to what is worth observing, describing and hence including in a scientific analysis of the world. Value judgements, therefore, stand at the root of all knowledge.

Theoretical knowledge and religious knowledge nonetheless employ different types of value judgements. Science proceeds on the basis of *concomitant* value judgements, which, Ritschl notes, 'are operative and necessary in all theoretical cognition, as in all technical observation and combination' (1900: 204–5). Another type of value judgement is the *independent* value judgement, which is moral in character in that it produces moral 'pleasure or pain' by encouraging or repelling moral action (1900: 205). We can see how Ritschl has derived this distinction from Kant, since concomitant value judgements investigate empirical facts and construct causal relations between them, whereas independent value judgements analyse the moral will and its effects. Ritschl, however, goes further than Kant by classifying religious knowledge as a type of independent value judgement, not restricted to moral judgements, since not every religion unites feelings of pleasure and pain with a moral will or ethical action. Although the religious value judgement involves feelings of pleasure or pain, it does so just in relation to the level of success it achieves in controlling natural forces, or as Ritschl puts it, 'in which man either enjoys dominion over the world vouchsafed him by God, or feels grievously the lack of God's help to that end' (1900: 205). In other words, religious cognition is evaluated on the basis of the satisfaction it produces in the individual's quest to attain the highest good as defined by a religion's particular view of the world and as expressed by its modes of worship and extraordinary experiences or moods. Christianity, because of its fundamental moral basis in the idea of the Kingdom of God, unites the moral with the religious value judgement. It remains religious, nevertheless and not strictly moral, in the sense that it 'deals with the relations between the blessedness which is assured by God and sought by man', but unites this with a perfect moral will because the blessedness achieved obtains to 'the whole of the world which God has created and rules in harmony with His final end' (1900: 207). Here again we can note Kant's influence, who, although excluding happiness as the motivation for moral behaviour, envisaged in the immortal progress of the soul an є which combined perfectly happiness with the moral will.

It is this close connection to the Kantian analysis which, perhaps more than any other, led Walter Capps (1995: 19) to conclude that Ritschl 'was opposed to what he considered Schleiermacher's romanticizing of religion'. Capps adds: 'In contending against Schleiermacher's formulations, he sought to revive and update Kant's association of religion with a practical, ethical, or moral ideal'. The distinction within the independent value judgement between moral and religious knowledge demonstrates that this conclusion is too severe and one-sided. When taken as a whole, religion focuses on feelings of pain or pleasure, satisfaction or the lack thereof, in achieving control over forces outside of human control. When the doctrines of a religion help believers to achieve feelings of satisfaction, religion fulfils its purpose. By extension, we can conclude that for Ritschl Christianity alone possesses an elevated sense of blessedness which unites morality with feeling and thereby achieves a perfect unity between the ethical and religious good. Schleiermacher's emphasis on religion as feeling undoubtedly informed Ritschl's interpretation of the religious value judgement, which at its most common or universal level seeks to attain a feeling of satisfaction or blessedness.

With respect to our later analysis of key phenomenologists of religion, it is important to note that for Ritschl religion was not comprised primarily or exclusively of doctrines, but, alongside beliefs, he included the phenomena associated with religious experience, ritual and that which separates the religious from the non-religious within various cultures. Moreover, although as a theologian he established criteria for evaluating religions and for asserting the superiority of Christianity, his method of religious epistemology applied universally. In the final analysis, for Ritschl, the religious value judgement could be performed only by religious adherents, not by outsiders. Through beliefs, worship and moods, believers themselves evaluated the capacity of their religion to foster the highest good as defined by their particular view of the world. In this way, religion establishes its own unique and distinctive way of knowing, quite distinguishable from theoretical and moral knowledge. ↖ Sui generis

Wilhelm Herrmann: The communion of the Christian with God

Wilhelm Herrmann (1846–1922) is less widely known than Ritschl, but his influence on liberal Protestant theology has been significant, as evidenced by the importance for later theological developments of two of his students, the noted theologians, Rudolf Bultmann and Karl Barth. Herrmann, who almost always is included among those belonging to the 'school' of Ritschl, did not study under Ritschl, but while pursuing a theology degree in the University of Halle, he met Ritschl at the home of Professor F. Tholuck, with whom he lived as a student for two and a half years (Stewart in Herrmann, 1904: vi). Following this meeting, Herrmann engaged in regular correspondence with Ritschl. After concluding his studies at Halle with a period of intense study of Kant's philosophy, Herrmann began teaching religion at a gymnasium, or classical high school, where he developed for his students a primer on Ritschl's theology. Robert W. Stewart observes that Herrmann's correspondence with Ritschl during this period 'had the most powerful influence on his development' (1904: vii). In 1889, Herrmann was appointed Professor of Theology in the University of Marburg, a post he held until his retirement in 1916.

In two of his most important publications, *Faith and Morals* and *The Communion of the Christian with God*, Herrmann expanded Ritschl's idea of religious cognition by defining religion as relational, that is, founded on a personal experience between humans and supersensible realities. Herrmann sought through this approach to demonstrate that the Christian's communion with God through the historical personality of Jesus Christ represented the highest quality or purest form of religion humans were capable of experiencing. In this way, as Robert Voelkel (Herrmann, 1972: xxii) notes in his introductory comments to *The Communion of the Christian with God*, 'Herrmann, instead of looking to the broader dimensions of world history and cultural relativism, focused upon the real significance of the Ritschlian method for distinctively Christian theology'. In particular, Herrmann led a lifelong attack against Christian orthodoxy, as expressed in rationalistic metaphysics, which, as Voelkel notes, served to 'perpetuate orthodoxy's errors in the name of enlightenment' (1972: xxii). By creating levels of abstraction, Herrmann believed that metaphysics leads humans away from the true meaning of religion, which can be found only, again in Voelkel's words, 'by specific men in specific situations in such a way as to define their specific self-expression' (1972: xxv).

At the core of Herrmann's interpretation of religion is the notion 'faith'. In the first part of *Faith and Morals*, which he delivered as a lecture in 1890 a few months after Albrecht Ritschl's death, Herrmann examined faith 'as Ritschl defined it'. It is clear, however, that the ideas in the address constitute Herrmann's own definition of faith, a view supported by Voelkel (Herrmann, 1972: xxii) who claims that it was Herrmann 'who taught Ritschl in his later years to direct his guns also against metaphysics' (1972: xxii). In *Faith and Morals* Herrmann (1904: 35) defines faith as that which 'places the person in whom it arises in a condition which is the beginning of the blessed life'. This condition does not derive from doctrines, or rational explanations about God, but from two inward feelings: a sense of blessedness experienced in material and spiritual matters and a sense of self-denial in response to moral imperatives. 'Evidently', writes Herrmann, 'the man who can feel both these things is inwardly cut free from the world and brought into a life in that which is eternal' (1904: 36). Faith, on such a view, cannot be reduced to or equated with an abstract idea, since it becomes manifest only within concrete and specific religious experiences. 'A man can only say he has found God when it has become clear to him from some event in his own life to which he can assign a definite date that God has therein sought him out and touched him' (1904: 37). Such an unambiguous experience defines the core of religion, since there is 'no religious thought which does not express this direct relation of God to the particular man who cherishes the thought' (1904: 37). For example, we can formulate an idea of an Almighty God, expressed in thoughts or doctrines, but this does not equate with an experience of God. A genuine faith experience of the Almighty occurs only when a person experiences, in Herrmann's words, 'a Power which at this particular moment is for our sake causing the whole reality in which we stand' (1904: 37). The first principle within Herrmann's theology thus distinguishes sharply between religious faith as an experience of blessedness and abstract ideas concerning the object or objects of religion.

Concrete religious experiences for the Christian, although producing feelings of blessedness and moral obedience, are always related to the historical facts surrounding the life, death and resurrection of Jesus Christ. In this way, Herrmann asserted the unique character of Christian faith as having been produced from the human interaction with the personality of the historical Jesus. In the historical Jesus, the Christian witnesses one who fulfilled human blessedness, both material and spiritual, and at the same time embodied the moral imperative through ultimate self-denial. Jesus, however, does not produce faith by illustrating blessedness and morality. Faith occurs as one engages with and enters into an experience of Jesus through the pages of the New Testament, where the life and personality of Jesus draw people into the deepest and most satisfying experience possible. Near the end of his lecture on 'Faith as Ritschl Defined It', Herrmann describes precisely what such an experience entails.

> The same Man who becomes judge and conscience to the person who comes face to face with Him, interests Himself in him with a patient and unparalleled love. At the same time that He makes the sinner insecure by the simple power of His personal life, He sets him on his feet by His kindness. Therefore those who have been led by Him to feel the bitterness of their plight, yet felt themselves for that very reason drawn to Him ... By looking back at His Personality all men after Him can be freed from their inward dispeace and from the burden of their sin. (Herrmann, 1904: 42–4).

Christian faith thus results from a direct encounter between an individual and the living personality of the historical Jesus, a personality which shines through the pages of the New Testament to produce a direct, concrete and fully satisfying experience of religious blessedness. Herrmann's position is summarized succinctly by Robert W. Stewart, who translated Herrmann's essay, 'Faith as Ritschl Defined It', into English:

> Faith ... is not belief in any Church or in any doctrine; it is neither assent to the truth of the narratives of Scripture, nor is it acceptance of propositions in theology. It is a spiritual experience of an overpowering revelation of God, and this revelation comes to man through the circumstances in which he realizes, in his own life, contact with the inner life of Jesus Christ. (Herrmann, 1904: 3)

In *The Communion of the Christian with God* (1895), Herrmann addressed the question of faith that is not produced by a person's encounter with Jesus Christ, or what can be called religious faith in general. Faith in all religions, he argues, bears the same characteristics as does faith in Christianity. It is relational, best described as a communion with God, and it produces a vague sense of blessedness. Even among the most primitive savages, such as those of 'New Holland', 'we by no means wish to assert, even for a moment' that they have 'no pulsations of true religion' (1895: 53). Every religious person everywhere obtains some measure of a communion with God. Herrmann admits that in the cases of those who have had no contact with civilization, 'we do not know through what medium' faith as communion with God reaches them (1895: 53). Since religion equates to faith and not doctrine, however, to deny that religious people throughout the world have achieved faith would constitute at worst a 'contradiction', but at the very least 'an exaggeration' (1895: 53).

Herrmann's resolution to the problem of faith outside knowledge of Christ is in one sense not a resolution at all, since he admits he cannot explain it. Rather he asks his reader to consider the significance of historical facts. Jesus Christ has changed history and thus has changed the experience people have of God. It is no use seeking to go back to a period before Jesus because that belongs to an entirely different historical context. For example, however hard one seeks to gain an understanding of the kind of faith that belonged to pious Israelites before Jesus, we can never do so since 'the revelation which was given to Israel can no longer satisfy our need' (1895: 53). Since the coming of Jesus, the religious landscape has altered radically making 'the knowledge of God and the religions which have been and which are possible to men placed in other historical conditions ... impossible to us' (1895: 54). In particular, after the coming of Jesus, people feel themselves separated from God and thus in need of intimate communion with him in a way that people with no knowledge of Jesus can feel. Likewise, since the coming of Jesus, the moral consciousness has been so deepened that 'we cannot go back to simple indifference to a moral demand' (1895: 54). Nor can we try to find the experience of faith in nature. Such attempts belong to a period 'that has now been outlived' (1895: 54) since we now know ourselves to be intertwined with historical and social contexts. For these reasons, the historical fact of Jesus cannot be ignored when considering the nature of religious faith. 'To overlook Him is to deceive ourselves as to the best treasure which our own life possesses. For He is precisely that fact which can make us certain, or which can make us see it to be a reality that God communes with us' (1895: 55).

Herrmann's contribution to an understanding of religion now becomes evident. In a creative reinterpretation of Ritschl's independent value judgement, he synthesized Schleiermacher's core of religion as feeling with Kant's understanding of religion as moral obligation and set both within a Hegelian emphasis on the operation of the spirit in history. By so combining these influences, Herrmann separated religion from intellectual abstractions, rooted it in a feeling of satisfaction that can be produced only under the power of the historical Jesus and assigned its value within human existence to its capacity for securing an intensely intimate communion with God under total obedience to the moral law.

Rudolf Otto in the Ritschlian tradition

Most historians of comparative religions regard the German theologian Rudolf Otto (1869–1937) as having played a critical formative role within the development of religious studies as an academic field in its own right. Eric Sharpe (1986: 161–7), for example, devotes an entire section of his history of comparative religion to Otto, rating Otto's landmark book *The Idea of the Holy* (first published in 1917 as *Das Heilige*, with its English translation first appearing in 1923) as having achieved 'near-canonical status as one of the books which every student of comparative religion imagines himself or herself to have read' (1986: 161). Russell McCutcheon (1999: 69) traces Otto's importance directly to the influence he exerted over thinkers in what I will later describe as the 'Chicago school' of phenomenology, particularly over the theories of Joachim Wach and Mircea Eliade.

Otto's first teaching post was in 1904 at Göttingen, where, as Sharpe notes, the history of religions school 'was in its heyday', and where he was a colleague of Husserl (1986: 162). In 1917, he took a post in the University of Marburg, following Herrmann's retirement, where he stayed for the remainder of his career (Sharpe, 1986: 162). McCutcheon (1999: 69) notes that Otto followed in the tradition of Schleiermacher by emphasizing 'the private and emotional quality of religion', which led Otto to submit that students of religion needed to have experienced religion personally if they were to achieve an understanding of anything other than the 'externals of religion'. This is demonstrated near the outset of the *Idea of the Holy*, where Otto invites his readers to direct their minds 'to a moment of deeply-felt religious experience' and requests, if they cannot summon such a memory, that they 'read no further' (1926: 8).

The most important concept in Otto's thinking for later phenomenologists derives from his description of the core of religion as an unknown and unknowable 'holy', which is expressed and hence becomes observable in religious experience. In his discussion of the 'elements of the "numinous"' in *The Idea of the Holy*, Otto (1926: 10) refers to the experience of 'creature-feeling', which he defines as 'the emotion of a creature, abased and overwhelmed by its own nothingness in contrast to that which is supreme above all creatures'. The numinous is objective, outside the self, and responsible for the intensity of feelings produced within an individual who responds to it. Otto contrasts this with Schleiermacher's principle, where a feeling of dependence 'is merely a category of self-valuation or self-consciousness', and thus bears no testimony to the objective reality of the holy (1926: 10). The 'nature of the numinous', described by Otto as the '*mysterium tremendum*', is manifested in the feelings it produces in humans, which vary in intensity and mood depending on how it is apprehended by the individual. Sometimes, the feeling it produces 'comes as a gentle tide'; at other times, it is experienced as 'thrillingly vibrant and resonant'. It may even 'burst in sudden eruption up from the depths of the soul with spasms and convulsions' (1926: 12–13). The tremendous mystery has a dark side, in which its 'crude, barbaric antecedents' appear, but it 'may be developed into something beautiful and pure and glorious' (1926: 13). By calling the source of these experiences a mystery, Otto implies that the cognitive content of the numinous remains 'hidden' and lies beyond the ability of humans to conceptualize it or put it into words. In this sense, the mysterious numinous reality is couched in negative terms as something unknown in itself, but it can be seen positively in the feelings it produces in human experience, such as awe, majesty, dread, fear, urgency to be in relation with it and concurrent emotions of fascination and revulsion.

Following Kant's distinction, Otto indicates that the *mysterium tremendum* is not just an analytic term, but also a synthetic one. The mystery is 'wholly other', unknown in itself, but because it is revealed in feelings such as awe, majesty, fear, urgency and fascination, the *tremendum* adds knowledge to the meaning of mystery, after experience, not simply by definition (1926: 25). The student of religion thus discovers not so much the concepts necessary for understanding religious experience through observing the way humans apprehend the tremendous mystery, but they see directly the transcendent in operation, so to speak, through the ways humans respond to the numinous. Otto explains:

> It is through this positive feeling-content that the concepts of the 'transcendent' and 'supernatural' become forthwith designations for a unique 'wholly other' reality and quality, something of whose special character we can *feel*, without being able to give it clear conceptual expression. (emphasis his) (Otto, 1926: 30).

The essence of religion, its core irreducible element, thus is presented by Otto as the inexplicable and inexpressible mystery, which is 'wholly other'; its reality is confirmed in its manifestations which are made explicit in the way humans respond to it through 'creature feeling'.

With this structure firmly in place, we can now ask in what way Otto used his analysis of the numinous as a basis for comparing religions. In the first instance, Otto notes that 'creature feeling' does not result from a moral sense. The depreciation of the self in the face of the overwhelming mystery occurs not because an individual has transgressed the moral law, but results instead from 'the feeling of absolute "profaneness"' (1926: 53). A sense of 'profaneness' is 'natural' to human beings, since the person who experiences the tremendous mystery does so, not as a result of guilt in response to particular actions or deeds, but simply because of 'his own very existence as creature before that which is supreme above all creatures' (1926: 53). Yet, the feeling is one of 'appreciation' for the 'holy', which stands in sharp contrast to the 'profane': in the former, the person values the numinous; in the latter, the individual 'treats the profane as "unworth"' (1926: 53). Religions develop from this fundamental core experience of the holy. In 'the lowest and earliest level of religion of primitive man', the sense of the numinous is experienced as 'the stupor before something "wholly other", whether such an other be named "spirit" or "daemon" or "deva", or be left without any name' (1926: 27). In what Otto calls 'highly-developed' religions, 'the appreciation of moral obligation and duty, ranking as a claim of the deity upon man, has been developed side by side with the religious feeling itself' (1926: 53). This means that the transcendent is fundamentally an ontological category, and not a moral one. The human can experience the overwhelming power of the numinous without ever having developed a sense of sin and the need for deliverance from evil (1926: 54–5).

For Otto, Christianity demonstrates its superiority over other religions just at this point. In Christianity, the feelings of awe result not from the need for protection against an unknown mysterious force, but from a fuller moral understanding of the relationship between the sacred and the profane. The Christian's awareness of the critical need for atonement with God means that 'mere awe, mere need of shelter from the "tremendum" has ... been elevated to the feeling that man in his "profaneness" is not *worthy* to stand in the presence of the holy one' (emphasis his) (1926: 56). For Christians, the need for atonement, to be put right with the transcendent, results from the desire to enter into an intimate communion with the numinous. This desire escalates to a 'craving', in Otto's words, for 'the *summum bonum*', which deepens and strengthens the religious feeling (1926: 57). When comparing religions on the basis of the quality of experience they are capable of engendering in their adherents, it can be seen that 'no religion has brought the mystery of the need for atonement or expiation to so complete, so profound, or so powerful expression as Christianity. And in this ... it shows its superiority over others' (1926: 58).

That Otto follows in a line from Kant and Schleiermacher through Ritschl is made clear from the way he emphasizes that religion is not primarily defined by concepts, doctrines or intellectual ideas, but by judgements of value, both experiential and moral. In a way consistent with the thinking of Ritschl and Herrmann, Otto stressed that morality within religion represents advanced stages in religious understanding, characteristic of higher religions like Christianity, but that the fundamental core of religion is associated with the feeling of the creature for the numinous, a tremendous, fascinating mystery which both attracts and repels at the same time. This core of religion cannot be reduced to thoughts or ideas, since it displays the human response to the utterly inexpressible, wholly other. Religions, nonetheless, can be compared, largely by the quality of the experience they produce in response to the numinous, the most superior of which is expressed in Christianity, where believers understand profanity not simply as a natural condition universally experienced by virtue of being a creature, but as a moral separation from God whose act of redemption in Christ combines an inherent creature-feeling with an intimate relationship with the holy based on respect and love. Otto's emphasis on religion as securing the *summum bonum*, or highest good, also lends support to his connection to Kant through the Ritschlian theological tradition, where true religion is described ultimately as securing blessedness for humanity, interpreted as an intimate communion with God and experienced through an elevated moral consciousness.

The faith–faiths distinction: A. G. Hogg

The argument that Albrecht Ritschl and Wilhelm Herrmann directly influenced the development of the phenomenology of religion finds further support from a somewhat unexpected source: the thinking and writing of the Scottish missionary theologian Alfred George Hogg (1875–1954). Although he is more often associated with missionary theology and philosophy than with religious studies, A. G. Hogg made an important contribution to the understanding of Christian–Hindu relations in India during the first half of the twentieth century, and through this to methodologies in the study of religions generally. As a Professor of Philosophy, and later Principal, at the Madras Christian College, Hogg's students included many who subsequently became leaders in Indian social, political and religious life, including the celebrated Hindu philosopher and former Vice-President of India, Sarvapalli Radhakrishnan. Hogg's importance has been noted by many Indians, both Christian and Hindu, and his significance has been documented in numerous studies by Eric Sharpe (1971, 1994, 1999; see also Cox, 1979: 241–56).

Hogg was born in 1875 in Egypt of Scottish missionary parents. In 1893, he obtained a place at Edinburgh University to read philosophy under the noted idealist Andrew Seth Pringle-Pattison. In 1900, he entered New College in Edinburgh, then the theological college of the United Free Church of Scotland, and completed his studies in Halle, Germany under the prominent Ritschlian, Arthur Titius. In January 1903, Hogg arrived at the Madras Christian College in India under an appointment from the Foreign Mission Board of the United Free Church, where he stayed for the remainder of his career until his retirement in 1938. During his years in Madras, he became well known in his own right as an idealist philosopher both in India and back in his Scottish homeland. In 1919 he became an honorary president of the Nagpur Philosophical Society, after delivering a series of lectures on ethics and metaphysics. In 1923, the University of Edinburgh awarded him the degree of D.Litt., which the editor of the *Madras Christian College Magazine* described as 'a further recognition of his erudite study and literary distinction' (Madras Christian College, 1924: 70). By 1935, his reputation as a philosopher had reached such a high status in India that he was elected President of the Indian Philosophical Congress. In 1938, he was requested by John R. Mott, Chair of the International Missionary Council, to remain in India following his retirement to participate in the international conference of the Council which was being held at the new Tambaram campus of the Madras Christian College. Hogg's chief contribution to the deliberations resulted from his debate with the Barthian theologian, Hendrik Kraemer, whose book *The Christian Message in a Non-Christian World* (1938) was written specifically for the conference. Hogg returned to Scotland after the conclusion of the Tambaram meeting early in 1939, and in 1945 made his final contribution to the Christian understanding of Hinduism by delivering the Alexander Duff lectures in the Universities of Edinburgh and Glasgow, which were published in 1947 as *The Christian Message to the Hindu.*

During his early years at the Madras Christian College, Hogg published several articles in the *Madras Christian College Magazine* in which he constructed his vital distinction between faith and faiths, derived largely from Wilhelm Herrmann. Hogg defined faith as the simple trust or assurance in the highest and best that humans in any religion are capable of experiencing. Faiths, or beliefs, refer to thoughts, ideas, cognitions and doctrines, in other words, the intellectual ways that humans achieve and maintain the faith experience. Beliefs are best regarded as opinions that change according to developments in history, knowledge and scientific advancements, whereas faith remains a core experience of communion with the highest and best that people know. Although every religion enables its adherents to achieve faith, the quality of faith varies according to the beliefs through which it is achieved and maintained. This analysis reflects Herrmann's insistence that trust in the historical Jesus produces the highest religious experience possible, and corresponds to a Ritschlian judgement of value.

In his earliest published article, 'Agnosticism and Faith' (1903), Hogg credited the school of Ritschl with opening new horizons for contemporary religious thinking by demonstrating the superior value of personal religion over theological and metaphysical speculation. Although he does not mention them by name, it is clear that Hogg is referring to the followers of Ritschl when he commends contemporary thinkers for injecting into theology 'a new hope, a new insight and a new reverence' (Hogg, 1903: 83). The next year, he published an article entitled 'The Christian Interpretation of Mediation', in which he specifically praised the Ritschlians for 'at once exhibiting the greatest apparent contrast with dogmatic tradition and the greatest real accordance with Christian sentiment' (1904: 360). The central principle of Ritschlianism, 'the confession that Jesus himself is the sole final revelation of God, being *literally* God manifest in the flesh' (emphasis his) demonstrated for Hogg that truths or theories of any kind (faiths), do not define the *content* of religious revelation (1904: 360). Rather, revelation occurs in personal experience (faith), which in its most satisfying form 'consists in the overmastering and authoritative impression Jesus makes on the hearts which He has compelled to submit to Him' (1904: 360). Hogg's first book, *Karma and Redemption* (1909), applied this distinction to what he regarded as the cardinal beliefs through which adherents to Hinduism and Christianity achieve faith. As the title suggests, he identified the point of significant contrast between Hinduism and Christianity as the Hindu doctrine of karma and the Christian belief in redemption. Hogg contended that although karma enables the Hindu to achieve faith, its potential quality is diminished when contrasted with the Christian teaching on redemption. By redemption, Hogg meant that humans can be delivered in the present moment by faith from any spiritual or physical affliction which causes them to doubt the goodness of God.

In 1911, Hogg published his second book, a series of daily Bible studies and meditations entitled *Christ's Message of the Kingdom*, which was read widely both in India and in Britain. A careful study of the book shows that Hogg was interpreting biblical texts to advance his argument that God wills to bring his kingdom to earth at any moment, but that he has been restricted in this by humanity's lack of faith or trust. The arrival of the kingdom would mean the eradication at once of all spiritual and physical suffering. God's supernatural power to redeem humans from evil, however, does not await the consummation of the kingdom. God responds to personal faith by releasing individuals from every evil that is truly an evil and not a blessing in disguise. This view was developed after a close reading of another Ritschlian theologian, Julius Kaftan, who interpreted Ritschl's emphasis on the Kingdom of God as defining the focal point in the teaching of Christ. This message provided Hogg with a positive, and as he conceived it, a uniquely relevant, message to the Hindu. He believed that the Hindu's deepest religious need had stemmed from a dissatisfying experience of the natural world. He considered that the development of what he called 'Higher Hinduism', by which he meant primarily Advaita Vedanta, as disclosed by the early-ninth-century philosopher Sankara, represented a Hindu response to the pervasive Indian belief in karma. The law of karma posits that what a person experiences in life corresponds to a strict moral system of ultimately fair rewards or punishments. Of course, the moral law operates only at the larger cosmic scale and thus must be accompanied by the transmigration of souls.

For Hogg, karma and transmigration have produced throughout India a vision of endless cycles of birth and rebirth with no escape from a closed system. Each new deed produces its own consequences and these are perpetuated eternally. According to Hogg, Sankara's philosophical monism regarded the vast karmic system as *maya* (illusion) and thus constructed a way of release (*moksha*) through the awareness that the highest world force or universal spirit (*Brahman*) is identical with the individual self (*atman*). Knowledge of the essential *Brahman-Atman* unity thus results in deliverance from the endless cycles of rebirth. Hogg argued that Christianity also views the world as dissatisfying. However, the Christian response is to emphasize not emancipation out of this world, but the possibility that the present world can be redeemed from evil. It is at this point that Hogg found the central point of contrast between Hinduism and Christianity most effective for communicating the Christian message in India. The belief in karma has produced the universal Indian desire for escape; redemption culminates in the Christian conception of God as one who incarnates himself into human history to transform evil into good. It is clear that this represents Hogg's application of the Ritschlian judgement of value to the Hindu–Christian encounter. Because the doctrine of karma fails to ensure an enduringly satisfying faith experience, the Christian belief in redemption should be urged on Hindus as that alone which can overcome their fundamental disaffection with the world.

In 1917, Hogg elucidated this view fully in one of his major works on Christian theology through a four-part series printed in the *International Review of Missions* entitled, 'The God that must needs be Christ Jesus'. He followed this in 1921 by delivering the Cunningham Lectures in New College in Edinburgh in which he expounded the philosophical and scientific justification for a world which can be regarded not as bound by unchanging laws of nature, but as one which is plastic and malleable to the perceptions of faith. These lectures, later published as *Redemption from this World or the Supernatural in Christianity*, drew heavily on Kantian formal categories of the mind, reinterpreted as faith, in which the world is shown to respond positively to the mental conceptions humans form about it. Ultimately, Hogg argued, the universal experience of suffering, and even death, can be eliminated if humans perceive the world through the category of the mind which Hogg called faith.

At the Tambaram meeting of the International Missionary Council in 1938, Hogg used his distinction between faith and beliefs to criticize Hendrik Kraemer's thesis that stressed a radical discontinuity between the Christian message and what Kraemer called the merely human apprehensions of reality expressed in the 'non-Christian' religions. Hogg offered an alternative to Kraemer by affirming continuity between the *faith* of Christians and adherents of other religions while maintaining that a fundamental discontinuity exists between the *beliefs* of Christianity and those of other religions. This distinction, which Kraemer dismissed as pure subjectivity akin to the liberal theology of Schleiermacher, was based on Hogg's lifelong conviction, influenced more by the Ritschlians than Schleiermacher, that, although Christian beliefs potentially produce the highest quality of faith, adherents of all religions testify to a deep and profound spirituality. It is clear from his contribution to the Tambaram debate that Hogg maintained to the end of his life the critical distinction between a universal religious faith and a quality of faith derived directly from religious beliefs. He interpreted this distinction in his 1945 Duff Lectures in a Ritschlian way in terms of the Kingdom of God and thus demonstrated that the kingdom idea had been motivated uniformly by his desire to select the most relevant contrasting beliefs between Christianity and Hinduism in order to show how Christian faith leads humanity to the highest good, the *summum bonum*. He consistently argued that if humanity could learn to trust God unconditionally by viewing reality through the spectacles of faith, the world could be transformed from its predominant experience of suffering to one of universal beneficence.

Any analysis of Hogg's contribution to Christian–Hindu understanding must take into account that his approach was developed when the pervasive tendency of Western missionaries was to treat Hinduism with condescension, as exemplified in J. N. Farquhar's theory of 'fulfilment' (see Sharpe, 1965). Hogg's frank commitment to Christian beliefs, accompanied by his unshakeable respect for Hindu faith, anticipated a view later adopted by many Christian theologians who were involved in inter-religious dialogue (see Whaling, 1999: 251–2). Outside theology and philosophy, in the field of religious studies, Hogg's influence can be shown to have had both direct and indirect implications. For example, his critical distinction between faith and faiths was adapted by Wilfred Cantwell Smith, a key scholar whom we will consider in the section on North American phenomenologists of religion. In his groundbreaking book, *The Meaning and End of Religion* (first published 1962), like Hogg, Smith defined 'faith' as the core of religion while sharply distinguishing it from doctrines or beliefs. This same distinction, as we will see in our discussion of North American phenomenologists, relates theoretically, and through Otto historically, to Mircea Eliade's highly influential thesis that the universal 'sacred', akin to Hogg's concept of 'faith', is apprehended and expressed through its particular manifestations (Eliade, 1987: 11).

Conclusion

Important connections have been made in this chapter between German romantic and idealistic philosophy and the development of Ritschlian theology in the nineteenth and early twentieth centuries. The innovative combination made by the Ritschlians between religion as morality and feeling, expressed in the ultimate culmination of history in the Kingdom of God, brought together Kant, Schleiermacher and Hegel into a synthesis that would have important implications for the development of the phenomenology of religion. As we have noted, the Ritschlian theological influence within religious studies has received relatively little attention in prior academic research, but even a brief review of Ritschl and Herrmann in relation to Otto's thinking, as I have done in this chapter, demonstrates that such an influence is far more significant than has been acknowledged previously. Otto, like the Ritschlians, emphasized the importance of feelings for understanding the core element within all religions, but stressed that Christian experience, since it was based on the atonement, induced both a sense of moral estrangement from and reconciliation with the transcendent. Nowhere can the Ritschlian influence be seen better than in A. G. Hogg's analysis of Hindu–Christian relations, in which the fundamental categories of faith and their expressions through beliefs were clearly delineated, and which in broad outline came to define one of the principal assumptions operating within most later phenomenological interpretations of religion.

References

Benét, W. R. (1977), *The Reader's Encyclopedia* (London: Book Club Associates, 2nd edn).

Braun, W. and McCutcheon, R. T. (eds) (2000), *Guide to the Study of Religion* (London and New York: Cassell).

Busse, R. P. (1995), 'Religious cognition in light of current questions', in D. Jodock (ed.), *Ritschl in Retrospect: History, Community, and Science* (Minneapolis: Fortress Press), pp. 166–85.

Capps, W. H. (1995), *Religious Studies: The Making of a Discipline* (Minneapolis: Fortress Press).

Cox, J. L. (1979), 'Faith and faiths: The significance of A. G. Hogg's missionary thought for a theology of dialogue', *Scottish Journal of Theology*, 32, 241–56.

Eliade, M. (1987) [1959], *The Sacred and the Profane: The Nature of Religion* (San Diego, New York and London: Harcourt).

Fitzgerald, T. (2000), *The Ideology of Religious Studies* (New York and Oxford: Oxford University Press).

Flood, G. (1999), *Beyond Phenomenology. Rethinking the Study of Religion* (London and New York: Cassell).

Hefner, P. (1972), 'Albrecht Ritschl: An Introduction', in A. Ritschl, *Three Essays* (translated by P. Hefner) (Philadelphia: Fortress Press), pp. 3–50.

Hegel, G. W. F. (1910) [1807], *The Phenomenology of Mind* (translated, with introduction and notes by J. B. Baillie) (London: S. Sonnenschein).

Hegel, G. W. F. (1977) [1807], *The Phenomenology of Spirit* (translated by A. V. Miller) (Oxford: Clarendon Press).

Hegel, G. W. F. (1969) [1812–1816], *Science of Logic* (translated by A. V. Miller) (London: George Allen and Unwin Ltd).

Herrmann, W. (1895), *The Communion of the Christian with God: A Discussion in Agreement with the View of Luther* (translated by J. S. Stanyon) (London: Williams and Norgate, 3rd edn).

Herrmann, W. (1904), *Faith and Morals* (translated by D. Matheson and R. W. Stewart) (London: Williams and Norgate).

Hogg, A. G. (1903), 'Agnosticism and faith', *Madras Christian College Magazine*, New Series 3 (2), 75–84.

Hogg, A. G. (1904), 'The Christian interpretation of mediation', *Madras Christian College Magazine*, New Series 3 (7), 357–69.

Hogg, A. G. (1909), *Karma and Redemption. An Essay Toward the Interpretation of Hinduism and the Re-Statement of Christianity* (London, Madras and Colombo: Christian Literature Society).

Hogg, A. G. (1911), *Christ's Message of the Kingdom. A Course of Daily Study for Private Students and for Bible Classes* (Edinburgh: T. and T. Clark).

Hogg, A. G. (1917), 'The God that must needs be Christ Jesus', *International Review of Missions*, vols 21–4, 62–73, 221–32, 383–94, 521–33.

Hogg. A. G. (1922), *Redemption from this World or the Supernatural in Christianity* (Edinburgh: T. and T. Clark).

Hogg, A. G. (1947), *The Christian Message to the Hindu* (London: SCM Press).

James, G. A. (1995), *Interpreting Religion. The Phenomenological Approaches of Pierre Daniel Chantepie de la Saussaye, W. Brede Kristensen, and Gerardus van der Leeuw* (Washington, DC: The Catholic University of America Press).

Jodock, D. (ed.) (1995), *Ritschl in Retrospect: History, Community, and Science* (Minneapolis: Fortress Press).

Kant, I. (1909) [1785], *Fundamental Principles of the Metaphysic of Morals*, in T. K. Abbott (translator), *Kant's Critique of Practical Reason and other Works on the Theory of Ethics* (London, New York and Bombay: Longmans, Green, and Co., 6th edn), pp. 1–84.

Kant, I. (1909) [1788], *Critical Examination of Practical Reason*, in T. K. Abbott (translator), *Kant's Critique of Practical Reason and other Works on the Theory of Ethics* (London, New York and Bombay: Longmans, Green, and Co., 6th edn), pp. 87–262.

Kant, I. (1929) [1787], *Critique of Pure Reason* (translated by N. K. Smith) (London: Macmillan and Company, 2nd edn).

Kant, I. (1977) [1783], *Prolegomena to Any Future Metaphysics That Will Be Able to Come Forward as Science* (the Paul Carus translation extensively revised by J. W. Ellington) (Indianapolis: Hackett Publishing Company).

King, R. (1999), *Orientalism and Religion. Postcolonial Theory, India and 'The Mystic East'* (London: Routledge).

Kraemer, H. (1938), *The Christian Message in a Non-Christian World* (London: Edinburgh House Press).

McCutcheon, R. T. (ed.) (1999), *The Insider/Outsider Problem in the Study of Religion: A Reader* (London and New York: Cassell).

Mackintosh, H. R. (1937), *Types of Modern Theology: Schleiermacher to Barth* (London: Nisbet and Co. Ltd).

Mackintosh, H. R. and Macaulay, A. B. (eds) (1900), 'Editors' Preface', in A. Ritschl, *The Christian Doctrine of Justification and Reconciliation* (Edinburgh: T. and T. Clark), pp. v–vi.

Madras Christian College (1924), 'College Notes', *Madras Christian College Magazine*, Quarterly Series 4 (1), 70.

Morris, B. (1987), *Anthropological Studies of Religion: An Introductory Text* (Cambridge: Cambridge University Press).

Otto, R. (1926), *The Idea of the Holy: An Inquiry into the Non-Rational Factor in the Idea of the Divine and Its Relation to the Rational* (translated by J. W. Harvey, 4th impression, revised with additions) (London: Humphrey Milford and Oxford University Press).

Popkin, R. H. and Stroll, A. (1986) *Philosophy: Made Simple* (London: Heinemann, 2nd edn).

Richmond, J. (1978), *Ritschl: A Reappraisal* (London and Glasgow: Collins).

Ritschl, A. (1900), *The Christian Doctrine of Justification and Reconciliation* (translated and edited by H. R. Mackintosh and A. B. Macaulay) (Edinburgh: T. and T. Clark).

Ritschl, A. (1972), *Albrecht Ritschl. Three Essays* (translated with an introduction by P. Hefner) (Philadelphia: Fortress Press).

Schleiermacher, F. (1922) [1821], *The Christian Faith in Outline* (translated from the German by D. M. Baillie) (Edinburgh: W. F. Henderson).

Schleiermacher, F. (1988) [1799], *On Religion: Speeches to Its Cultured Despisers* (Introduction, translation and notes by R. Crouter) (Cambridge: Cambridge University Press).

Scruton, R. (1982), *Kant* (Oxford and New York: Oxford University Press).

Scruton, R. (1984), *A Short History of Modern Philosophy: From Descartes to Wittgenstein* (London and New York: Ark Paperbacks).

Sharpe, E. J. (1965), *Not to Destroy but to Fulfil. The Contribution of J. N. Farquhar to Protestant Missionary Thought in India before 1914* (Uppsala: Gleerup).

Sharpe, E. J. (1971), *The Theology of A. G. Hogg* (Madras: Christian Literature Society).

Sharpe, E. J. (1986), *Comparative Religion: A History* (London: Duckworth, 2nd edn).

Sharpe, E. J. (1994), 'A. G. Hogg', in G. H. Anderson (ed.), *Mission Legacies: Biographical Studies of Leaders of the Modern Missionary Movement* (Maryknoll, New York: Orbis), pp. 330–8.

Sharpe, E. J. (1999), *Alfred George Hogg, 1875–1954: An Intellectual Biography* (Chennai, India: Christian Literature Society).

Smith, W. C. (1964) *The Meaning and End of Religion. A New Approach to the Religious Traditions of Mankind* (New York: Menton Books).

Stewart, R. W. (1904), 'Biographical note', in W. Herrmann, *Faith and Morals* (translated by D. Matheson and R. W. Stewart) (London: Williams and Norgate), pp. v–xii.

Stewart, R. W. (1904), 'Introduction by the Translator', in W. Herrmann, *Faith and Morals* (translated by D. Matheson and R. W. Stewart) (London: Williams and Norgate), pp. 3–5.

Voelkel, R. T. (1972), 'Introduction', in W. Herrmann, *The Communion of the Christian with God: Described on the Basis of Luther's Statements* (London: SCM Press, 2nd edn), pp. iv–lxviii.

Whaling, F. (1999), 'Theological approaches', in P. Connolly (ed.), *Approaches to the Study of Religion* (London and New York: Continuum), pp. 226–74.

Chapter 3

Ideal Types and the Social Sciences: The Contributions of Troeltsch, Weber and Jung to Phenomenological Thinking

Many of the key phenomenologists of religion I will consider in Chapters 4, 5 and 6 can be characterized as reacting against developments in the late nineteenth and early twentieth centuries which witnessed the increasing impact of the new disciplines of anthropology, sociology and psychology on the study of religions. Under the influence of Darwin's theory of biological evolution, the philosopher Herbert Spencer's notion of the evolution of ideas and Ludwig Feuerbach's projection theory, late nineteenth-century social scientists, such as the anthropologists Edward Burnett Tylor and J. G. Frazer, the sociologist Émile Durkheim and the psychoanalyst Sigmund Freud, obtained a theoretical framework through which they could interpret the developments of cultures according to levels of complexity and sophistication. This enabled them to study religions as representing stages of cultural evolution, from the lowest, most primitive forms found in simple societies, such as those in Africa and Australia, to the highest, most complex examples, notably those religions of Western civilization. The anti-evolutionary tendency of key phenomenologists is reflected by their refusal to reduce religion to any explanation emanating from the emerging social sciences and by their insistence that believers' perspectives must retain a central place within any truly scientific interpretation of religion.

Although in many ways, through the course of the twentieth century, the phenomenology of religion has been identified by its opposition to the reductionist tendencies of the social sciences, not all formative influences on phenomenology need to be seen in such a negative light. An important figure who has contributed to a study of religions in the twentieth century by combining theological, and hence pro-religionist sentiments, with sociological interpretations of religion, is Ernst Troeltsch. Troeltsch is particularly important as a formative influence because he followed in the line of Ritschlian theologians, but in certain important respects, he restated Ritschlian ideas in historical terms and thereby contributed to a new science of religion that was grounded in empirical methods without undermining the religious commitments of adherents. Troeltsch also modified the typological classifications developed by his friend and colleague, the sociologist Max Weber, whose sociological method contributed significantly to a dynamic understanding of religious processes within history. Both Troeltsch and Weber, therefore, must be seen as important figures who, from perspectives within the social sciences, influenced many of the central ideas that later characterized the positions maintained by phenomenologists of religion. Another leading thinker from the social sciences who adopted a positive attitude to religion is the psychoanalyst, Carl Gustav Jung, who played an important contributing role in the phenomenological interpretation of symbols, particularly in the writings of Mircea Eliade. Jung developed an early association with Sigmund Freud, but later broke with him on interpretations of psychoanalysis, including the relative significance of the sexual drive in the human psyche. Unlike Freud, Jung regarded religion as therapeutic, since the universal archetypal images reflected in religious symbolism helped reconcile many of the psychological conflicts that had produced neurotic behaviours and attitudes among his patients, particularly those in the second half of life.

In this chapter, I will outline some of the significant formative influences on the phenomenology of religion that emerged out of the burgeoning social sciences in the nineteenth century by focusing primarily on the thinking of Troeltsch and Weber in the history and sociology of religions, and on Jung in psychoanalysis, each seen as a positive contributor to later phenomenologies. I begin by providing a brief overview of some of the key trends within nineteenth-century thinking as a backdrop to understanding the development of the phenomenology of religion generally, but more specifically to place Troeltsch, Weber and Jung into their social and historical contexts.

Philosophical and contextual background of the nineteenth century: From Feuerbach to Müller

As we saw in the last chapter, following the philosophy of Fichte, the German philosopher G. W. F. Hegel sought to demonstrate that all human intellectual development up to now has resulted from the logically necessary working out of Mind coming to know itself through a dialectical process. Hegel believed that human consciousness, when viewed in the light of history, was moving towards an eventual completion or fulfilment in a final unifying synthesis in the Absolute Spirit, which generates no further theses or antitheses. Hegel's philosophy had important implications for the subsequent development of the social sciences, partly through the influential writings of his student Ludwig Feuerbach (1804–1872), who eventually became a leading critic of Hegel's idea of the transcendence of an Absolute Mind. For our purposes, the key idea in Feuerbach's philosophy is the concept of 'projection'. If, for Hegel, all historical processes ended in the Absolute Spirit, which was in fact human consciousness of an Absolute consciousness, Feuerbach wanted to know why we need the Absolute Spirit at all. In fact, consciousness really is the consciousness of the self, projected on to something we call Absolute. Feuerbach outlined this argument in his influential book, *The Essence of Christianity* (first published in 1841) in which he contended that God represents the essence of the human perfected, but

68

abstracted from individual, embodied humans and objectified and worshipped as a distinct entity. Humanity therefore attributes to God its own highest feelings, thoughts and hopes and thereby reverses the proper subject–predicate order. Rather than affirming the biblical formulation, 'Man is made in the image of God', Feuerbach argued that its proper rendering should be: 'God is made in the image of man'. In *The Essence of Christianity*, Feuerbach put it this way: 'Consciousness of God is self-consciousness, knowledge of God is self-knowledge' (1893: 12).

This reversal of the customarily understood subject–predicate order defined for Feuerbach the basis for a genuine humanism. So long as humans project on to an ideal being the attributes most desired by humans themselves, humanity will be plagued with self-deprecation, guilt and helplessness. However, if humans can understand that the very ideals that are projected on to the divine being are in fact *human* ideals, then humans can concentrate on developing these ideals within themselves. The subject–predicate reversal shows this clearly. For example, if we say, 'God is love', the very attribute we most admire in ourselves (the capacity to love) is attributed to a being, who, because of its transcendent perfection, makes all human efforts to practise love appear impotent and flawed. However, if we begin to see that the perfect love projected on to God is in fact that which originates in human situations, we can reverse the subject–predicate order and declare: 'Love is God'. This would allow humans to cultivate, develop and enlarge the love we already know within experience towards creating its perfection within human history.

Since for Feuerbach 'all the attributes of the divine nature are ... attributes of the human nature' (1893: 14), this principle can be applied to any of the attributes commonly projected on to God, such as omnipotence, omniscience and omnipresence. It is clear that power, knowledge and space result from human experiences in the world. That humans possess limited power, knowledge and spatial extension should not thereby result in the devaluing of human efforts in favour of a being who is postulated to have these very human experiences, but without limitation. The subject–predicate reversal should be applied in every case. Instead of declaring that 'God is all-powerful', we should affirm that power, used responsibly and ethically, is divine. Or, rather than calling God all-knowing, we should elevate knowledge, wisdom and reason to the level of ideal human aspirations. Instead of calling God present everywhere, we should affirm that humanity is capable of filling space with meaning and purpose. These examples show abundantly that belief in God dehumanizes all legitimate human efforts by deflecting the human imagination away from this world to a postulated other world ruled over by a divine being of our own creation. The idea of God is thus shown by Feuerbach not only to be dehumanizing, but in sharp contrast to Hegel, historically destabilizing.

With Hegel's philosophy of history having been transformed into a humanistic project by Feuerbach, the time was ripe for an all encompassing naturalistic interpretation of history to appear. This was supplied by Charles Darwin (1809–1882), who published *On the Origin of the Species* in 1859 (Darwin, 1964 [1859]).

Darwin's theory of biological evolution, in a general sense, explained the emergence of life forms, including humans, as resulting from competition among and within species through a process which eventually weeds out the less fit. This is accomplished by the selective retention of desirable genes, which are then transmitted from generation to generation, ensuring that some species obtain an advantage over others. Throughout the course of history, this dictates that only the fittest survive. As environmental conditions change, some characteristics which had proved valuable for survival were rendered less advantageous allowing other, better adapted species to take their place. Eventually, humans would emerge as the species most suited to survival and thus as the one form of life that could dominate all other life forms.

Darwinism provided much more than a biological explanation for the emergence and evolution of life forms. It provided a comprehensive theoretical framework within which human history, including human cultures and their religions, could be interpreted. When accompanied by the Hegelian concept of history as the working out of the dialectical process turned back upon itself by Feuerbach's reversal of the subject–predicate order, it provided a firmly empirical basis for how lower forms of life, cultures and religions evolve historically into higher forms. In his seminal work, *Comparative Religion: A History*, Eric Sharpe (1986: 48) argues that when viewed in Darwinian terms, religion became something it had never been before: 'From becoming a body of revealed truth, it became a developing organism'. Quoting the Oxford anthropologist R. R. Marett, Sharpe entitles a chapter of his book, 'Darwinism makes it possible', implying that before the Darwinian interpretation of history, although there had been various descriptions of the religions of the world, and in particular of savage peoples, they lacked a suitable method by which to systematize their material. Descriptions were recorded in rather haphazard fashion and the most exotic and strange behaviour of primitive people was emphasized. With the Darwinian thesis, a certain order could be inserted into the data. Sharpe explains:

> The coming evolutionism changed the picture radically. In the new evolutionary perspective, the mind of 'primitive' man was removed from the lower rungs of the hierarchical ladder, and could be seen to be human, if childish, and therefore worth studying. His religion could also be seen to be eminently worthy of attention, if only as a means of demonstrating the earliest stages through which the faiths of mankind had passed on their way to the heights of ethical monotheism, or the heights of agnosticism, whichever was preferred. (Sharpe, 1986: 48–9)

The Darwinian interpretation of history and culture obtained a powerful voice through the philosopher Herbert Spencer (1820–1903), whom Morris (1987: 2) associates among the founders of social science, along with Durkheim and Weber. Spencer articulated an evolutionary interpretation of history, even before Darwin's theories became widely known. In his *First Principles*, he defined evolution as 'a universal process', which, 'under its primary aspect, is a change from a less coherent form to a more coherent form' (1880: 327). Based on a model derived from physics, where the law of evolution entails 'the integration of matter and concomitant dissipation of motion' (1880: 285), Spencer held that, in its biological and social outworkings, evolution always displays a development 'from a homogeneous state to a heterogeneous state' (1880: 330). Forces that create diversity also foster complex organisms which have intricate and specialized functions. These, when unified, produce highly efficient forms of life. Spencer applied this process to religion, and, not unexpectedly, found that the religion of 'primitives' was marked by a certain 'incoherence' and 'homogeneity', which through time became both more coherent and diverse in the complex structures of the great religions.

The evolution from homogeneity to heterogeneity becomes apparent when the simple organization of primitive societies is shown to progress into more specialized forms in higher civilizations. Spencer explained: 'The developmental process by which these traits are turned into those of the adult European ... is a continuation of that change from the homogeneous to the heterogeneous' (1880: 342). For example, all cultures draw close connections between law, manners and religion. In advanced societies, these fall to the responsibility of diverse individuals and/or institutions. In primitive societies, however, they usually are combined in one person, the chief, since 'very early ... in the process of social evolution, we find an incipient differentiation between the governing and the governed' (1880: 343). For primitives, the chief held ultimate responsibility for the law and its enforcement. At customary functions, the chief acted as the 'Master of Ceremonies'. Because of the ubiquitous primitive preoccupation with ancestors, the Dead Chief became what Spencer called 'the aboriginal God' (Sharpe, 1986: 33). Humankind's first deity, Spencer speculated, was the one who in life was given the deepest respect by enforcing the law, and the one who embodied customary practice and tradition. He outlined this process in the following way:

> As all ancient records and traditions prove, the earliest rulers are regarded as divine personages. The maxims and commands they uttered during their lives are held sacred after their deaths, and are enforced by their divinely-descended successors; who in their turn are promoted to the pantheon of the race, there to be worshipped and propitiated along with their predecessors; the most ancient of whom is the supreme god, and the rest subordinate gods. (Spencer, 1880: 343–4).

For Spencer, therefore, the evolution of cultures, which had been confirmed by the Darwinian theory of biological evolution, was based on what he regarded as wide ranging scientific evidence, beginning with the physical sciences. His conclusions, which influenced philosophy, sociology and religion, were so far-reaching that it is no exaggeration to conclude that his views contributed significantly to the widely held assumption in nineteenth-century thinking that religion, along with every other area of human culture, ought to be treated in evolutionary terms.

Another key figure in nineteenth-century evolutionary social theory is Auguste Comte (1798–1857), whom J. G. Platvoet (1988: 6) calls the 'father of sociology'.

The movement he founded, positivism, came to define in precise terms for nineteenth-century thinking what constitutes a social *science*. Comte's major contributions to thought, like those of Spencer, strongly influenced nineteenth-century theories in philosophy, history, sociology and religion. His theory was based on the evolutionary assumption that societies go through three stages: the theological, the metaphysical (philosophical) and the positive. The overall effect of these stages is progressive, from lesser to more advanced levels of intellectual development. The theological stage, which he associated with the medieval period in the West, is characterized by beliefs in gods and spirit forces. Humans explain all events by personified supernatural causes of successively, a fetishist, polytheist and monotheist type. He did not regard this as wrong, but as childlike, immature, and thus at a lower level of development. The metaphysical stage is associated with the scientific revolution, but was not science in the true sense, because metaphysical philosophers try to explain reality with the aid of abstract notions about the essence of things. Hence, philosophers ask: What is the substance of the universe? What is the nature of ultimate reality? Such questions lead to speculation without recourse to the actual world in which we live. By the positive stage, Comte meant 'positive science' through which humans attempt to discover the laws by which empirical reality is governed. At the time he was writing, Comte believed that the stage of positive science had just begun in Western Europe. This phase, which Comte endorsed and which he saw himself as a key figure in producing, confines explanations of events in the world to verifiable and measurable correlations between phenomena.

In his discussion of Comte, Brian Morris (1987: 51–2) lists four characteristics of positivism:

1. It is opposed to metaphysical thinking. Only logical and scientific propositions are considered valid; all other forms of understanding, such as art and religion, are seen as meaningless.
2. Science deals with empirical facts that are independent of the human subject. Scientific knowledge consists of analytical–inductive generalizations based on sense impressions.
3. A distinction is made between judgements of value and judgements of fact. Value judgements or morality have no empirical content that can be tested. Scientific facts, including social facts, are value-free and neutral.
4. The natural and social sciences share an essential unity. The same methods employed within the natural sciences are applicable within the social sciences. Social sciences, like all sciences, identify causes and their necessary effects.

Comte's positivism thus assumed a progressive movement in human societal and intellectual development in stages from childish (religious) to philosophical (speculative) to scientific (testable and realistic), thereby influencing both how religions were regarded (childlike) and how they should be studied (scientifically, according to fact, not value).

It was against the backdrop of evolutionary theory, but from an entirely different perspective, that the foundational thinking of F. Max Müller (1823–1900) emerged. A philologist and student of Sanskrit, Müller is frequently called the 'Father of Comparative Religion' (Sharpe, 1986: 35). He was born in Germany, studied at Leipzig, but spent most of his life in England in Oxford. He introduced into the field the concept of the study of religion as a science, and sought to investigate texts as evidence of his theory that the religious impulse in humanity began as a quest for the infinite. The quest for the infinite, however, does not make a religion. It must be accompanied by a sense of morality. Hence, Müller defined religion as 'the perception of the Infinite' under 'such manifestations as are able to influence the moral character of man'. He added: 'If there are perceptions of the Infinite unconnected as yet with moral ideas, we have no right to call them religious' (1891: 294–5).

To understand Müller as a foundational figure in the study of religions, we need to place him first as a philologist and through this as a scholar in comparative mythology and religion. By studying the emergence of language, he believed he could uncover how mythology developed and what role it played in religious thinking. In the Vedic literature of India, he believed he could analyse myths as expressed at a very early phase in human history. By studying the *Rg Veda*, in particular, he discovered a pattern beneath apparently disorganized textual material that displayed an order in nature by referring, for example, to the seasons, the regularity of the lunar cycle and normal movements of stars and the sun. The regularity of natural phenomena thus explained the earliest form of mythological thinking and indicated the universal human quest for the infinite. In his Gifford Lectures, delivered in the University of Glasgow in 1890, Müller explained: 'The Veda has now become the foundation of all linguistic, mythological, and religious studies ... Nay, religion itself ... assumes, when watched in the Veda, a character so perfectly natural and rational, that we may boldly call it now an inevitable phase in the growth of the human mind' (Müller, 1891: 18).

This analysis led to Müller's own take on the evolution of humanity. According to Laurens P. van den Bosch (1999: 31), Müller did not share the same view of evolution as had been advocated by many social scientists and ethnologists under Darwinian influence. By studying the ancient Vedic texts, van den Bosch (1999: 31) contends, Müller believed he could trace 'the development of ancient Indian religion through its various stages' beginning with its 'childhood'. The place of nature in the earliest form of religious thinking among the people of India by this form of analysis becomes apparent, and it serves as a pattern for the 'childhood' stage among all peoples. In his two volumes outlining world mythology, entitled *Contributions to a Science of Mythology* (1897), Müller writes: 'If the ancient Greeks or Aryans of India began to ask, whence came the rain and lightning, whence sprang hail and snow, heat and cold, day and night, coming and going in regular and irregular succession ... this arose not only from a necessity of thought but at the same time a necessity of language' (cited by van den Bosch, 1999: 31). Just as language formation moves from concrete to abstract expressions as humans develop from childhood to adulthood, so too does this occur in human cultural development, as for example in giving a name to the wind or the sun (as a child would) to more philosophical or abstract ways of speaking. On this assumption, Müller argued that there is truth in all religions, even those regarded as the lowest developmentally. Even primitive religions, which in many ways are the most childlike, possess a quest for the infinite with all its moral implications.

Because he regarded the study of religions to be a science of how humanity had perceived the infinite and how that had influenced morality, Müller was an early exponent of the idea of natural revelation, whereby the yearning for God could be traced to a divinely implanted predisposition within humanity as a whole. For Müller, religion originates from these innate human responses to natural phenomena which 'arouse in the human mind the idea of something beyond, of something invisible, yet real, of something infinite or transcending the limits of human experience' (Müller, 1898: 90). Although in this sense Müller could be regarded as an evolutionist of sorts, he maintained that religions change throughout history, just as languages, concepts and expressions change. This led to his assertion that religion may have originated not as a childlike idea or a primitive concept, which was expressed as the worship of objects or in the performance of magical acts, but had resulted instead from the innate concept of the infinite in human thinking processes. This original idea of seeking the infinite, in some situations, could even have devolved, as opposed to evolved, into forms of magic and superstition (Morris, 1987: 93). As a result, if we want to understand 'pagan religions' in a true and fair way, we must evaluate them in terms of their 'progress towards truth' and regard them as 'a yearning after something more than finite' by acknowledging within them 'a growing recognition of the Infinite' (Müller, 1891: 328). For Müller, then, religious development, because of the innate religious instinct in humanity, could just as well be classified by stages of decay as by stages of progress, giving a quite different interpretation to evolution than was being expressed, for example, by Comte or Spencer. On the basis of his positive evaluation of the religious impulse in humanity, Müller condemned narrow, one-dimensional and confessional approaches to the study of religion as unscientific. He is famous for applying Goethe's oft-quoted statement to religion: 'He who knows one, knows none.'[1] By this, Müller, speaking from his own context in Oxford, meant primarily that if a scholar of religions were to know only Christian history and theology, the scholar would know nothing of religion. In this sense, for Müller, all religions contain truth and all are worth studying comparatively and scientifically, a process for him that must be founded on sound philological knowledge.

This brief survey demonstrates that the pervasive intellectual climate throughout nineteenth-century thinking, from Hegel to Müller, was dominated by concepts of development, progress and evolutionary theory. With Darwin, the speculative progression of Hegel seemed to be founded in humanistic and empirical facts, carried to their most extreme in Feuerbach and Comte. Spencer's philosophical analysis of social processes, accompanied by Müller's emphasis on language and texts, seemed to confirm that it was possible to study the religions of humanity, scientifically and comparatively. The phenomenology of religion cannot be understood apart from these overriding trends that emerged out of nineteenth-century intellectual formulations, largely, as I noted above, in reaction against their reductionistic tendencies. The social sciences, however, also included positive approaches to religion, as noted in the case of Müller's philological method, that were embraced by phenomenologists of religion, if not explicitly, at least implicitly. In Ernst Troeltsch and Max Weber, specifically, we find social scientific interpretations of development that eschew evolutionary thinking without denying causative relations between dynamic historical and social forces. In C. G. Jung, religious myths and symbols are described as embedded in the very nature of the human psyche.

The significance of Ernst Troeltsch

In the preface to his edited volume, *Ernst Troeltsch and the Future of Theology* (1976), John Clayton Powell (1976: ix) observes that in recent years 'there has been in Germany and North America especially, and more recently in Britain as well, a renewed interest in nineteenth-century religious thought'. Powell adds that this is due in part to the fact that 'the mainly methodological issues which are central in contemporary religious studies were either raised first or refocused significantly during that century'. Powell argues that the clarification of such methodological issues 'will be advanced only after a thorough reassessment of certain developments in nineteenth-century religious thought, including some which are closely associated with the name of Ernst Troeltsch (1865–1923)'. Walter Capps (1995: 168) contends that the main contribution Troeltsch made to religious studies was to take 'Max Weber's insights' and apply them 'both to identify the makeup of the Christian religion as well as to help explain why it is able to assume a variety of cultural shapes and social expressions'.

In this section, I shall argue that Troeltsch's contribution to the study of religions falls into three categories: 1) his interpretation of the history of religions as the data for the study of all religions, including Christianity; 2) his formulation of a religious *a priori*; 3) his use of ideal types, which he devised under the influence of Max Weber. In this sense, I contend that Troeltsch's importance for the phenomenology of religion extends beyond his application, in Capps' words of 'a revised Weberian project' in support of the normativity of Christianity, to his constructing a comprehensive theory of the role of history in the overall academic study of religions.

Troeltsch first read theology at the University of Erlangen, but later moved to the University of Göttingen to study under Albrecht Ritschl (Drescher, 1992: 18–19). Ritschl's influence can be seen throughout Troeltsch's writings, particularly in his evaluation of Christianity in relation to other religions. Nevertheless, Troeltsch differed from Ritschl by emphasizing the importance of understanding Christianity as an empirical phenomenon within the overall history of religions, an approach that ran the risk of making Christian truth relative among all other truth claims. As Hans-Georg Drescher (1992: 160) notes, 'If as a result of the application of historical thinking to Christianity, Christianity is in principle recognized as a historical phenomenon, then it is a conditioned, relative expression of religion and not the realization of the "essence" of religion'.

Troeltsch and the history of religions

In a series of articles Troeltsch wrote between 1897 and 1916, but not translated into English until 1991 under the title *Religion in History*, Troeltsch spells out clearly how he interpreted the history of religions, and the place of Christianity within it. In the earliest of the articles, entitled 'Christianity and the History of Religion', Troeltsch acknowledges that the modern, scientific approach to history has created 'entirely new' problems for Christianity. Specifically, 'the rise of a comparative history of religion has shaken the Christian faith more deeply than anything else' (1991: 77). By demonstrating that people everywhere employ 'myths and traditions', which bind 'cultures and religious laws to the whole of life', the unique nature of Christian faith appears to have been disproved (1991: 78). Moreover, ethnographers and anthropologists have discovered that so-called primitive peoples, 'peoples-without-history', share many of the same features as 'the oldest traces of the cultural and religious development of the civilized nations' (1991: 77). Troeltsch concludes: 'As a result, Christianity lost its exclusive-supernatural foundation' and became 'perceived as only one of the great world religions, along with Islam and Buddhism' (1991: 78).

The implications of this extend beyond Christian theology for Troeltsch to challenge the concept of religion as a whole. Religion, as something unique within human experience, seems undermined by the vast diversity of beliefs and practices. Historical study seems to suggest that even belief in God, which, in Troeltsch's words 'manifests itself in a thousand different forms', depends on highly specific social and cultural contexts (1991: 78). Many religious beliefs, which claim some sort of revelation from a supernatural source, contradict one another. Troeltsch asks, if religion is understood as humanity communing with the Deity, how can there be *religion* at all when there are so many *religions?* In this sense, historical study seems to have entirely relativized religion and made any search for a common core within religious life impossible.

Troeltsch resolves this problem initially by arguing that historical study uncovers basic tendencies within human nature that have operated universally over the generations. This leads to a new interpretation of religion founded fully on 'a thorough knowledge of the empirical history of religion' (1991: 79). He suggests that in the first instance we must learn to view religion more sympathetically by ridding ourselves of 'doctrinaire, rationalistic, and system-atizing presuppositions' in order that we might 'focus more intently on the characteristic, distinctively religious phenomena' (1991: 79). This implies that the historian should not study 'average people', since these may not reflect the core values of the religion. Through this method, the scholar achieves an understanding of that which cannot be analysed or dissected further, 'an ultimate and original phenomenon that constitutes ... a simple fact of psychic life'. What appeared to be dissolved into myriad diversity now reconstructs itself as a unifying principle. 'Everywhere the basic reality of religion is the same: "an underivable, purely positive, again and again experienced contact with the Deity" ' (1991: 79). Rather than undermining universal religious faith, therefore, the history of religions discloses 'a common dynamism of the human spirit which advances in different ways as a result of the mysterious movement of the divine Spirit in the unconscious depth of the human spirit, which is everywhere the same' (1991: 79).

The core of religion, that without which religion would not be religion, thus is the same everywhere. The more the different religions recognize this, the more will be their commitment to its expression, accompanied at the same time by a reduction of dogmatism and misunderstanding, which Troeltsch refers to as a tendency within religions to 'self-seclusion'. In this way, the history of religions fosters tolerance because it encourages members of a particular religion to comprehend 'the truth-moments of all others' (1991: 79). Moreover, this means that religion, as the expression of 'exceptional profundity, power and lucidity' resulting from 'a great sense of awe before the mystery of a supersensible world' cannot be reduced to mere cultural, social, ethical or even aesthetic factors (1991: 79). Religion remains an irreducible, core experience of the human with the divine, which is experienced in its essential character everywhere.

The core element of religion, as Troeltsch has already noted, expresses itself in concrete historical movements. This means that the manifestations of the core of religion are multiple and that they can be evaluated in terms of how adequately they satisfy the essential religious quest of humanity found in religion's core. As a Christian theologian, it is not surprising, therefore, when we find Troeltsch employing comparisons that invoke values through which he ultimately defines Christianity as the religion supremely capable of expressing the deepest, most profound inner meaning of the human communion with the Deity. Troeltsch argues that Christianity alone 'addresses itself to every individual without exception, presupposing this essential core in everyone', and thereby attains 'full inwardness and purely human universality' (1991: 79–80). Buddhism, for example, does not appeal to a universal experience, since it 'originates as merely the religion of a monastic order' and Islam 'represents a regression from Judaism and Christianity and has never been able to conceal its characteristic ties to the Arab nation and war' (1991: 80).

Above all, what the history of religions makes clear is that human existence comprises a search for meaning, peace of the soul with God, and the need to overcome what Troeltsch calls 'the anguish of the world and all pains of conscience' (1991: 84). Only a religion of redemption can satisfy these deepest needs of humanity, which 'Christianity alone among the religions completes' (1991: 84). The empirical study of history thus demonstrates that a prophetic, redemptive religion constitutes the 'high point' of religion. Troeltsch concludes: 'It is because of its empirical uniqueness and the inner coincidence of what it is with what it demands that we recognize in the Prophetic-Christian religion ... a new point of departure ... in the history of religion; not a conclusion and end calling for rest but the beginning of a new day for the world, with new work and new struggles' (1991: 84).

Although he used historical methods for comparative evaluation, Troeltsch's first lasting contribution to the study of religions resulted not from his efforts to prove the superior nature of Christianity following this method, but from his assertion that the study of religions, including Christianity, must proceed by examining the facts disclosed by history. That he discovered a unifying principle in these facts led him to the problem of relating religions one to the other, but it nevertheless ensured for the history of religions the *sui generis*, or non-reductive nature of its subject matter. Despite his insistence on evaluating religions according to what he claimed were empirically verifiable principles, the relativity of historical facts continued to challenge him, and increasingly it became clear that his assertion that Christianity reflects a 'high point' within the universal human apprehension of religion could not be confirmed by empirical studies alone. For this reason, in his book *The Absoluteness of Christianity and the History of Religions* (1972), first published in German in 1901, Troeltsch moved in a Kantian direction by positing the existence of a religious *a priori*.

The religious *a priori* in Troeltsch's thinking

In his discussion of the significance of Troeltsch's *The Absoluteness of Christianity and the History of Religions* (first published in 1902 as *Die Absolutheit des Christentums und die Religionsgeschichte*), James Luther Adams (1976: 211) notes that Troeltsch addressed the problem of the absoluteness of Christianity by 'proposing a means for the validation of religion ... at the normative level' by discerning a universal 'idea' of religion, which finds its expression in many forms throughout the world's religions. According to Adams, Troeltsch maintained that all religions develop in an evolutionary manner, culminating in Christianity, but they are not evaluated in historical terms following a Hegelian philosophical theory. Rather, they are judged with respect to a Kantian *a priori* 'idea' of religion, which forms the basis on which all historical manifestations of the religions occur. By the religious *a priori*, Adams explains, Troeltsch asserted 'a mystical immediacy with the transcendent', which could overcome 'Kantian metaphysical agnosticism' (1976: 211).

In an essay entitled, 'On the Question of a Religious A Priori' (1991) (see *Religion in History: Essays*), first written in 1907, Troeltsch attributes his use of this term squarely to the problem of relativism that has resulted from the historical study of religions. The notion that historical factors operate differently in the case of Christianity than in other religions has been entirely discredited, and yet, the sense of the Christian that there is something unique in Christianity is so strong that 'it is impossible to relinquish this whole area to a mere psychologism' (1991: 34). Moreover, if the sense of normativity in truth and morals is surrendered, we are left with 'the relativising of everything actual'. Such relativism in turn generates 'an increased yearning for the absolute', which Troeltsch defines precisely as a yearning for religion (1991: 34).

As we have seen, Kant distinguished between phenomenal and noumenal types of knowing, the former referring to the world of experience and the latter to a world of ideas. Science is able to construct laws that operate within the phenomenal world precisely because the mind possesses *a priori* categories that structure the world identically for every person. The categories of the mind thus create the synthetic *a priori*, that which is known formally before experience, with the content filled into the forms after experience. Troeltsch applied this Kantian formulation to religion, whereby he asserted 'the autonomy of reason', that is the primary place given to the rational process, the results of which can be evaluated precisely according to the validity of the rational argument, not to emotions, feelings or psychological states, which are relative to the person experiencing such states. This analysis had the effect of modifying in some ways Kant's distinction between pure and practical reason, in Drescher's (1992: 146) words, by identifying a process for interpreting culture according to a 'pre-understanding' founded on a 'rational critical basis'.

Troeltsch admitted that he had gone somewhat beyond Kant by stressing 'the synthetic function ... of a unified personality, and deriving this unified personality from a rational core that lies behind the working of the soul and its foundational unity' (1991: 36). His aim clearly was to give 'equal recognition' to the experiential and the rationally necessary in order to overcome the twin problems of what he called 'psychologism', which corresponds to relativism, and 'dogmatic supernaturalism', which makes its appeal to a 'miracle-apologetics'. The challenge for Troeltsch was to 'extract that which is valid and normative from the psychological and historical' (1991: 37). Troeltsch thus defined his primary task as emancipating historically conditioned religions from a simple cultural relativism and subjecting them to laws of universal validity derived from the *a priori* of reason. Intuition plays an important role in this process, since the evidence obtained by penetrating into the inner core of religion is inexact and subject to error. The process nonetheless is based on 'serious comparison, deliberation and absorption' of particular values which must be thought of as 'approximations to an objective whose general direction is known' (1991: 38). Moreover, such intuitive insight involves an 'act of the will', since the scholar must 'dare to believe that the right path has been found'. The objectivity sought, therefore, is intuited as an *a priori* principle and cannot be derived from experience itself. In this sense, Troeltsch is following an entirely Kantian principle wherein the 'thing-in-itself', that which really exists behind a phenomenal appearance, cannot be known through pure reason.

It is fundamental to note that Troeltsch proposed the religious *a priori* in support of the *sui generis* character of religion itself. He opposed reducing religion to a psychological process precisely because this would make religion appear 'solely as a product and not as a normative principle that posits requirements of its own' (1991: 39). Troeltsch contended that to argue that this kind of assertion is inconsistent with so-called scientific explanations of religion is nonsense, since the assumption, for example, that religion is a product of the struggle for existence in a Darwinian sense is based equally on a presupposition which involves intuition and will. 'The certitude of their validity is not derived from any science' since science itself rests upon 'the spontaneous creative power of reason' (1991: 39). Neither can the religious *a priori*, as a function of reason, be defined as identical with a scientific *a priori*, any more than it is identical with intuitions about morality or aesthetics. The religious *a priori* stands on its own and in certain important respects is different from other types of *a priori* ways of thinking, but it shares with them a similar commitment to reason, 'the power working from the absolute and the unified', as opposed to psychological flux, which is 'atomistically and relativistically divided' (1991: 40).

unique

Up to this point, Troeltsch has avoided defining precisely what constitutes the religious *a priori*, but since he distinguishes it from other types of *a priori*, he is forced to make some comment on its function and actual content. He asserts in the first instance that 'it is the nature of every *a priori*' to be unconditioned, absolutely prior, and thus 'to point to an active presence of the absolute spirit in the realm of the finite' (1991: 41). The religious *a priori*, however, seems to demand a further stage, that is, a 'surrender of the finite spirit to the absolute spirit'. This act of will, freely undertaken, 'demands the formation and the attainment ... of a personality imbued with absolute values' (1991: 41). Thus, the religious *a priori*, which asserts 'an interrelationship of nature and spirit, of the cosmic process and productive freedom' without ever being able to make this fully comprehensible, depends on the prior assumption of the reality of God and the soul. This cannot be proved by appeals to pure reason or to irrational claims to a supernatural authority. It can only be posited as an *a priori* condition for religious knowledge.

On the basis of the religious *a priori*, Troeltsch claims to arrive at the superiority of Christianity over any other world religion. This cannot be derived on the basis of a mere confession of faith, but must result from an analysis of religious reasoning. In *The Absoluteness of Christianity and the History of Religions*, he applies this principle by asserting that the concepts of God and the soul, which as we have seen are critical elements within the religious *a priori*, are inferiorly conceived by every other religion when compared to Christianity. Religions of the law, by which he means primarily Judaism and Islam, 'proclaim the divine will, but they leave the natural man to overcome the world in his own strength' (1972: 114). The non-Christian religions of redemption, primarily Hinduism and Buddhism, 'dissolve man and the world in the divine essence but in the process forfeit all positive meaning and content in the divine nature' (1972: 114). By contrast with all religious expressions, therefore, only Christianity is validated according to the rationality prescribed by the religious *a priori*, since only Christianity 'has disclosed a living deity who is act and will in contrast to all that is merely existent, who separates the soul from the merely existent and this separation unites it with himself' (1972: 114). The religious *a priori* dictates that Christianity represents the 'convergence point of all the developmental tendencies that can be discerned in religion' (1972: 114). Troeltsch's second major contribution to religious studies, the religious *a priori*, although used to engage in 'evaluative comparison', which was rejected by later phenomenologists, nevertheless retained important principles inherent within all phenomenologies of religion: intuitive insight, choice concerning the object of religion and the non-reductive nature of religion itself.

Troeltsch's understanding of ideal types

As I noted above, Walter Capps credits the primary significance of Troeltsch for the study of religions to have been his development of ideal types of religious organization, which he derived from a sociological model in close association with Max Weber. Indeed, Troeltsch and Weber were close friends and colleagues. In Heidelberg, where Troeltsch lived and worked for 20 years as a theologian, Weber and Troeltsch shared a house. After Weber died in 1920, Troeltsch wrote that he had 'experienced the infinitely stimulating power of this man in daily conversation', and admitted, 'I am conscious of owing to him a large part of my knowledge and skill' (1991: 362). Typologies became a critical component within all phenomenologies of religion, and in this sense, the Weberian influence through Troeltsch rightly counts as a positive contribution to the phenomenology of religion originating from sociology as a discipline.

It is important to note another connection between Troeltsch and Weber, that of Wilhelm Dilthey, the late nineteenth and early twentieth-century German philosopher, whom Troeltsch refers to as 'my teacher' (1991: 33) and whom Brian Morris asserts strongly influenced Weber's emphasis on the ethical dimensions within religions (Morris, 1987: 57). In particular, Dilthey distinguished between the natural sciences (*Naturwissenschaft*) and human or cultural sciences (*Geisteswissenschaft*). This distinction had important consequences for all scientific approaches to the study of human behaviour in that those who emphasized the positivist, experimental function of the natural sciences over the sometimes intuitively interpretative role of a social scientist tended to explain human activity by reference exclusively to measurable or quantifiable data. Those who advocated the development of the study of human activity as a social science, as Robert Farr (1996: 21) notes in the case of social psychology, 'analysed mind in terms of its external manifestations, i.e. in terms of culture'. Morris (1987: 57) explains that Dilthey 'advocated a philosophy of life stressing the significance of lived experience and the need for the intuitive understanding of specific cultures or world views'. This idea was particularly important both for Troeltsch's and Weber's interpretations of ideal types.

In his essay comparing what he called Stoic-Christian natural law with modern secular ideas of natural law (1991) [1911], Troeltsch clarified how sociological typologies could be employed to interpret the rise of the early Christian community. He argued that the early church, because of its 'religious and liturgical unity', constituted a new sociological entity based on its ability to combine 'radical religious individualism with an equally radical religious socialism' (1991: 323). This produced a conflict with the social ideals that dominated late antiquity, which in the process of resolving, produced what Troeltsch calls 'three major types' and 'which in later history become increasingly accentuated and mutually antagonistic' (1991: 324). The three types, which can only be understood in relation to the 'sociological structuring of the Christian religious idea', are the church-type, the sect-type and the enthusiasm/mysticism-type.

By the church-type, Troeltsch refers to a community founded on the theological conviction that salvation constitutes part of a divine decree, and in that sense has already been realized. Salvation in this view necessarily involves the religious community, but it also requires individuals to claim that which has been given freely and which has already been accomplished by the death of Christ. Although individual responsibility is apparent in this scheme, what Troeltsch calls 'objective salvation' is embodied in the Church, which he says has been 'established by Christ, with its apostles and their priestly successors, and with the sacraments instituted by Christ and permeated by his living presence' (1991: 324). This basic form has been carried forward in history by Catholicism, 'with its constant reinforcement of the objective principle' and Protestantism, which like Catholicism 'is the institution of salvation and grace'. The church-type accepts that neither the church nor the world is yet perfected, and thus as an organization within society 'it can compromise with the existent structures of the world' and even acknowledge that, although they are 'less than Christian' they are 'very useful (for the time being) in the discipline and organization of the sinful world' (1991: 324).

The sect-type operates very differently from the church-type. Theologically, the sect-type demands the actual overcoming of sin and living up to the divine commandments. In this sense, in sociological terms, it refuses to compromise with the world by fostering 'a state of holiness based on a conscious act of the will' (1991: 325). Many examples of this type are found in Christian history: Montanist and Donatist movements, medieval Waldensians and even early forms of the Franciscans. The sect-type is characterized by a 'strict Christian radicalism', which, Troeltsch says, makes it 'bound to clash with natural sociological laws and with competing social ideals'.

Enthusiasm or mysticism constitutes Troeltsch's third religious type. This type is characterized by its aim to attain immediate union with God through an inward religious experience. The organizational forms of religion are quite secondary to this aim and are useful, according to Troeltsch, 'merely as a stimulus and a means towards the inner, timeless intercourse with the divine'. In Christian history, mysticism finds its seeds in the experiential enthusiasm of the Apostle Paul, which when fused with neo-Platonism, produced the idea that the sparks of God can be found in the finite and which through contemplation can elevate the soul towards its union with the divine. Troeltsch argues that Christian mysticism flourished particularly when 'inward feeling was left unsatisfied by the objective character of cult, dogma, and institution' (1991: 326). The mystical-type differs from the sect-type in that it is concerned strictly with 'the immediacy of union with God' whereas the sect-type 'is at pains to stay close to the letter of the historical word ... and to the organization of the holy community' (1991: 327). The mystic, by contrast, is largely indifferent to the world, and limits contact with others to those who share the same orientation in life towards the transcendent. As a result, mysticism need not produce a social organization, and is often associated with monastic orders or the life of a hermit. For Troeltsch, 'the relationship of mysticism to the secular world is essentially that of indifference to, and abstention from, all concerns and forms of life incompatible with the mystical life of God' (1991: 328).

By identifying these three typical social groupings that arose in history as a result of particular theological constructs, Troeltsch was able to analyse various responses from within church history to what he called 'non-Christian ideals of social life'. He then applied this to his interpretation, based on a consideration of the idea of natural law in late antiquity and in modern times, of the proper role of the church within the social conditions defined by the contemporary secular state. It is important to note that for Troeltsch the ideal types could only be understood in historical contexts, and that their usefulness was purely analytical. He was employing sociological categories, which he derived from theological positions, to demonstrate the way the dynamic forces of history interact with social conditions. The ideal types thus were pragmatic constructs, and in no way can be regarded as essential categories or ontological realities.

The third contribution of Troeltsch to the phenomenology of religion thus finds its greatest significance in his sociological and historical uses of religious typologies. When this is considered in the light of Edmund Husserl's eidetic vision, which I outlined in Chapter 1, it will be clear that phenomenologists of religion consistently walked a fine line between employing socially and historically contingent categories while constructing universally valid types with a timeless and essential meaning. I will return to this issue frequently in the next chapters, but I raise it here to indicate the importance of Troeltsch, who was writing from within a religious position, for those who contend that a sympathetic approach *to* religion need not result in decontextualized studies *of* religion. It is precisely at this point that the thinking of Max Weber becomes most significant.

Max Weber: Sociology and religion

Eric Sharpe (1986: 177) argues that the work of the sociologist Max Weber 'was of an astonishing breadth and penetration', but that much of it 'is of only indirect significance for the comparative study of religion'. Usually, Weber is discussed in relation to religion in terms of his own use of ideal types, which in many ways were similar to those of Troeltsch (although from a non-theological point of departure), and with respect to his analysis of the impact of the Protestant ethic on the rise of capitalism. In both cases, the importance of Weber may be more extensive than is implied by Sharpe, since we see in Weber a way in which religion can be interpreted by using social categories without reducing religion to those categories. Unlike other earlier scholars writing from within the social sciences, Weber's influence thus was regarded positively by phenomenologists of religion and at least indirectly invoked in their own analyses of religious types and social processes (James, 1995: 178).

In his text, *The Sociology of Religion*, Malcolm Hamilton discusses Weber under the heading 'Religion and Rationality', arguing that Weber identifies developments within religions through a process of 'ethical rationalisation' (Hamilton, 2001: 158). This results in Weber's thinking, according to Hamilton, from his psychological interpretation of the role of religion in human existence. In Hamilton's view, Weber regards religion as 'a response to the difficulties and injustices of life' at varying levels of complexity (2001: 156). At the most basic level, humans have employed magic to try to manipulate forces to overcome problems related to survival and well being. As religious thinking develops, humans begin to assign personal power to unseen beings, which then must be persuaded to assist in overcoming human difficulties and ensuring justice. This of necessity produces a priesthood, which is a natural development from magicians, but priests, unlike magicians, manifest the ethical dimension within religion, since they act on behalf of the community to ensure its best interests.

Hamilton notes that this process corresponds to Weber's well-known analysis of religious personalities, and particularly to the way charismatic religious figures give way eventually to the institutionalization of religious traditions. In Hamilton's analysis, Weber places great importance on the charismatic prophet as one who produces religious innovation by 'indulging in practices which are extraordinary or by undergoing some extraordinary experience' (Hamilton, 2001: 159). The 'non-routine' nature of the charismatic figure challenges established religious practices, creating a new religious movement or at least initiating substantial change within traditional institutions. When the charismatic figure gathers followers, however, the followers begin to routinize beliefs and practices surrounding the prophet, resulting eventually in the re-establishment of institutions, out of which new charismatic personalities may emerge, thus launching a new cycle from routinization to non-routinization to re-routinization (see Chryssides, 2001). The important point in this description in relation both to Troeltsch and to later phenomenologies of religion is to observe the way Weber employs 'ideal types'. These can be seen in the above examples through his analysis of the historical emergence of priestly classes, by his application of religious ideas to economic systems, and by his interpretation of how typical social patterns produce new religions.

Brian Morris (1987: 59) argues that Weber introduced distinct methodologies into the sociological theory of religion by analysing collective social types as if they were, in Weber's words, 'particular acts of individual persons'. This approach, which Morris labels 'methodological individualism', interprets broad social and collective movements as particular instances with specific historical contexts as opposed to a Durkheimian 'holistic' system which reified collectivities, in Morris's words, 'as organisms or cultural totalities' (1987: 60). Morris contends that this aligned Weber with Hegel and Spencer as a 'historical sociologist, not a systems analyst' (1987: 60). The same point has been made by the phenomenological sociologist Alfred Shutz (1972: 6), who argues that Weber 'reduces all kinds of social relationships and structures, all cultural objectifica-tions, all realms of objective mind, to the most elementary forms of individual behavior'. Schutz adds that Weber's sociological method analysed individual actions in order to identify 'ideal types' that could be used to interpret collective behaviour (1972: 6–7). A second methodological emphasis found in Weber's writings is the concept of *Verstehen*, which Morris (1987: 60) says is 'best described as comprehending social action through an empathetic understanding of another person's values or culture'. This led to Weber's rejection of 'objective meaning' in any sense that removes interpretation from its relation to the subjective perspective of the interpreter. Morris argues that this means that Weber, who was concerned primarily to explain social facts, insisted that there could be no disjuncture between objective and subjective meaning in the interpretation of social reality (1987: 60).

Stephen Sharot (2001: 20) insists that Weber's primary contribution to a 'comparative sociology of world religions' is to have created an action-orientated model for understanding ideal types. By this, Sharot refers to the 'dimensions of any action: goals, means and conditions'. Following Weber, Sharot argues that religious action needs to be understood from 'the subjective viewpoint of the actors' (2001: 60–1). Hence, a Weberian interpretation would define the goals (or ends) as 'the future state of affairs toward which actors understand their actions to be oriented' (2001: 61). The means refer to 'those objects, persons, groups, or processes that actors understand can be used or manipulated in the realization of their ends' (2001: 61). The conditions, on the other hand, represent precisely those factors which the actors themselves understand as incapable of being manipulated to foster their desired ends, but which 'must be taken into account or addressed' (2001: 61). Sharot's analysis underscores the point made by Morris that an 'insider's' perspective was paramount for Weber's method for obtaining understanding, and it confirms Shutz's argument that social processes as interpreted paradigmatically through individual behaviours were integral to Weber's construction of the ideal types.

Because of their central importance to the thinking of key figures in the phenomenology of religion, Weber's application of 'ideal types' as religious action or dynamic processes within history, and his theory of *Verstehen* as a non-reductive empathetic methodology, need to be considered by examining his own writings. I propose to do this primarily by analysing an influential early essay Weber wrote on methodology entitled 'Critical Studies in the Logic of the Cultural Sciences' (1949: 113–88). I will then show how Weber exemplified these methods by relating religious beliefs to economic systems in his collection of essays which later became famous as *The Protestant Ethic and the Spirit of Capitalism* (1992; first published 1930).

Weber's statement on the methodology of ideal types

Weber's essay, 'Critical Studies in the Logic of the Cultural Sciences', carries the sub-title, 'A critique of Eduard Meyer's "methodological views"' (Weber, 1949:

113–88). Although the article focuses on historical interpretation in the thought of Eduard Meyer, who held the Chair of Ancient History at the University of Berlin, Weber's primary aim in the article is to analyse history as revealing causal processes that can be classified into typologies.[2] At the beginning of the essay, he asserts, 'The most significant achievements of specialist methodology use "ideal typically" constructed conceptions of the objectives and methods of the special disciplines' (1949: 114). Weber proceeds on the assumption that was widely held by European sociologists, as I noted above, that the natural sciences and the cultural sciences can be distinguished in terms of aims and results. This is particularly important for understanding human, as opposed to natural, actions, and thus for an understanding of historical methodologies. In the case of human actions, history cannot be reduced to rules that apply to cause and effect in the same way that natural actions can. This is because humans, both the interpreter of actions and the actors themselves, are subject to errors in judgement on one level, but, even more significantly, because humans are affected by what Weber calls 'temperaments' and 'moods'. This would seem to negate the possibility of forming general 'action' typologies, which adhere to historical patterns or rules, since, Weber argues, 'reality is constituted only by the concrete and particular' (1949: 128–9).

This would indeed be the case were it not for the fact that the scholar always operates on the assumption that cultures themselves have pre-selected certain points of significance, which the historian then isolates for comparison with other points of cultural significance in different contexts. This does not imply that the same points of cultural significance become dominant in every culture, but that the scholar compares 'analogous' events, in Weber's terms, 'as a means' of imputing 'causal agency'. For this reason, it is incorrect to conclude that the 'particular as such is the subject matter of history' (emphasis his) (1949: 130). This is particularly important to remember when attempting to isolate causal factors under the broad heading of 'development'. By way of example, Weber refers to research on the development of the concept of the 'state' among kinship groups, as noted in recent studies on the Tlingit and Iroquois peoples of North America, both of which moved from kinship-based structures to conceptions of a 'state'. In themselves, the studies of development in Tlingit and Iroquois societies are 'of no significance', but as 'heuristic devices in the causal interpretation of other historical developments', such analyses are 'uncommonly significant' (1949: 132–3).

The particular event and the ideal type thus are inextricably connected, the former providing the real context for analysis and the latter 'typical' instances of an abstract concept. The scholar must use the device of discovering the connections and meanings within the particular (heuristic process), that is to construct an ideal type or pattern, while still upholding the principle that the particular fact represents the actual or real causal event. The ideal type cannot be regarded as real in the same way as the particular event, but the significance of the particular can only emerge from the comparison or analogous reasoning that produces the ideal type. Nevertheless, in order to guard against transforming the ideal type into an ontological or philosophically essentialist category, Weber emphasizes that 'the meaning of history as a *science of reality* can only be that it treats particular elements of reality not merely as heuristic *instruments*, but as the *objects* of knowledge, and particular causal connections not as premises of knowledge but as *real* causal factors' (emphases his) (1949: 135). Only in this way can the ideal type illuminate the significance of particular facts while, at the same time, safeguarding the empirical nature of historical studies. The ideal type, as an heuristic device, thus operates ahistorically or atemporally, but can be confirmed only by reference to actual historical or temporal facts.

In the process of constructing the ideal type in relation to concrete historical facts, the scholar employs the skill of interpretation, which Weber says is like an 'art' and which involves a high level of 'subjectivizing'. This is why by interpreting events the scholar reaches 'the outermost edge of what can still be called the "elaboration of the empirical by thought"' (1949: 145). Interpretation involves 'value analysis' on the part of the scholar, the aim of which is to produce understanding (*Verstehen*). Weber notes, by way of example, that one may 'interpret' the meaning of Goethe's *Faust* or 'investigate' the writings of Karl Marx with respect to their 'intellectual content' rather than speaking about historical events as such, and thus could be regarded as going 'beyond history' (1949: 147). Yet, this would not be entirely true since to understand Goethe or Marx requires an appreciation for the social milieu behind the writing of *Faust* and an awareness of the historically given problem situation that produced *Das Kapital*. In this sense, a 'value judgement' is closely connected with empathetic understanding of the concrete object of research while at the same time it reflects the individual preferences of the scholar which result from 'a thoroughly concrete, highly individually structured and constituted "feeling"' (1949: 150). This does not reduce interpretation to pure subjectivity, since, Weber explains, 'in constructing historical individuals, I elaborate in an explicit form the focal points for possible "evaluative" attitudes which the segment of reality in question discloses and in consequence of which it claims a more or less universal "meaning"' (1949: 151).

To return to the example of Marx's *Das Kapital*, Weber observes that as a literary product, it shares with other books a common format (its structure) and substance (paper and ink), but these characteristics (typologies of sorts) do not define its meaning as a philosophical work. Rather, it is the intellectual content and the circumstances it addresses that make it at the same time specific and general, referring to and emerging from historical conditions and yet falling within a broader type of intellectual discourse. Which specific conditions the scholar emphasizes within the historical milieu of the mid-nineteenth century and which intellectual components emerge as important for the eventual interpretation, nevertheless, are selected by the scholar out of his or her informed interest (1949: 152). Moreover, many constructs that scholars take to be value-free and historically given have resulted from the very same process of selection and interest, such as the idea of culture, the concept of 'Christianity', and even the notion of a nation-state like 'Germany'. 'These', Weber concludes, 'are individualized value concepts formed as objects of historical research' (1949: 160).

Thus far in his argument, Weber has only imprecisely articulated the fully dynamic nature of the ideal type. Towards the end of 'Critical Studies in the Logic of the Cultural Sciences', he discusses the 'driving forces' or 'reverse obstacles' to development, and offers capitalism as an example, in which the complex relationships between causes, patterns, rules of action and linkages between events become evident. One 'law' of causation may overrule another 'law'; a 'certain rule' may be 'transcended by certain causal linkages'. These can all be fitted together if they are understood as abstractions of components within a 'real causal chain'. The abstractions define the possible objective or real occurrences, which help scholars create 'a causal complex with a certain structure' (1949: 188). It is no coincidence, of course, that Weber selected capitalism as his example, since this essay was written at nearly the precise time he was working on his theory linking Protestantism with capitalism, a connection that demonstrates vividly how the ideal type, as a structural pattern, emerges as an abstraction from the scholar's interpretation of particular facts, all in aid of achieving understanding. In his comments on 'Critical Studies in the Logic of the Cultural Sciences', Edward Shils (1949: iv) observes that 'the intricate task of explaining causally the emergence of an "historical individual" (in this instance, modern capitalism) finds its methodological reflection in this essay which treats of the nature of explanation of particular historical events in its relationship to general or universal propositions'. To fully appreciate Weber's argument in the article we have been considering, therefore, we need to analyse the thesis he advanced in *The Protestant Ethic and the Spirit of Capitalism* (1992 [1930]).

Weber on Calvinism and capitalism

It is important to note at the outset that Weber's analysis of the relationship of Protestantism to the rise of modern Western capitalism forms just one part of his overall discussion of the relationship of world religions to economic systems. In other places, he also discusses Hinduism, religions of China (particularly Confucianism) and Judaism in relation to the types of economies they produced, partly to indicate why Western capitalism could not have emerged under the world views of India and China (see Weber, 1958, 1968). Judaism, because of its relation to Christianity and the prophetic tradition, presents a different case altogether (see Weber, 1952). For our purposes, however, it will be important to note that the 'religious action' characterized by the Protestant ethic serves as a model for the dynamic ideal typology, which can be applied elsewhere in other religions with far different historical results. For Weber, moreover, although the Reformation spawned numerous forms of Protestantism, the closest causal link to the rise of modern Western capitalism can be seen in Calvinism, particularly due to its teaching on predestination, which Weber calls 'its most characteristic dogma' (1992: 56). In support of this, Weber cites the Westminster Confession of 1647 where (Chapter III, No. 3) this doctrine is most clearly articulated: 'By the decree of God, for the manifestation of His glory, some men and angels are predestined unto everlasting life, and others foreordained to everlasting death' (1992: 57). Weber indicates that he is not interested in evaluating this teaching, but rather in uncovering its 'historical significance' (1992: 58).

In the Protestant perspective, as opposed to Roman Catholic theology, the doctrine of predestination meant that humans were entirely dependent on the grace of God for salvation. Nothing they could do would make any impact on their ultimate salvation. No priests could intervene; no interpreters of God's word could help; no sacraments could administer grace to assist with salvation; no Church could make any impact on the inscrutable will of God (even members of the Church included those who were destined to damnation). Even Christ is understood as dying only for those who have been chosen. Weber concludes: 'This, the complete elimination of salvation through the Church and the sacraments ... was what formed the absolutely decisive difference from Catholicism' (1992: 61). With the dominance of the doctrine of predestination, Weber reasons, so too the problem of who is elected to salvation and how those so elected can obtain certain knowledge that they are chosen by God, came to the forefront. 'So, wherever the doctrine of predestination was held, the question could not be suppressed whether there were any infallible criteria by which membership in the *electi* could be known' (1992: 66). Paradoxically, despite the radical doctrine that absolutely nothing a person can do can influence his or her ultimate salvation, according to Weber, the evidence of grace, that which alleviates anxiety about whether one is chosen or not, results from 'intense worldly activity' (1992: 67). This follows precisely from Calvinist doctrine because for the Calvinist there was no atonement for sin by the accumulation of good works. All the Calvinist did for good was in order to magnify the glory of God and to promote the will of God on earth. This meant, according to Weber, that 'the God of Calvinism demanded of his believers not single good works, but a life of good works combined into a unified system' (1992: 71). Participation in such a life-consuming effort was the only proof of grace available to the believer.

The effect of this world view was to create an ascetic attitude far different from that of individual ascetics within monastic orders during the Middle Ages. The Calvinist elect formed a 'spiritual aristocracy' which the damned were forever forbidden from joining. This created an attitude towards one's neighbour 'not of sympathetic understanding based on consciousness of one's own weakness, but of hatred and contempt for him as an enemy of God bearing the signs of eternal damnation' (1992: 75). Although this extreme perspective emerged out of Calvinism and its Puritan offspring, Weber argues that it applies much more generally and broadly to Protestantism as a whole, although much less so to Lutheranism. The direct connection to capitalism, nevertheless, cannot be applied uncritically to Protestantism as a whole, but more specifically to ascetic Protestantism and its close association with Calvinism and 'the side of English Puritanism which was derived from Calvinism' (1992: 102).

The moral attitude of the ascetic Protestant is defined by a corresponding attitude towards work. Only in the next life is one permitted to rest, to enjoy for the sake of enjoyment, to luxuriate in the eternal presence of God. In this life, the member of God's elect must avoid all temptations of the flesh, never waste time, or even engage in what Weber refers to as 'inactive contemplation' (1992: 104). This attitude towards idleness bears directly on the concept of wealth, particularly the accumulation of wealth. According to Weber, 'if that God, whose hand the Puritan sees in all the occurrences of life, shows one of His elect a chance of profit, he must do it with a purpose' (1992: 104). Indeed, individual profit for the glory of God, not for personal indulgence and pleasure, becomes a Christian duty, and that as a direct result of a life of work. And, yet, the greater one obtains profit, the heavier is that person's responsibility before God and the more becomes the duty to increase profit, in Weber's words, 'by restless effort' (1992: 115).

As a prelude to exposing finally the connection between Protestantism and modern capitalism, Weber summarizes his argument thus far:

> This worldly Protestant asceticism, as we may recapitulate up to this point, acted powerfully against the spontaneous enjoyment of possessions; it restricted consumption, especially of luxuries. On the other hand, it had the psychological effect of freeing the acquisition of goods from the inhibitions of traditionalistic ethics. It broke the bonds of the impulse of acquisition in that it not only legalized it, but (in the sense discussed) looked upon it as directly willed by God. (1992: 115)

Thus, a class of bourgeois entrepreneurs was created under the guise of faithfully performing the duties of God's elect in order to magnify the glory of God and promote his will on earth. Moreover, those who worked for the entrepreneurs were infused with an equally ascetic duty to work hard, never to lie idle and thereby to contribute to the productivity of the collective enterprise. 'The treatment of labour as a calling became as characteristic of the modern worker as the corresponding attitude toward acquisition of the business man' (1992: 121).

What Weber calls 'the spirit' of modern capitalism thus has its roots fully in the form of Protestant asceticism he has been describing. The main difference is that, in its contemporary expressions, the religious content has faded away with the inevitable consequence that whereas the duty-bound Puritan freely engaged in such a life of grinding devotion to work, the modern person does so out of necessity. With the rise of modern technology and production, the capitalist economic order has begun to dominate 'worldly morality' and can rightly be described as a 'tremendous cosmos' (1992: 123). Weber likens the resulting impact on the 'idea of duty in one's calling' which is so characteristic in modern life, to 'the ghost of dead religious beliefs' prowling about in our lives (1992: 124). At the end of his argument, Weber admits that he is now treading very near the 'world of judgments of value and faith', which his purely historical analysis has attempted thus far to avoid. He claims much less for his conclusion, that of identifying causal affects in one, 'though very important point'. The larger task remains: assessing 'the quantitative cultural significance of ascetic Protestantism in its relation to the other plastic elements of modern culture' (1992: 125).

The Protestant Ethic and the Spirit of Capitalism demonstrates precisely how the idea of 'religious action' works. The ideal type is presented in social, historical and cultural specificity, but at the same time it operates as a paradigm for understanding the causal relationships between religious beliefs and socio-economic systems. By identifying patterns operating in history and relating them to the specific content of historical periods, Weber demonstrates how deeply embedded religious ideas foster overriding cultural systems. Involved in this entire process, of course, is a hermeneutic principle that recognizes the role of the interpreter in the eventual outlining of meaning. The end of the process is understanding in the deepest sense of the word (*Verstehen*), but the meaning disclosed is entwined neither within value judgements nor metaphysical (non-historical) propositions. The resulting patterns remain fully accountable to socio-historical data and thus comprise a genuine contribution to the social and cultural sciences (*Geisteswissenschaft*). An ideal type, therefore, far from being conceived as an essential, atemporal, ontological reality, is shown by Weber, not only to be dynamic, but as having emerged out of the creative hermeneutical analysis by the scholar of concrete and particular socio-historical facts.

Carl Gustav Jung: The psychological advantage of religion

The theories of C. G. Jung (1875–1961) constitute the third major formative influence on the phenomenology of religion from within the social sciences that I am considering in this chapter. Jung's thought emerges from the psychoana-lytical school founded by Sigmund Freud, but, unlike Freud, Jung regarded religion as a positive force within the development of the human personality. For this reason, whereas many phenomenologists of religion reacted against Freud's concept of religion as an 'infantile neurosis', they incorporated some of Jung's main ideas into their thinking, particularly the universal archetypal images embedded within human consciousness. Jung cannot be connected directly to the social sciences and to history in the same way as Troeltsch and Weber, and he had no direct connection with either. Nevertheless, certain similarities can be noted between Jung's archetypal images and the ideal type, particularly as they came to be employed later in the process of individuation, which Jung regarded as emblematic of sound psychological health. As we will see in Chapter 5, certain of these key ideas are translated directly into the study of religions by Mircea Eliade. At this point, however, I simply want to outline briefly Jung's main ideas, particularly since his positive contribution to the phenomenology of religion stands in stark contrast to those scholars writing from within the social sciences who were rejected by phenomenologists as reductionists, among whom Freud was regarded as a chief culprit (see, for example, Thrower, 1999: 143–50; Preus, 1987: 178–204; Pals, 1996: 54–87).

Jung was born in 1875 in northern Switzerland in a village on Lake Constance, where his father was serving as a pastor of the Swiss Reformed Church. As a young boy, Jung's family moved to a town near Basel, where Jung later studied medicine, specializing in psychopathology. In 1906, Jung sent a manuscript he had written on word association to Freud, and later sent him a copy of his book, *The Psychology of Dementia Praecox* (1936 [1907]). Subsequently, Freud invited Jung to visit him in Vienna. Michael Palmer (1997: 85), who has written a substantial and important book comparing Freud and Jung on religion, notes that their first meeting lasted 13 hours without interruption! Between 1908 and 1911, Freud and Jung were close colleagues, but in 1911, a severe break occurred between them over Jung's book *The Psychology of the Unconscious* (1915 [1911]), in which Jung disagreed with Freud's views on the central place of sexual urges in the unconscious. In 1912, during a lecture tour to the United States, Jung publicly criticized Freud, after which their relationship deteriorated and eventually became acrimonious.[3] Anthony Storr (1997: 90) records that after 1913, Jung entered a period of intense psychological difficulty. His last meeting with Freud had been at the Fourth International Psychoanalytic Congress, held in Munich in September 1913. Storr describes the period that followed the next few years in Jung's life as evidence that he was suffering from a psychosis. Jung wrote in October 1913:

> I saw a monstrous flood covering all the northern and low-lying lands between the North Sea and the Alps. When it came to Switzerland I saw that the mountains grew higher and higher to protect our country. I realised that a frightful catastrophe was in progress. I saw the mighty yellow waves, the floating rubble of civilisation, and the drowned bodies of countless thousands. The whole sea turned to blood. This vision lasted about one hour. (Cited by Storr, 1990: 90)

Jung regarded these visions as prophetic, since they seemed to be speaking about the events of World War I. Storr notes: 'This interpretation may have convinced Jung that he was not suffering from a psychosis, but it obviously demands belief in precognition ... Jung seems to have believed that he was the vessel of a higher power which granted him insight into the future' (1990: 91).

After World War I, Jung recovered from what Storr describes as his mental illness, largely with the aid of religion. He later held that for all those over the age of 40, mental health depended on gaining a religious outlook on life. After 1921, his reputation grew largely as a result of his book published in that year, entitled *Psychological Types* (1923) [1921], which Palmer (1997: 86) describes as his 'principal contribution to the psychology of the unconscious mind, and in which he distinguishes between two attitudes to life, the introverted and the extraverted, and to some of his other key ideas: the collective unconscious, archetypal images and the individuation process'. In 1948, Jung and some of his close associates founded the C. G. Jung Institute in Zurich, which became a centre for the practice of Jungian psychotherapy.

The unconscious and the universal archetypes

Freud believed that humans inherit at birth certain instincts embedded in what he called the 'id' – leftovers from the earliest primal instincts where the sons killed their father in order to have access to the women controlled by their father.

For Freud, therefore, all humans inherited an unconscious consisting primarily of repressed individual instincts, remaining collectively in the human psyche from this earlier period in human evolution (see Freud, 1938, 1949). Jung also believed that mental functions could be understood in terms of conscious and unconscious processes and, like Freud, he believed that humans possess a conscious level (Freud's ego), a personal unconscious and a collective unconscious. The personal unconscious varies from individual to individual; it is unique to each person, since it is acquired and developed during a lifetime. Palmer (1997: 99) writes: 'Here we find ... all psychic material that has yet to reach the threshold of consciousness'. These include, in Jung's words, 'lost memories, painful ideas that are repressed (i.e. forgotten on purpose), subliminal perceptions by which are meant sense-perceptions that were not strong enough to reach consciousness, and finally, contents that are not yet ripe for consciousness' (cited by Palmer, 1997: 99). In his essay, 'Archetypes of the Unconscious', first published in 1934, Jung (1972: 4) calls the personal unconscious 'a more or less superficial layer of the unconscious', superficial because 'it rests upon a deeper layer, which does not derive from personal experience and is not a personal acquisition but is inborn'. This deeper layer is the collective unconscious, which refers to that store of images, myths, religious ideas and dreams that are inherited and common to all humanity. Jung explains: 'I have chosen the term "collective" because this part of the unconscious is not individual but universal; in contrast to the personal psyche, it has contents and modes of behaviour that are more or less the same everywhere and in all individuals' (1972: 3–4).

The term 'unconscious' conveys in many ways an ambiguous concept, since what is not conscious would seem to be incapable of being spoken about without bringing its contents to consciousness. For this reason, Jung explains that the contents of the unconscious must be 'capable of consciousness', and thus it is legitimate to speak of the unconscious 'only in so far as we are able to demonstrate its contents' (1972: 4). This applies both to the personal and to the collective unconscious. In the latter case, the content consists of 'archetypes' in the sense that the archetype is 'an unconscious content that is altered by becoming conscious' (1972: 5). In a lecture delivered in 1938 on 'Psychological Aspects of the Mother Archetype', Jung (1972: 10–11) described the universal collective unconscious as determining the patterns within which all human thought processes necessarily proceed, much like Kant's categories of the mind. He explained that 'thinking, understanding, and reasoning' operate within the mind as *psychic functions coordinated with the personality and subordinate to it'* (emphasis his). The content of the collective unconscious is manifested through myths, visions, religious ideas and dreams, such as hero and saviour myths, resurrection stories and 'the concept of the Great Mother'. Each of these themes recurs in all human cultures and in every historical period, and thus conveys fundamental and shared psychological experiences which confirm the existence of universal archetypal images.[4] In this sense, the *content* of the collective unconscious consists of the archetypal images, which represent the pre-existent forms or primordial types that have existed since the remotest times of humanity. They reside in the unconscious of all humans, in Jung's words, as 'unconscious but nonetheless active-living dispositions ... that perform and continually influence our thoughts and feelings and actions' (1972: 13). Their existence is perfectly normal, and thus when they appear in dreams, myths, stories, fantasies and imagination, they simply reflect the accumulated mental imagery which has become part of the human psyche during the process of evolution. Of course, the archetypal images also appear in psychological disturbances, such as schizophrenia, but this does not make the archetypes themselves aberrant, since they persist as universal human phenomena.

Particularly important archetypes, for Jung, included images of the Mother or of child gods. The Mother, for example, often appears in myths or dreams in the form of goddesses, or in symbols of fertility such as mother earth, or even in natural symbols like a ploughed field or a garden. The universal character of the Mother image is capable of numerous permutations. Thus, Jung explains, 'it can be attached to a rock, a cave, a tree, a spring, a deep well, or to various vessels such as the baptismal font, or to vessel-shaped flowers like the rose or the lotus' (1972: 15). Many archetypal symbols are repeated in religious stories and texts, particularly the child god, notable in Christianity and Buddhism, and in Judaism through the miraculous rescue of the baby Moses. Even in Islam, stories abound concerning the miracles of Muhammad as a child. The child god archetypal image is replicated in many European fairy tales and in the visionary poetry of Blake. Closely related to these are hero images, sometimes expressed in myths describing how the hero overcomes a monster, a powerful figure, or even death. Numbers also contain symbolic meaning, and, for Jung, the number four or the quaternary symbolizes completion or wholeness, as exemplified in such symbols as the mandala (1968: 138).

Jung also placed a strong emphasis on the coincidence of opposites. For example, he argued that each person is characterized by the forces of *anima* and *animus*; that is, every human displays both male and female personality characteristics. The *anima* is the feminine side of the male; the *animus* the masculine side of the female. In the male, the *anima* is first encountered in the mother, just as in the female the *animus* is first met in the father. In a man, the *anima* he acquired from his mother will be projected on to those females who awaken positive or negative characteristics within him. A similar pattern occurs in the projection by the female of her father's animus on to males. The *anima/animus* coincidence often appears in dreams, and is evidence of the collective unconscious. Jung (1968: 99) explains: 'The collective unconscious as a whole presents itself to a man in feminine form. To a woman it appears in masculine form'. In religion, such coincidence of opposites often occurs, such as in the Yin-Yang convergence in Chinese religious philosophy.

For Jung, psychological health results from humans achieving what he called individuation through which a person becomes whole or complete. Jung (1923: 561) calls it 'a *process of differentiation*, having for its goal the development of the individual personality' (emphasis his). It is a natural process within the psyche, similar to any physical development, such as growth or ageing. Michael Palmer (1997: 143) explains: 'The personality is thus destined to individuate just as surely as the body is biologically destined to decay'. Jung's description is based on Freud's distinct phases in human psychological development from infancy through puberty to adulthood. The individual crossing from one stage of development to another can become stuck and never move to the next phase, resulting in mental disturbances or neuroses that persist into adult life. For Jung, factors which can cause the process of individuation to become thwarted include both inherited tendencies and environmental factors. In the latter case, these include such things as the influence of parents, education and the social conditions in which one grows up.

Clearly, then, individuation is the goal of human psychological development, which will be achieved if intervening factors do not halt its progression. Individuation could be called the 'archetype of wholeness', which entails the individual attaining a proper balance between seemingly contradictory impulses, such as a balance within mental and physical experience and a stability between conscious and unconscious factors. It also produces the union of opposites within the person, such as harmony between *anima* and *animus*, between extravert and introvert tendencies. It entails cooperation between the mental activities expressed in terms such as intuition, rational thinking and emotional feeling. Individuation also reconciles opposing powers in human nature, specifically between what Jung called the shadow side, 'the subjective components of conscious functions' (1968: 22), which often entail hostile or aggressive impulses, and the capacity for spontaneous affection. In other words, individuation entails the realisation of the self, or what Jung calls 'individual completion or wholeness', a 'healing' process which involves a 'descent into the depths' (1968: 137).

Jung's understanding of religion as therapeutic

Since individuation is a lifelong process, it can be divided into distinct phases. During the first half of life (approximately), the self is initiated into external reality. The self must learn to live with the outer world, to accommodate to social demands and to set out life goals that are determined by circumstances and forces beyond its control. The second half of life is distinguished by initiation into inner reality during which philosophical and religious questions come to the fore and people begin to concern themselves with the ultimate meaning of life. This is why individuation is frustrated in later life if a religious resolution is not achieved. Jung (1960: 399) explains: 'We cannot live the afternoon of life according to the programme of life's morning; for what was great in the morning will be little at evening'. The young person cannot 'be too preoccupied with himself ... but for the ageing person it is a duty and a necessity to give serious attention to himself'. Religion and individuation thus are closely connected within the latter half of life. Palmer (1997: 150–2) suggests that Jung even went so far as to associate individuation with recognizing or experiencing (almost mystically) the unity of the self with God. In Jung's words: 'The extraordinary difficulty in this experience is that the self can be distinguished only conceptually from what has always been referred to as "God", but not practically. Both concepts apparently rest on an identical numinous factor which is a condition of reality' (1963: 546).

One of Jung's last published works was entitled *The Undiscovered Self* (2002), published in German in 1957 and in English in 1958. It was written at the height of the cold war and in some ways is fiercely anti-Soviet, but it is equally opposed to the rationalism and materialism of the West. A large portion of the book focuses on Jung's attitude towards religion. I draw attention to this because it underscores the positive regard Jung had for the power of religion in producing psychological health and well being in humans. In Chapter Three on 'The Position of the West on the Question of Religion', Jung refers to the traditional and collective convictions of the churches, the creedal approach, as 'unreflecting belief', and hence as opposed to a faith resulting from inner experience. When religious faith does not reach to the inner self, but remains on the surface only, it is 'apt to disappear as soon as one begins thinking about it' (2002: 26). This is because in the cold light of reason, religion appears irrational. What is needed if religion is to endure is to recognize that faith has been instilled into us through the experiences of 'trust and loyalty' (2002: 26). Deep religious faith thus forms part of the human collective unconscious, which operates at the level of symbol and cannot be destroyed by rationality or scientific knowledge. To attain and maintain psychological well being, each person needs an individual faith and experience of an 'extramundane principle', which Jung claims is 'capable of relativizing the overpowering influence of external factors' (2002: 16). He adds: 'The individual who is not anchored in God can offer no resistance on his own resources to the physical and moral blandishments of the world' (2002: 16). Religious faith, which Jung defines as 'the evidence of inner, transcendent experience', is that alone in the modern world which can protect humans against 'the otherwise inevitable submersion in the mass' (2002: 16).

Since the psychological advantage of religious faith is so strong, Jung is not concerned with the illusory nature of the object of the religious consciousness, nor even with magical thinking, which Freud had so closely associated with the neuroses resulting from belief in transcendent beings (see Cox, 1993: 119–25). In sharp contrast to Freud, Jung affirms the psychological benefits derived from religious thinking, even if religious beliefs seem to contradict scientific evidence.

> The performance of a 'magical' action gives the person concerned a feeling of security which is absolutely essential for carrying out a decision, because a decision is inevitably somewhat one-sided and is therefore rightly felt to be a risk. ... When the rationalist directs the main force of his attack against the magical effect of the rite as asserted by tradition, he has in reality completely missed the mark. (Jung, 2002: 18–19)

The essential point for Jung, therefore, in any consideration of religion is not to enter into a debate concerning the ontological reality of transcendental objects or spiritual entities, but to affirm their positive psychological effects on feelings of security and well being, which are attached in many traditions to images of God as a beneficent father. Michael Palmer (1997: 195) argues that Jung's interpretation of religion demonstrates how thoroughly he rejected Freud's reduction of religion to a psychic projection of the father image and hence to an infantile neurosis.[5] Of course, religion contains father images for Jung, but rather than seeing God as a fantasy substitute for an earthly father, the belief in which perpetuates childhood dependence, God is fitted into the archetypal imagery. The importance of belief in God for psychological health thus becomes substantiated as the universal, archetypal father within the collective unconscious. Jung concludes that the 'religious function cannot be disposed of with rationalistic and so-called enlightened criticism' since it plays a fundamental role in fostering the human progression towards individuation (2002: 19).

Conclusion

George James (1995: 57) argues that among what he calls the 'familial traits' held in common by phenomenologists of religion can be found a strongly 'anti-reductionistic' penchant. By reductionism, James means 'the tendency to treat the subject matter of religion in such a way as to deny it the status of a distinctive object of inquiry'. In addition to marginalizing the study of religion as a legitimate field within academic contexts, reductionism has the further complication of explaining religious phenomena away by making them aspects of other more primary modes of analysis, such as psychology, sociology or economics. James notes: 'To study religion as purely economic behaviour or as psychosis would clearly be reductive' (1995: 59–60). I have argued in this chapter, that, although we will find phenomenologists of religion reacting against forms of social scientific reductionism, many of which directly resulted from Feuerbach's projection theory and were rooted in a positivist philosophy, the contribution of the social sciences to the phenomenology of religion, as illustrated in the cases of Troeltsch, Weber and Jung, cannot be understood strictly in such a negative sense. Rather, key ideas emerging from the social sciences have helped shape the way phenomenologists have delineated their field of study and how they have developed methods they regarded as unique to their discipline.

In this chapter, we have seen that Ernst Troeltsch, although primarily a theologian in the Ritschlian tradition, developed a hermeneutical method that placed the historical data of religion squarely within social contexts. His use of ideal types, although intended in the end to provide evaluative criteria for comparing Christianity with other religions, emphasized the dynamic character of religious movements in history, thereby underscoring the close connection between the history of religions and the social sciences. At the same time, the ideal type was conceived by Troeltsch as forming a structure of the religious consciousness, which was embedded in all human thought processes, a structure he called the 'religious *a priori*'. We have also noted that Troeltsch was a close friend and colleague of Max Weber, who in his own right demonstrated how social processes, modelled on the idea of individual action, demonstrate universal patterns that can be applied rationally to assist in understanding the causative role religious ideas play in forming socio-historical and economic systems. Weber's emphasis on the dynamic model of religious forces in history was exemplified in many of his writings, but perhaps most famously in his analysis of how the Calvinist form of Protestantism resulted in the dominance of the capitalist economic system. This twist in the concept of ideal types aided in attaining genuine understanding (*Verstehen*) of how religious elements are interwoven into social and economic processes, without reducing them to such courses of action. From an entirely different perspective, C. G. Jung also contributed to a positive understanding of religion by identifying the foundational, universal archetypes within the human collective unconscious, which are expressed in forms closely related to phenomenological typologies in the study of religions, such as myth, ritual and symbol. That the archetypes play a constructive role in human psychological health in some ways must be regarded as a side-effect of Jung's analysis that is not directly relevant to constructing phenomenological typologies, but it should be noted that the positive role Jung attributed to the archetypes fostered at the very least an empathetic attitude towards religious belief and practice. Moreover, it should be emphasized that Jung's theory justified the claim that the universal structure of the consciousness is religious at its core, implying that the study of religious myths, rituals and symbols discloses precisely how humans think.

After having reviewed the formative influences on the phenomenology of religion through the philosophical movement founded by Edmund Husserl in Chapter 1, through the theological school of Albrecht Ritschl in Chapter 2, and in this chapter by analysing the positive influences of Troeltsch, Weber and Jung, I am now ready to consider the key thinkers in the phenomenology of religion itself. I begin with the Dutch school of phenomenology in the next chapter and move iff subsequent chapters through the British school to key thinkers in North American phenomenology. Throughout these discussions, the formative influences I have identified in the first three chapters will reappear, usually in the background and often implicitly, but nevertheless as persistent threads which, when bound together, help to clarify the significance of each as contributing to the phenomenological movement in the study of religions as a whole.

References

Adams, J. L. (1972), 'Introduction', in E. Troeltsch, *The Absoluteness of Christianity and the History of Religions* (London: SCM Press Ltd), pp. 7–20.

Bosch, L. P. van den (1999), 'Friedrich Max Müller: His contribution to the science of religion', in E. R. Sand and J. P. Sorensen (eds), *Comparative Studies in History of Religions* (Copenhagen: Museum Tusculanum Press, University of Copenhagen), pp. 11–39.

Capps, W. H. (1995), *Religious Studies: The Making of a Discipline* (Minneapolis: Fortress Press).

Chryssides, G. (2001), 'Unrecognised charisma? A study of four charismatic leaders'. Paper presented at the International Conference, 'The Spiritual Supermarket: Religious Pluralism in the Twenty-first Century', organized by INFORM and CESNUR. www.cesnur.org/london2001/chryssides.htm.

Clayton, J. P. (ed.) (1976), *Ernst Troeltsch and the Future of Theology* (Cambridge: Cambridge University Press).

Cox, J. L. (1993), *Changing Beliefs and an Enduring Faith* (Gweru, Zimbabwe: Mambo Press).

Darwin, C. (1964) [1859], *On the Origin of the Species* (with an introduction by Ernst Mayr) (Cambridge, Massachusetts: Harvard University Press).

Drescher, H. G. (1992), *Ernst Troeltsch: His Life and Work* (London: SCM Press).

Farr, R. M. (1996), *The Roots of Modern Social Psychology: 1872–1954* (Oxford: Blackwell).

Feuerbach, L. (1893), *The Essence of Christianity* (translated from the second German edition by Marian Evans) (London: Kegan Paul, Trench, Trübner, and Co).

Freud, S. (1938) [1919], *Totem and Taboo* (Harmondsworth: Penguin Books).

Freud, S. (1949) [1940], *An Outline of Psycho-Analysis* (translated by J. Strachey) (London: Hogarth Press).

Hamilton, M. (2001), *The Sociology of Religion. Theoretical and Comparative Perspectives* (London and New York: Routledge, 2nd edn).

James, G. A. (1995), *Interpreting Religion. The Phenomenological Approaches of Pierre Daniel Chantepie de la Saussaye, W. Brede Kristensen, and Gerardus van der Leeuw* (Washington, DC: The Catholic University of America Press).

Jung, C. G. (1915) [1911], *The Psychology of the Unconscious* (translation with introduction by B. M. Heindel) (London: Kegan Paul).

Jung, C. G. (1923) [1921], *Psychological Types or the Psychology of Individuation* (translation by H. G. Baynes) (London: Routledge and Kegan Paul).

Jung, C. G. (1936) [1907], *The Psychology of Dementia Praecox* (translated by A. A. Brill) (New York: Nervous and Mental Disease Publishing Company).

Jung, C. G. (1960), *The Structure and Dynamics of the Psyche* (*Collected Works*, Volume 8, translated by R. F. C. Hull) (London: Routledge and Kegan Paul).

Jung, C. G. (1963), *Mysterium Coniunctionis. An Inquiry into the Separation and Synthesis of Psychic Opposites in Alchemy* (*Collected Works*, Volume 14, translated by R. F. C. Hull) (London: Routledge and Kegan Paul).

Jung, C. G. (1968), *Analytical Psychology: Its Theory and Practice. The Tavistock Lectures* (London: Routledge and Kegan Paul).

Jung, C. G. (1972), *Four Archetypes: Mother, Rebirth, Spirit, Trickster* (London: Routledge and Kegan Paul).

Jung, C. G. (2002) [1958], *The Undiscovered Self* (London: Routledge).

Meyer, E. (1894), *Geschichte des Alterthums* (Stuttgard: J. G. Cotta).

Meyer, E. (1902), *Zur Theorie und Methodik der Geschitchte* (Halle a.S.: M. Niemeyer).

Morris, B. (1987), *Anthropological Studies of Religion: An Introductory Text* (Cambridge: Cambridge University Press).

Müller, F. M. (1893) [1873], *Introduction to the Science of Religion* (London: Longmans, new edn).

Müller, F. M. (1891), *Physical Religion* (London: Longmans, Green, and Co).

Müller, F. M. (1897), *Contributions to the Science of Mythology. Volumes I and II* (London: Longmans, Green, and Co.).

Müller, F. M. (1898), *Theosophy or Psychological Religion* (London: Longmans, Green, and Co.).

Palmer, M. (1997), *Freud and Jung on Religion* (London and New York: Routledge).

Pals, D. L. (1996), *Seven Theories of Religion* (New York and Oxford: Oxford University Press).

Platvoet, J. G. (1988), *A Concise History of the Study of Religions* (Harare: University of Zimbabwe, Department of Religious Studies, Classics and Philosophy; internal publication).

Preus, J. S. (1987), *Explaining Religion. Criticism and Theory from Bodin to Freud* (New Haven and London: Yale University Press).

Schutz, A. (1972) [1932], *The Phenomenology of the Social World* (translated by G. Walsh and F. Lehnert) (London: Heinemann Educational Books).

Sharot, S. (2001), *A Comparative Sociology of World Religions. Virtuosos, Priests, and Popular Religion* (New York and London: New York University Press).

Sharpe, E. J. (1986), *Comparative Religion: A History* (London: Duckworth, 2nd edn).

Shils, E. A. (1949), 'Foreword', in M. Weber, *The Methodology of the Social Sciences* (translated and edited by E. A. Shils and H. A. Finch) (New York: The Free Press), pp. iii–x.

Spencer, H. (1880), *First Principles* (London and Edinburgh: Williams and Norgate, 4th edn).

Storr, A. (1997), *Feet of Clay: A Study of Gurus* (London: HarperCollins).

Thrower, J. (1999), *Religion: The Classical Theories* (Edinburgh: Edinburgh University Press).

Troeltsch, E. (1972) [1902], *The Absoluteness of Christianity and the History of Religions* (London: SCM Press).

Troeltsch, E. (1991), *Religion in History: Essays* (translated by J. L. Adams and W. F. Bense, with an introduction by James Luther Adams) (Edinburgh: T. and T. Clark).

Weber, M. (1949), *The Methodology of the Social Sciences* (translated and edited by E. A. Shils and H. A. Finch) (New York: Free Press).

Weber, M. (1952), *Ancient Judaism* (translated and edited by H. H. Gerth and D. Martindale) (Glencoe, Illinois: Free Press).

Weber, M. (1958), *The Religion of India: The Sociology of Hinduism and Buddhism* (translated and edited by H. H. Gerth) (Glencoe, Illinois: Free Press).

Weber, M. (1968), *The Religion of China: Confucianism and Taoism* (translated and edited by H. H. Gerth) (New York: Free Press).

Weber, M. (1992) [1930], *The Protestant Ethic and the Spirit of Capitalism*

(translated by T. Parsons, with an introduction by A. Giddens) (Lc
New York: Routledge).

Notes

[1] Müller used this phrase in his *Introduction to the Science of Religion* (1873).
Sharpe thought it was so important that he coined it as the title of his second
chapter in *Comparative Religion: A History* (1986).

[2] Meyer's most important work is often regarded as *Geschichte des Alterthums*
(1894), which outlines a sociological theory of antiquity with direct
relevance to religious history. Weber's article, however, responds directly to
Meyer's earlier book on historical method and theory entitled *Zur Theorie
und Methodik der Geschitchte* (1902), in which Weber (1949: 116) contends
that Meyer defines 'the proper objects of scientific knowledge' as: a) mass
phenomena in contrast to individual actions; b) the typical in contrast to the
particular; and c) the development of social classes in contrast to the
political actions of individuals.

[3] In the Tavistock Lectures delivered in 1935, Jung observed: 'I started out
entirely on Freud's lines. I was even considered to be his best disciple. I was
on excellent terms with him until I had the idea that certain things are
symbolical' (Jung, 1968: 140).

[4] The collective unconscious is what Storr (1997: 98) defines as 'a level of
mind responsible for producing myths, visions, religious ideas and certain
varieties of dream which are common to many cultures and to many periods
of history'.

[5] Palmer concludes, perhaps unexpectedly, that Jung must be considered in
the long run as far more dangerous to religion than Freud despite the fact, or
perhaps because of the fact, that in popular perceptions Jung is regarded as
more friendly to religion than Freud. For Palmer, Freud's opposition to
religion makes him an identifiable enemy, whereas Jung's complicity with
religion disguises that he subtly undermines it. 'To agree with Jung', Palmer
argues, 'is to adopt a conception of God as an innate human disposition,
indistinguishable as anything other than a particular psychic state'. By
limiting God to that which proceeds from the psyche, Jung, in Palmer's view,
has made faith into a purely subjective experience, 'guaranteed in much the
same way as, say, a belief in flying saucers' (Palmer, 1997: 195).

Chapter 4

The Decisive Role of Dutch Phenomenology in the New Science of Religion

What became known generally as the phenomenological movement in the study of religions resulted from the writings of three seminal thinkers working in Dutch universities: W. Brede Kristensen, who for 36 years was Chair of the History of Religion in Leiden University; Gerardus van der Leeuw, Kristensen's student, who was Professor of the History of Religion in the University of Groningen; and another of Kristensen's students, C. Jouco Bleeker, who was Professor of the History of Religion in the University of Amsterdam. Throughout the first two-thirds of the twentieth century, these scholars almost single-handedly defined the basic tenets associated with the phenomenology of religion. By so doing, they claimed to have identified the epistemological tools necessary for students of religion to attain objective knowledge of religious life and practice while at the same time reflecting accurately and fairly the perspectives of believers. The end-product of this method was the phenomenology of religion, whose advocates claimed to have combined the empirical investigation of religious phenomena with an in-depth understanding of the faith of adherents. This contention became the cornerstone supporting the persistent argument that religion constitutes a unique category, *sui generis*, within human experience making the discipline which studies it, phenomenology, irreducible to any other field of scholarly investigation. It is perhaps this contention, at once its most controversial and most lasting, that caused Arie Molendijk (2000: 26) to refer to the 'Dutch science of religion' as 'notorious for its contribution to the phenomenology of religion'.

This chapter focuses primarily on the thought of Kristensen, van der Leeuw and Bleeker as key figures in the Dutch school of phenomenology. It is important to note, however, that their theories of religion were developed within a larger framework of theological studies and the burgeoning field of comparative religion in the Netherlands at the close of the nineteenth century. Before exploring in some depth the theories of Kristensen, van der Leeuw and Bleeker, I place them into context by outlining briefly some of the principal ideas that were developed by their predecessors in the comparative study of religions in Holland, C. P. Tiele and Pierre Chantepie de la Saussaye.

The early Dutch school

The founder of a comparative approach to the study of religions in The Netherlands was C. P. Tiele (1830–1902), whom Eric Sharpe (1986: 35) regards as so important that he rates him as a competitor with Max Müller for the title 'Father of Comparative Religion'. Molendijk (2000: 26) adds that 'the international reach of early Dutch science of religion ... is probably best exemplified by the work of Tiele'. In 1876, the Dutch government enacted the Higher Education Bill which established special chairs in the history of religions in Leiden and Amsterdam within theological faculties (Molendijk 2000: 21), with Tiele being appointed one year later as the University of Leiden's first professor in this field. Tiele was a scholar of ancient Egyptian religions and a minister in the Arminian wing of the Dutch Church, which emphasized free will, as opposed to predestination as taught by its dominant Calvinist wing. In academic settings, Tiele adopted what he called a 'Science of Religion', a term he had employed as early as 1866, to distinguish his area of study from 'confessional theology', which up until the passage of the Higher Education Act had occupied a dominant position in Dutch universities. Tiele's 'Science of Religion' included both the history of religions and the philosophy of religion, because in Molendijk's words, he 'did not want to do "just" history, but to analyze and evaluate religions and religious phenomena as well' (2000: 21). As a Christian minister, Tiele did not oppose confessional interpretations of theology, but he wanted to replace them in the university with non-confessional, scientific studies of religion, both in the comparative study of religions and in theology proper. Under Tiele's method, confessional theology would surrender its traditional authoritative role in Dutch universities by itself becoming subject to the methods of comparative research. Nevertheless, a theological motive remained in his method, as Molendijk notes: 'Science of religion was expected to fulfil (most of) the tasks of the old theology and to show the superiority of Christian religion' (2000: 22–3).

In his Gifford Lectures delivered in the University of Edinburgh in 1896 and 1898, and later published as *Elements of the Science of Religion* (1897), Tiele outlined in successive stages what he regarded as essential tasks for the science of religion. He first defined religion as 'those manifestations of the human mind in words, deeds, customs, and institutions which testify to man's belief in the superhuman, and serve to bring him into relation with it' (Tiele, 1973: 97). To investigate religion, the scientist begins with an accurate description of 'all religions of the civilised and uncivilised world, dead and living', including 'all the religious phenomena which present themselves to our observation' (Tiele, 1973: 97). At the next stage, the scholar compares religious phenomena in order to determine their origins and to classify them according to their place in the evolution of human religious development. Tiele explained that the scientist of religion 'judges, in so far as his task is to compare the different manifestations of religious belief and life, and the different religious communities, in order to classify them in accordance with the stage and direction of their development' (1973: 98). Finally, the scientist of religion discovers the inner core of all religions through careful comparisons made at each stage of the process. Tiele believed this step-by-step methodology would show that religions develop in phases from nature religions through mythological religions through doctrinal religions and finally to the world religions. Key to the transition from the nature stage to the mythological stage is the capacity for humans to think symbolically. Throughout this progression, the scholar looks for the abiding core of religion as it is expressed at every level of development. Tiele believed that discovering the essence of religion could only occur at this final stage in the process. Otherwise, preconceived notions of religion would render the entire method unscientific. He explained that by building on 'established facts', the scholar of religion seeks to discover the 'necessary manifestations of religious consciousness' or what he calls the 'constant elements' operating within all religious phenomena (1973: 102).

[handwritten margin notes: "Surely "nature" religions" "See Symbols in nature"]

As Walter Capps (1995: 119) observes, Tiele concluded that the inner core of religion is best described by the word 'adoration', which refers to a universal essence that is expressed through the cultural values of one generation and then handed down to the following generation, which in turn reformulates it into more sophisticated levels of understanding. The enduring quality of 'adoration' is ensured by its uncanny ability to bind together the seemingly opposing tendencies in religion, one which stresses transcendence and the other which values immanence.

> In adoration are united these two phases of religion which are termed by the schools 'transcendent' and 'immanent' respectively, or which, in religious language, represent the believer 'as looking up to God as the Most High', and as 'feeling himself akin to God as his Father'. (Cited by Capps, 1995: 119)

The act of adoration unites the believer immanently with the transcendent, precisely because 'to adore' involves, in Tiele's words, 'the elements of holy awe, humble reverence, grateful acknowledgment of every token of love', while, at the same time, it fosters a feeling in the religious person that 'he belongs to the adored one forever, in joy and in sorrow, in life and in death' (cited by Capps, 1995: 119). Although Tiele may be understood as advocating a form of 'historical evolutionism' (Platvoet, 1998: 116), he did not, ostensibly at least, argue for the superiority of one form of religion over the other, since the abiding core of religion was the same everywhere. He acknowledged that religions change and develop, grow, recede and sometimes die out, but 'this transitoriness is precisely one of the strongest proofs of the development of religion'. This is because, although the religions of humankind undergo persistent change, the essence remains constant. 'Though ever changing in form, religion lives like mankind and with mankind' (cited by Capps, 1995: 120).

The first thinker to employ the term 'phenomenology of religion' was another Dutch scholar, Pierrre Daniel Chantepie de la Saussaye (1848–1920), who became Professor of the History of Religion in the University of Amsterdam from 1879, a post he surrendered in 1899 to take up a professorship in theology in Leiden. Molendijk (2000: 25) asserts that 'compared to Tiele, the influence of

Chantepie de la Saussaye was smaller' because 'he did not possess Tiele's zeal to fight for the new discipline, and in the course of his career he was drawn ever more to ethics and traditional theology'. In his *Manual of the Science of Religion*, first published in English in 1891, Chantepie defined the phenomenology of religion as a method for classifying and comparing religious beliefs and practices, which would produce a new discipline, falling midway between the history of religions and the philosophy of religion (1891: 67). He divided the history of religions into ethnographical and historical sub-disciplines, with ethnography providing details of the religions of the 'savage' tribes while history recounted the development of the religions of the 'civilized' nations (1891: 8). The philosophy of religion also consisted of two parts: a psychological one focusing on the subjective elements of religion and a metaphysical emphasis which outlined the objective side of religion (1891: 8).

Chantepie regarded the phenomenology of religion primarily as a method for creating typologies out of the vast material produced from historical and ethnographic studies. He described 'religious acts, cult, and customs', particularly as they are expressed in rituals, as the 'richest material' for the phenomenology of religion (1891: 69). He admitted that every religious act must be preceded by a thought or a doctrine, but he argued that doctrines often attempt to explain rituals. For this reason, the phenomenology of religion, although concerned with religious doctrines, must regard 'ritual customs' as more stable than doctrines. Rituals carry us 'back to the most distant times', whereas doctrines 'tend to adapt themselves to new requirements' (1891: 68). Religious sentiments, which Chantepie regarded as the inner states of the religious person, give rise to the outward expressions of religion and thus vary according to historical periods, social groupings and individual predispositions. Since religious sentiments cannot be observed directly by the scholar, the most important outward expression of a religion is found in its rituals, which, because they are practical, show how the religious person 'desires in his religion ... an attainment of certain benefits' (1891: 70). In some cases, these benefits are entirely material, as often appear in the customs of 'so-called savages', but in other instances 'what is desired consists in more general spiritual objects, as when for instance a ceremony is meant to preserve a certain cosmic order' (1891: 70).

Chantepie defined religion as 'a belief in superhuman powers combined with their worship'. For him, this meant that religion everywhere directs attention to 'the living God who manifests Himself among all nations as the only real God' (1891: 71). When scientists of religion observe, describe and classify religious phenomena, they actually are recording the revelations of God. For Chantepie, the very existence of religion confirms that God has been at work everywhere and that God has implanted within humans the innate capacity to respond to the divine revelation. In other words, if God did not exist, religion would not exist, and no scientific study of religion would be possible (Platvoet, 1998: 123). This view derives from Chantepie's acceptance of the widely maintained theological distinction between natural and revealed religion, whereby divine revelation is attested universally in nature, in history and what Chantepie called 'the innermost being' of the human. He explained:

> Though by many He is only partially known, or not known at all, because divine honour is paid to His works and His powers rather than to Himself, yet in the end all worship is meant for Him, and man cannot conceive anything divine that is not really derived from Him. (Chantepie, 1891: 71)

By affirming the universality of natural revelation, Chantepie, like Tiele, separated the science of religion from confessional theology, the latter being confined largely to a study of the special revelation of God in Christ. This meant that for scholars to understand religion scientifically, they must note variations in the human responses to God, compile them historically and classify them for comparative purposes according to phenomenological typologies. Chantepie nonetheless rejected the evolutionary thesis advocated by Tiele, arguing that the Darwinian model of biological evolution cannot apply to human religious development. As Platvoet (1998: 124) notes: 'Chantepie ... refused to accept Darwinism ... as valid also for human spiritual evolution: humans, having religions, could not have evolved "from a lower species of beings without religion" since "a religious being cannot evolve from a non-religious being"'. Platvoet (1998: 124) suggests that Chantepie accepted instead, 'albeit cautiously', the theory of primitive monotheism, later expounded fully by Wilhelm Schmidt, that humans were originally monotheistic and that this early belief had been corrupted through history (Sharpe: 1986: 182–4). This view enabled Chantepie to affirm that even the religions of the most primitive tribes reflected a divine origin despite their having degenerated into polytheism and superstition.

We see in Tiele and Chantepie two pioneers in the academic study of religions who broke new ground necessary for subsequent thinkers in the Dutch school of phenomenology. Taken together, their main contribution was to have distinguished between confessional theology and academic theology, thereby framing the study of religions within scientific categories. Both clearly were theologians in the broadest sense, believing that religion has its roots in a universal, natural revelation from a divine source, although Tiele cast his analysis into evolutionary terms rejected by Chantepie. Both also related the study of religion to philosophy, which they regarded as providing the analytical tools necessary to confirm the reality of God in the historical data of religion and the phenomenological typologies derived from the data. The components of the foundational Dutch school of phenomenology thus are already present in the nascent ideas expressed by Tiele and Chantepie: the emphasis on accurate and fair descriptions of religious beliefs and practices through history, a framework for scientific analysis through the creation of typologies and a proper role for philosophy as the discipline dedicated to demonstrating the reality of God. In the words of Walter Capps (1995: 120), these themes became 'paradigmatic' for later phenomenologists who shared with Tiele and Chantepie the 'conviction that all concerns about essences and origins must be suspended until after careful and painstaking phenomenological analysis has been conducted'.

W. Brede Kristensen: The primary tasks and purpose of the phenomenology of religion

Although he was born in Norway and earned his Ph.D from the University of Oslo in 1896, I have included W. Brede Kristensen (1867–1953) in the Dutch school since between 1890 and 1892 he studied under Tiele, and for 36 years, beginning in 1901, he held the Chair of the History of Religion in Leiden University as Tiele's successor. Sigurd Hjelde (2000: xix) notes that Kristensen's name 'is generally associated with the Dutch phenomenology of religion that came to play such a central role around the middle of the twentieth century, a position which was also shared by many of his students such as Gerardus van der Leeuw and C. J. Bleeker'. Kristensen became a Dutch citizen in 1917, although he continued to maintain ties with his native Norway. During his lifetime, Kristensen was known primarily as a scholar in the religions and languages of the Ancient Near East, with specific expertise in the historical religions of Egypt, Greece, Rome, Persia and Mesopotamia (Naguib, 2000: 105–6). His major work in English on theory and method in the study of religion, *The Meaning of Religion*, was not published until 1960, seven years after his death, and comprises lecture notes he delivered in Leiden, which were compiled and translated by John B. Carman of Harvard University. At a symposium held nearly 30 years after his translation of Kristensen's work, Carman (2000: 158) observed: 'It is perhaps unfortunate that the best known work of Kristensen's available in an international language of scholarship should focus on his general lectures on phenomenological topics rather than on more specific subjects, like the Egyptian boat or the Delphic tripod'. Even in continental Europe, his contribution to phenomenology, in Eric Sharpe's words, was 'curiously delayed' since his major work in this field was not published in Norwegian until one year after his death, under the title *Religionshistorisk Studium* (*The Study of the History of Religion*) (Sharpe, 1986: 227–8; Kristensen, 1954). Although his substantial contribution to methodological issues did not occur until after 1950, Kristensen is presented in this chapter as first in the line of the Dutch phenomenologists because he occupies a strategic position between Tiele's and Chantepie's theologically slanted views of comparative religions and van der Leeuw's definitive application of phenomenological principles to the study of religion.

In *The Meaning of Religion*, Kristensen defined 'phenomenology's way of working' as the 'grouping of characteristic data', thus making phenomenology a 'systematic discipline'. The phenomenologist organizes and classifies religious beliefs, cult activities and ritual performances 'according to characteristics which correspond as far as possible to the essential and typical elements of religion'. In this way, the task of phenomenology becomes that of 'illustrating man's religious disposition' (1960: 8). This interpretation resulted in the general characterization of phenomenology in academic circles following Kristensen primarily as a method for identifying religious typologies, by which the scholar organizes the vast diversity of human religious beliefs and behaviours into orderly categories for comparative purposes. George Alfred James (1995: 144) argues that

Kristensen is largely responsible for embedding this 'ahistorical trait' into the phenomenological method because, in Kristensen's words, phenomenology 'takes out of their historical setting the similar facts and phenomena which it encounters in different religions, brings them together, and studies them in groups' (1960: 2).

For Kristensen, phenomenological typologies consist of generalizations based on comparisons of specific or particular historical data. This makes the typologies in themselves concepts, not mirrors of reality. The ideas they convey are derived from real actions and beliefs, but in themselves they are not realities. For example, it is well known that sacrifices occur almost universally in specific religions. The phenomenologist thus studies and compares sacrificial acts in various situations under the general category 'sacrifice'. Sacred kingship provides another example of a religious typology. By comparing 'the sacredness of the Greek and Roman kings ... in the light of the ancient concept of kingship', a general category emerges. Kristensen admits that 'the Ancient conception of kingship or the religious essence of sacrifice is a concept and not historical reality (only particular applications are reality), but we cannot dispense with those concepts' (1960: 7). Typologies serve a distinct function for the scholar: 'They give the research direction and lead to the satisfying result of understanding the data' (1960: 7). Kristensen thus concludes that the fundamental aim of the phenomenological method is 'to come as far as possible into contact with and to understand the extremely varied and divergent religious data, making use of comparative methods' (1960: 11).

This relationship, which Kristensen described between particular data and phenomenological generalizations, dictated how he explained distinct, but closely connected, tasks for history and phenomenology. 'History provides the material for the research of Phenomenology' but phenomenology consists of 'the different types of religious thinking and action, of ideas of deity and cultic acts' (1960: 9). The relationship between phenomenology and history thus is mutually interactive. If the material for the pheneomenologist is obtained from particular historical studies, that which the historian describes is informed equally by the phenomenological classifications. For example, the category 'sacrifice' defines in a broad sense what the historian is looking for when identifying and describing particular sacrificial acts. In this sense, phenomenology provides the material for the 'History of Religion'. At the outset of The Meaning of Religion, Kristensen puts it this way: 'Phenomenology of Religion is the systematic treatment of History of Religion' (1960: 1).

A third factor in a general science of religion in Kristensen's analysis is reserved for philosophy, which has as its 'chief task' to formulate the 'essence' of religion (1960: 9). This can be easily misunderstood as the phenomenologist's task, that is, as a search for 'religion's core' based on the premise that some common idea stands behind all religious ideas. Kristensen argues that this idea is impractical since even the idea of 'God' is not universal in the major religions, as is well exemplified in Buddhism (1960: 8). If variations in belief about an ultimate reality can be demonstrated, how difficult would it be to find some common core in the myriad of other religious phenomena? To search for the 'common element' behind all religious practices thus results in 'nothing and everything', since 'seen more deeply, everything is held in common' (1960: 9). The philosopher, not the phenomenologist, searches for essences; phenomenology provides for philosophy the 'general picture' of religious thoughts and actions (1960: 9).

This point is extremely important in Kristensen's analysis. The typologies identified by the phenomenologist are generalizations, but they are not essences in a philosophical sense. Generalizations are necessary to speak comparatively about various religious acts; they make understanding possible. Typologies do not, however, reflect a core or essence of religion, although they provide necessary tools for the philosopher. A good illustration of a philosophical approach as opposed to historical and phenomenological methods, according to Kristensen, is found in Rudolph Otto's concept of 'the holy', which Kristensen regarded as thoroughly inadequate for an empirical study of religions. What is presented by Otto as a:

> study in Psychology of Religion is actually quite theoretical in nature and properly belongs to the field of Philosophy of Religion, which seeks to determine the essence of religion in general. But it is fatal for the rest of the exposition that Otto the philosopher and systematic theologian does not see that on this basis no translation is possible to the *historical* understanding which he sets as his goal. (Emphasis his) (Kristensen, 1960: 16)

In this light, Otto could act quite properly as a philosopher by defining the essence of religion as the holy, but he was incorrect methodologically when he sought to establish the idea of the holy as an historical or empirically verifiable fact.

Thus, very much like Tiele and Chantepie before him, Kristensen argued for a middle role for phenomenology of religion between history and philosophy. He saw each discipline as having specific tasks, but also each as closely linked to the other. 'None of the three is independent; the value and the accuracy of the results of one of them depend on the value and accuracy of the other two' (1960: 9). Phenomenology, nonetheless, plays a pivotal role, because it stands as the point of contact for the other two. It utilizes the data of history by making sense of its diversity and it provides the categories on which the philosophy of religion depends. In this sense, phenomenology links the particular with the universal: 'Phenomenology is at once systematic History of Religion and applied Philosophy of Religion' (1960: 9).

Phenomenology as a scientific method

Kristensen believed that for the study of religion to be worthy of the name 'science' a clear methodology was needed to ensure that the researcher's conclusions remained fully empirical and hence testable. The basic elements of Kristensen's approach later became synonymous with the phenomenological method in the study of religion. He began by clarifying the relationship between the scholar of religion and the adherents within religious communities. He then discussed what constitutes objective and subjective knowing, and finally introduced intuition as a key hermeneutical tool. In these ways, Kristensen anticipated later controversies in the study of religions, such as the insider–outsider problem, the issue of self-reflexivity on the part of the researcher and the problem of identifying meaning within religious data. He did this in the first instance by giving absolute priority to the perspectives of believers, arguing that adherents understand their own religion better than anyone from the outside ever could. This view led to Kristensen's assertion that the scholar of religion experiences a peculiar dilemma, in a way quite unlike researchers in other disciplines. On the one hand, a researcher is required to fairly represent a religion, which requires that the religious community must in some sense be described from the inside, that is, from the standpoint of its own practitioners. On the other hand, a scholar can never enter sufficiently enough into a religion of which he or she is not a part to appreciate it as a believer does. 'Every believer looks upon his own religion as a unique, autonomous and absolute reality ... and thus incomparable' (1960: 6). The resolution to this problem cannot require researchers literally to become believers in the religions they are studying. 'We cannot become Mohammedans when we try to understand Islam, and if we could, our study would be at an end: we should ourselves then directly experience the reality' (1960: 7).

The primary technique Kristensen advocated to overcome the problem of understanding from the inside is what he called in one place, 'empathy' (1960: 7) and later, 'an indefinable sympathy' (1960: 10). By these terms, he indicated that a scholar 'tries to relive in his own experience that which is "alien"' through 'an imaginative reexperiencing of a situation strange to us' (1960: 7). This engages the researcher's mind in an act of representation, but does not replicate the actual religious experience itself. 'The "existential" nature of the religious datum is never disclosed by research' (1960: 7). Empathy, nonetheless, involves more than play acting, since it is virtually impossible for a scholar of religion to employ this technique unless the scholar has some personal experience of religion. 'We make use of our own religious experience in order to understand the experience of others' (1960: 10). This implies that the principal way to overcome the distance created by being an outsider is not to convert to the religion one is studying, but to employ one's own religious sensibilities to attain a feeling for how adherents experience and understand the absoluteness of their own religion. The use of empathy deepens the personal faith of the researcher, since, Kristensen asserts, 'when religion is the subject of our work, we grow religiously' (1960: 10).

Whether or not a genuinely scientific methodology has been applied is disclosed only by the research findings. Kristensen admits that on one level, scholarly investigation is primarily concerned with theoretical issues and not with matters of practice. But the descriptions and interpretations that are provided in research bear evidence as to whether or not a scholar has employed an empathetic technique. If the results of theoretical research cannot be affirmed by believers themselves, the method has not been utilized and the results do not reflect scientifically the religious data. This is because researchers must be able 'to forget themselves' and 'surrender themselves to others' (1960: 13). This self-surrendering has the effect that those who are being studied will likewise surrender themselves to the scholar. To implant one's own preconceived notions and biases into the research findings is to contradict the evidence that comes from the believing communities. The sign that empathy has been employed and a degree of self-surrender has been accomplished is that the researcher gives full credit to the interpretations of the adherents, so much so that the researcher affirms the adage: 'the believers were completely right' (1960: 14). If a researcher cannot accede to such a testimony, empathy has not been utilized, the religion has been described and interpreted in ways that are foreign to the believers, and the religious reality has been negated. 'For there is no religious reality other than the faith of believers' (1960: 13).

Kristensen argues that the most glaring prejudice that has been foisted on the study of religions has resulted from the application of Darwinian biological evolution to an interpretation of religious development. This widespread idea has resulted in the misrepresentation and distortion of believers' perspectives. The scholar of religion must become aware of this and any other biased evaluations of religious data by refusing to use them as interpretative tools. In this regard, Kristensen anticipated how later phenomenologists would use Husserl's notion of *epoché*, although he did not actually employ the term himself. If one considers the implications of evolutionary theory when applied to religion, the distorting bias beneath its assumption becomes clear.

> The basic conviction is this, that the history of mankind has had just ourselves as its goal, and after frightfully great pains it has generated our civilization, as the result of all which had preceded it ... In religion as well as in the rest of our culture we stand on the apex of the historical pyramid. (Kristensen, 1960: 11)

Evolutionism comes in two forms: historical and idealist. C. P. Tiele provides an example of an historical evolutionist since he asserted, in Kristensen's words, that 'the results achieved in each historical period are handed down to the following generation by them and further developed' (1960: 11). As we have seen, Tiele contended that progress occurs through each transmission of values, which not only are preserved, but perfected, or at least developed to higher levels. Idealistic evolutionism has been articulated most forcefully in the Hegelian dialectic, where, as we noted in previous chapters, religion moves through a series of historical antitheses and syntheses to achieve finally the perfected state of the Absolute Spirit. In Kristensen's view, idealistic evolutionists interpret religion as detaching itself 'more and more from the undeveloped reality which is clothed in primitive forms and comes to light in full clarity in the highest civilization and the highest religions' (1960: 11–12).

Kristensen acknowledged that evolutionary interpretations can be defended on philosophical grounds, where the essence of religion is being described.

Philosophy has the right, indeed it is the duty of philosophy, to seek out essences. Some religious rites and practices can be said to reflect the essence in lesser or greater degrees or at lower or higher levels. But it must be remembered that the philosophical method is deductive; it begins with the essence and then interprets historical data. The historian and the phenomenologist by contrast operate inductively; they depend on facts which cannot be 'conjured up' by mere deduction. The historian and the phenomenologist thus do not begin with essences, but 'investigate what religious value the believers ... attached to their faith, what religion meant to them' (1960: 13). The empirical scholar of religion thus wants to understand the religion under study on its own terms, not on the basis of the scholar's prior assumptions about the essence of religion. For this reason, Kristensen concludes: 'All evolutionary views and theories ... mislead us from the start' (1960: 13).

Evaluative and informed comparisons

Kristensen considered finally in what ways a scholar of religion might evaluate religious phenomena legitimately. He rejected any form of evaluative comparison which presupposes an *a priori* ideal or standard whereby one religious act, belief or practice is judged as better, higher or more valuable than another act, belief or practice. This is what happens when the evolutionary thesis is imposed as an interpretative principle in the history of religions: evolutionary theorists tell the data what they want to hear and in this way arrive at predetermined results. Instead of judging by what the believers of the religion in question identify as the key to understanding their own beliefs and practices, the scholar who engages in evaluative comparison uses an alien interpretative key or prejudgement imposed from without. Kristensen argued that 'evaluative comparison' does not fall within the domain of the phenomenology of religion but instead 'belongs to the provinces of philosophy and dogmatic theology' (1960: 418). Over against evaluative comparison, Kristensen advocated the practice of an informed comparison based on historical data, a method which leads to understanding rather than judgement. At every stage in the comparative study of religions, the scholar must adhere to the overriding principle of phenomenology, which 'wishes to learn to understand the conception of the believers themselves, who always ascribe an absolute value to their faith' (1960: 418).

Intuition defines the way a scholar attains such an understanding in depth. This, as we have noted, draws on personal experience, the researcher's own religious awareness. Understanding in the fullest sense never occurs strictly from a rational or logical description of religious data. Kristensen explained that 'in Phenomenology we are constantly working with presumptions and anticipations' that do not 'take place outside our personality' (1960: 10). How then can we be assured that our predetermined biases are not causing us to fall into evaluative as opposed to informed comparisons? Here we return to Kristensen's idea of empathy through which one's own experience of religion creates an attitude fully disposed to understanding. Intuition depends on personal experience, which in turn leads to a sympathetic understanding of that which might otherwise seem strange, exotic or bizarre in the behaviour of others, and thus which we might tend to evaluate negatively. Intuition nonetheless does not provide an excuse for inventing, even through sympathetic imagination, what religious acts mean. Phenomenological insight works hand in glove with empirical studies. 'Sympathy is unthinkable without an intimate acquaintance with the historical facts – thus again an interaction, this time between feeling and factual knowledge' (1960: 10).

Of course, historians must select which facts to emphasize in order to build up a composite picture of any religious tradition. It is here that evaluation is employed legitimately, but it is informed by historical criteria which filter out what is important in the study of religions and what is trivial. 'That which is insignificant always proves to have no lasting existence in history' (1960: 14). The historian, as a result, does not need to pay attention to fleeting movements or to transitory events. 'As far as the Ancient religions are concerned, most of the data of this sort have disappeared without leaving us any trace' (1960: 14). Consequently, for the historian and phenomenologist of religion, working together, the cultic practices and dominant beliefs that have persisted sometimes for thousands of years comprise the phenomena of religion. These can be seen in such rituals as the 'worship of nature and spiritual beings' which continue unabated even to our own day. Other acts, such as sacrifice or divination, also can be traced throughout history into the current practices of religious communities. 'The enduring existence of all these religious data proves their religious value' (1960: 14). This is precisely why historical data provide the content for phenomenological typologies and why at the same time they help the scholar of religion avoid wasting time on 'passing movements and temporary phenomena' (1960: 14). In this way, history becomes the servant of intuition and sympathy: 'That which has been carefully weighed and approved by generations and has been able to serve as the basis of life has proved its inner value' (1960: 15).

Kristensen: A summary

Kristensen's main contributions to the phenomenology of religion were carried forward by his students, van der Leeuw and Bleeker, since, as we have seen, his own analysis of phenomenology did not make its international impact until after the publication of *The Meaning of Religion* in 1960. Nevertheless, primarily through van der Leeuw's writings during the 1930s, Kristensen's central ideas became accepted as foundational principles within the phenomenology of religion, the main tenets of which can be summarized under the following four points:

1. Kristensen clearly defined separate but inter-related roles for history, philosophy and phenomenology in the study of religions by associating history with empirical data, phenomenology with typologies and philosophy with essences.

2. He gave final authority for interpreting a religion to the believer that religion.

3. He employed intuitive insight into the meaning of religion by foste empathetic attitude towards religions of which he was not a men order to achieve understanding otherwise unavailable to him as an outsider.

4. He eliminated evaluative comparisons based on evolutionary or theological assumptions from the 'Science of Religion' by insisting on informed comparisons based on thorough empirical research.

In his analysis of Kristensen's legacy for the study of religions, particularly in light of recent critiques of phenomenology emanating from 'post-christian humanists', post-modernists and post-colonial scholars, John Carman (2000: 172) concludes that Kristensen must be credited with the central idea that the 'scholar's religious imagination' lies at 'the heart of his scholarship, without which he is only a collector of "facts"'. Carman notes that we may reject Kristensen's view, 'but we need to be aware of the implications for the history and phenomenology of religion of that rejection' (2000: 172). The implications, to which Carman refers, become even more explicit in the writings of the second figure in the Dutch school I am considering in this chapter, Gerardus van der Leeuw.

Gerardus van der Leeuw: Overcoming the subject–object dichotomy

In 1918, van der Leeuw was appointed Chair of the History of Religion in the University of Groningen, after which he developed his own interpretation of the phenomenology of religion in a way that, as Jacques Waardenburg (2000: 263) argues, although 'inspired' by Kristensen, 'took a different course'. Pettersson and Åkerberg (1981: 23) refer to van der Leeuw as 'the best known Dutch representative of phenomenology', although they contend that his thought has 'little influence in the professional scholarship among the phenomenologists of today'. His main contribution to the phenomenology of religion is found in his book, *Phänomenologie der Religion*, first published in German in 1933, and translated into English in 1938 as *Religion in Essence and Manifestation*. A letter written from Kristensen to van der Leeuw in 1933, which has been reproduced in an English version by Willem Hofstee (2000: 173–4) notes that Kristensen held reservations about van der Leeuw's methods in *Phänomenologie der Religion*, particularly the way he used essential categories, such as 'power'. Kristensen complained: 'Everybody thinks he understands the idea of "power" of these peoples ['negroes and indians'] – is it not an illusion? Are we not satisfying ourselves only with general terms and words?' Kristensen's objections underscore the fact that van der Leeuw, far more than his mentor, incorporated philosophical ideas into his interpretation of the phenomenology of religion, and can be seen as the first scholar of religion quite consciously to wed philosophical phenomenology in its Husserlian form to the phenomenology of religion.

Van der Leeuw introduces his long excursus into the phenomenology of religion, in *Religion in Essence and Manifestation*, by describing phenomenology as a method for outlining the proper relationship between subjects and objects, which in the study of religions indicates the relationship between the scholar, who seeks to understand the behaviours of religious practitioners, and the practitioners, who understand themselves in various ways as responding to supernatural forces. What the scientist of religion regards as the 'Object' of religion, religious people take to be the 'Subject' or 'active agent' of religion. In most cases, the active agent for believers is God, who becomes the source of their religion. In a statement that came to define one of the principal assumptions beneath the phenomenology of religion, van der Leeuw asserted that the sciences of religion cannot study God, but only what religious people say or do in relation to God: 'For Religion, then, God is the active Agent in relation to man, while the sciences in question can concern themselves only with the activity of man in his relation to God' (1963: 23).

Although for believers, the 'Subject' in religion is God, when the scholar identifies it as the 'Object' of religious experience, the term 'God' is shown to be too imprecise to convey what is actually meant as the 'active agent' for believers. Usually, in its various forms and meanings, God is understood by religious people as something 'Other' and is known through the 'power' it generates. If we study 'primitive religious experience' or even the vast majority of the religions of antiquity, we meet supernatural power as the primary component in the idea of God (1963: 23–4). This leads van der Leeuw to make an initial assertion about the nature of religion, which the remainder of his book is devoted to disclosing. He associates God, or the supernatural element in religion, with 'the simple notion of an "Other"', which he describes as 'something foreign and highly unusual', and, in a general line from Schleiermacher, argues that it produces in the religious mind 'the consciousness of absolute dependence' (1963: 24). This starting point is intended in van der Leeuw's phenomenology to provide sympathetic insight into the way devotees experience religion, while at the same time allowing scholars to study objectively the manifold religious apprehensions of the supernatural.

Van der Leeuw divides the sections of his book precisely according to this distinction between the 'Object' and the 'Subject' of religion. He calls part one 'The Object of Religion', which he devotes to a study of 'power' in various forms across cultures and in the history of religions. Part two switches to the 'Subject of Religion', which analyses the 'Object' from the perspective of believers, from the point of view of faith. In this sense, he explains, 'the Object of religion, to faith, is Subject in the sense of "the active and primary Agent"' (1963: 191). As Subject, the study for van der Leeuw turns towards the 'Sacred', because 'human life, in directing itself towards Power, "first" of all was touched by Power: in orientating itself to the sacred, that is to say, it participates in sacredness' (1963: 191). The analysis of the Subject in relation to the sacred is divided into three categories:

the sacred man; the sacred community; the sacred within man: the soul. Part three of the book is devoted to what van der Leeuw calls 'Object and Subject in their reciprocal operation' in which the main typological classifications are analysed in accordance with outward and inward action, outward referring to such typological categories as purification, sacrifice, divination and myth, with inward action analysing religious experience, including mysticism, love of God and conversion. Part four of the book examines the way religions relate to the world and part five classifies types of religions, such as religions of love or religions of nothingness, and concludes with a study of different types of founders of religions. In this way, van der Leeuw demonstrates how the phenomenology of religion works and how it produces an entire system for classifying and understanding religious communities and their practices.

Throughout this analysis, we are led to ask, with Kristensen, how van der Leeuw justifies from the outset making such seemingly sweeping assertions about supernatural power constituting the essential characteristic of religion. He claims to have reached such conclusions on the basis of scientific evidence. In his quite short chapter one (1963: 23–8), which runs to just six pages, he cites ideas of power as described by R. H. Codrington in Melanesia, where *mana* is understood as an 'influence', not physical, but somehow supernatural, then moves to descriptions of Native American ideas of power through the Iroquois concept of *Orenda*. In a few paragraphs, he then refers to a wide range of similar concepts in Borneo, ancient Germany, among the Arabic speaking peoples and even in the Christian celebration of the Eucharist. Van der Leeuw says that he has reached his conclusions 'empirically' by referring to this underlying concept which, when it is compared cross-culturally, is transformed into an idea of 'Universal Power'. Van der Leeuw's first component in a phenomenology of religion thus directly addresses the subject–object problem, which he seeks to overcome through the concept of supernatural power, defined as the insider's core experience of religion, and described from the outside, in its many forms and variations throughout history, in scholarly accounts that are subject to verification or falsification following sound scientific principles.

In chapter two of *Religion in Essence and Manifestation*, van der Leeuw argues that although the power he has described can be generalized, it is in reality experienced as 'specific power' manifested in 'particular instances' (1963: 29). Here we meet a second major element in van der Leeuw's phenomenology: the scholar of religion identifies universal or 'ideal' types hidden within particular instances. In one sense, only the particular occurrences are real, but the construction of ideal types is necessary if we are to make sense of the specific cases described. Thus, supernatural power is expressed in a wide variety of particular instances, such as the Indian stratification of classes according to a supernatural order, or an Estonian peasant's idea of 'luck' in the success of a good harvest (1963: 29). By themselves, they have no apparent similarity, until the scholar sees or intuits the common element in the idea of power. By themselves, particular instances do not disclose the meanings inherent within them. Only by comparative studies resulting in the building up of the ideal types can scholars make hermeneutical sense out of the myriad of individual situations they encounter.

Van der Leeuw's philosophical phenomenology of religion

The majority of van der Leeuw's 700 pages in *Religion in Essence and Manifestation* identify and exemplify the ideal types as seen from the perspectives of Subject and Object. It is not until the final section of his book, what he calls the 'Epilegomena', that we finally find van der Leeuw's method spelled out quite specifically in terms of a definitive phenomenological method. I want to devote considerable space to analysing this section, but it must be emphasized that in the Epilegomena he is telling his reader what he has already done, first by identifying power as the essence of religion and in the remainder of the book by defining what he means by the subject–object dichotomy in the study of religion and how it helps to produce the essential categories of phenomenology. Although, as Waardenburg notes (1978: 203), originally van der Leeuw intended to place this section at the beginning of *Religion in Essence and Manifestation*, as a kind of prolegomena, it is clear that he decided to place it at the conclusion in order that the methods he had employed throughout the book could be made explicit.

That van der Leeuw was concerned in the first instance with philosophical phenomenology is confirmed by his analysis found in chapter 107, which is the first chapter appearing in the Epilegomena. In this section, he defines a phenomenon as what appears to a subjective observer. The process works in a sequential fashion: 1) something exists; 2) that which exists 'appears'; 3) because it appears, it is a phenomenon (1963: 671). This apparently simple account of how a subject apprehends an object is complicated by the fact that what appears makes its appearance to someone, implying that ' "appearance" refers equally to what appears and to the person to whom it appears' (1963: 671). This classical Husserlian analysis of perception leads van der Leeuw to assert that what appears is not a pure object, since it results from the interaction between subject and object. This does not mean that the objective reality of the phenomenon that appears is determined or 'produced' by the subject. 'Its entire essence is given in its "appearance", and its appearance to "someone" ' (1963: 671). When this 'someone' begins to talk about what has appeared, to analyse it and to give it objective meaning, phenomenology, the study of phenomena, arises.

The appearance of a phenomenon can be analysed further according to a three-stage process. In the first instance, a phenomenon is concealed, but it gradually becomes revealed and finally it attains a relative transparency to the observer. The concealment of a phenomenon can be correlated to personal experience, since there always exists a distance between what one experiences and the construction of reality. In a way reminiscent of Husserl, van der Leeuw acknowledges that his experience of writing the words on the page a few seconds before 'is just as remote from me as is the "life" associated with the lines I wrote thirty years ago in a school essay' (1963: 671–2). In both cases, what he has written is completely past; the distance from the immediate past varies little from any reconstructed experience. 'In fact, the experience of the lines of a moment ago is no nearer to me than is the experience of the Egyptian scribe who wrote his note on the papyrus four thousand years ago' (1963: 672). This point is made to underscore that everything is reconstructed, even our most cherished thoughts. We objectify ourselves. In this way, the problem of understanding the 'other' as object, 'whether close by or in distant China, of yesterday or of four thousand years ago', is precisely the same (1963: 672). The interplay between subject and object is rooted fundamentally in our perceptions, a fact which cannot be avoided even in our most intimate personal experiences.

This means that we must reconstruct reality in order that what appears may gradually be revealed and thus be understood. The reconstruction begins by the observer attempting to make sense of otherwise 'chaotic' appearances by sketching an outline of it, that is, by giving it a structure. The structure of what is observed does not appear directly in experience, nor does it result from logical analysis. It is somehow 'understood' in its totality by the observer, in a way which draws us back to the primary relationship between the subject and the object. Structure is what has been 'significantly organized' by someone who thereby gives it an objective reality. The significance noted by the observer thus 'belongs in part to reality itself, and in part to the "someone" who attempts to understand it'. The subject–object distinction in this sense translates into a relationship between 'understanding' and 'intelligibility', which van der Leeuw asserts is connected in a way that is 'unanalysable' because it is an 'experienced connection'. The meaning that results from understanding and intelligibility again reflects the subject–object division. Understanding in a way 'dawns upon us', suggesting that meaning transcends either the subject or the object. The meaning of any structural connection at the same time is 'my' meaning and 'its' meaning, the subjective perception and the objective reality, 'which have become irrevocably one'. This results finally in the meaning of what appears becoming transparent through the 'act of understanding' (1963: 672–3).

In this way, understanding attains the deepest sense of comprehension, as implied in the German translation of the English word 'understanding' as *Verstehen*, a term Waardenburg argues is central to van der Leeuw's entire phenomenology of religion (1978: 224). Apprehending in depth means that what 'dawns upon' the observer is not momentary or even a single experience. It is a seeing into the meaning of a sequence of experiences, wherein the unity that binds them together is somehow intuited. This implies that the unity comprising understanding in depth allows the observer to see into the 'essential nature' of the experience, since it is a seeing into the connections that make it at once what it is as a component and what it is in its totality. Waardenburg (1978: 225) argues that van der Leeuw's use of *Verstehen* meant that the phenomenologist does not comprehend factual existence as such, but contemplates the essence of the phenomena through an analysis based on intuition rather than reason. Even if Waardenburg is correct on the point, it is clear that Van der Leeuw regarded the process of phenomenological understanding as fully scientific and that it could be outlined systematically.

The phenomenological method outlined

To demonstrate how such understanding is achieved, following Husserl, van der Leeuw drew attention to the relationships that subsist between appearances. By referring to 'backgrounds', 'nuances', similarities and differences, he concluded that these are 'perceptible relationships', not 'factual relationships', since 'they are valid only within the structural relations' (1963: 673). Drawing structural relationships leads to identifying 'types', or, as is now clear, following Weber, 'the ideal type', which he defines in detail in the Epilegomena, as 'timeless' in that it does not necessarily 'occur in history' but nonetheless 'possesses life, its own significance, its own law' (1963: 673). To explain how he arrived at an ideal type, which, as we have seen, at one level is what his entire book has illustrated, van der Leeuw resorts to phenomenology, which he argues falls into five stages: 1) assigning names; 2) interpolating the phenomena into our own lives; 3) observing the *epoché*; 4) clarifying what has been observed; 5) achieving genuine understanding (1963: 674–6). For purposes of his discussion, these are described as occurring in succession, but, he explains, 'in practice, they arise never successively but always simultaneously' (1963: 674).

What appears must be named in order that various phenomena can be identified, separated and classified. Certain practices fit under one classification and other practices do not, but each must be assigned a different designation for purposes of clarification. Hence, one act might be called 'purification' and another 'sacrifice', since they are distinguishable in kind. The process of assigning names is a necessary human act as old, says van der Leeuw, as Adam, but it carries the danger of reifying a particular term by transforming the name into an object or a thing, rather than experiencing it as a living entity which has appeared in a particular moment (1963: 674). To avoid this, van der Leeuw urges us to employ the technique of interpolating that which we have named and which may seem entirely alien to us into our own experience. The alien character of the 'other' is further complicated by the fact that what appears and that which we name comes to us in a mediated fashion, often as symbol through language, which we must interpret. Such an interpretation cannot be undertaken unless we experience it in our own lives 'intentionally and methodically'. This requires sympathetic understanding, much more akin to art than to pure logic, where one feels for that which is outside of oneself by bringing it into one's own experience. Yet, as van der Leeuw has already noted, even bringing what we have named into one's own experience does not remove entirely the separation between subject and object. Still, the idea of sympathetic interpolation is based on the fundamental principle, which to this day represents a hallmark of the phenomenology of religion, that 'the essentially human always remains essentially human, and is, as such comprehensible' (1963: 675).

A third component in van der Leeuw's outline of a phenomenological method introduces the idea of *epoché*, which he describes as 'a technical expression employed in current Phenomenology by Husserl and other philosophers', to indicate, 'that no judgment is expressed concerning the objective world, which is thus placed "between brackets"' (1963: 646). Van der Leeuw refers to *epoché* as observing 'restraint' by concerning itself only with phenomena and not by assuming what is 'behind' appearances. It is precisely all such prior assumptions which are restrained, or put within brackets, thereby suspending judgements about 'their real existence, or their value' (1963: 646). This method does not entail making the mind a blank tablet or empty space on to which the phenomena implant themselves. Rather, as employed by Husserl, as van der Leeuw notes, the 'phenomenological reduction' crosses through or brackets out what is accidental in appearances to obtain insight into their essence (1963: 676).

This leads to the fourth element in van der Leeuw's description of phenomenology, whereby the scholar clarifies what has been observed by placing phenomena into types or categories. In order that this not be confused with an evolutionary scheme, van der Leeuw underscores that the typologies do not imply 'causal connections' in the sense that 'A arises from B, while C has its own origin uniting to D'. The phenomenological method simply employs 'structural relations somewhat like a landscape painter combines his groups of objects, or separates them from one another'. The phenomenological clarification thus refers to structural relations and connections, 'and this means that we seek the ideal typical interrelation, and then attempt to arrange this within some yet wider whole of significance' (1963: 676). We see once again in this discussion that van der Leeuw understood the ideal types as emerging from a process of grouping religious phenomena into patterns of coherence, or as Gavin Flood (1999: 98) describes it, as 'the mapping of intersubjective networks of communication and the relating of diverse networks to each other'.

The end of this process, when stages one through to four are combined, produces understanding, as noted above, in the sense of comprehending deeply, thoroughly and intuitively. Van der Leeuw describes such a grasp of the phenomena as occurring when 'the chaotic and obstinate "reality" thus becomes a manifestation, a revelation'. Although this may sound entirely subjective, even mystical, for van der Leeuw, phenomenology represents a 'science', since the aim of all science is understanding that has resulted from an interpretative analysis of data. Following Dilthey, van der Leeuw concludes that science in the end is 'hermeneutics' (1963: 676). This method, moreover, applies equally to the interpretation of history as it does to present experiences, since all that has occurred in history reflects the same common humanity as that which is expressed in the actions of 'my nearest neighbour'. Van der Leeuw explains: 'Certainly the monuments of the first [Egyptian] dynasty are intelligible only with great difficulty, but as an expression, as a human statement, they are no harder than my colleague's letters' (1963: 677).

That phenomenology is a science is further corroborated by van der Leeuw's insistence that interpretations constantly must be tested against facts, in the cases of historical and textual studies, through archaeology and philology. Of course, no facts are pure, and hence must be subject to interpretation. When a scholar examines archaeological evidence or textual sources, the resulting interpretations are quite restricted and limited, thus necessitating further phenomenological analysis. Nevertheless, if the phenomenologist does not test interpretations of meaning against the data, the method falls into 'pure art and empty fantasy'. The aim of phenomenology thus is to obtain 'pure objectivity'. This does not mean attempting to disclose Kant's 'thing-in-itself', but 'to *testify* to what has been manifested' (van der Leeuw's emphasis) (1963: 677–8). As the phenomenological method demonstrates, this maps out an indirect route to objectivity since it involves 'a thorough reconstruction', but the scholar has no other choice. As Kant has demonstrated, in van der Leeuw's words, 'to see face to face is denied us', but, he adds, 'much can be observed even in a mirror; and it is possible to speak about things seen' (1963: 678). The ideal types, interpretations or meanings are not invented, but result from the interplay between hard fact and scholarly hermeneutics.

Religion as a category *sui generis*

The phenomenological method, as outlined by van der Leeuw, thus far has only indirect bearing on religion, since he was dealing almost exclusively with epistemological issues. In the next chapter of the Epilegomena, he considers religion as a category of its own and brings his reader back to the very first pages of his book by relating religion to power, as seen through the eyes of a religious person. Van der Leeuw asserts that the religious person 'tries to elevate life, to enhance its value, to gain for it some deeper and wider meaning' (1963: 679). This analysis looks at religion from the perspective of the human, what van der Leeuw calls the horizontal plane, and not as an attempt to interpret revelatory acts. Because religion has to do with asserting power by arranging life in some order, granting to it significance and meaning, religion is instrumental in producing cultures. In this sense, van der Leeuw explains in a footnote, 'Ultimately, all culture is religious; and, on the horizontal line, all religion is culture' (1963: 679).

Religious significance for a believer points towards an ultimate significance in life. It 'is that on which no wider or deeper meaning whatever can flow' (1963: 680). Religious meaning encompasses the 'whole'; it provides 'the last word'. Yet, for a religious person there is always something more, something inexplicable, something which remains 'eternally concealed'. Hence, by reaching the 'ultimate' meaning, the religious person discovers the limits of meaning. Van der Leeuw refers to the adherent as the '*homo religiosus*', one who seeks complete understanding and hence mastery over life in all cultural settings. For example, in agricultural societies, religion is associated with efforts to dominate the soil by making it fruitful. In hunting and gathering societies, animals are imbued with powers that the community must subdue in order to survive. Even in religions with profound philosophical systems, the philosophies are developed in order to subject the world to human control (1963: 680).

In addition to this horizontal plane, religion must also be understood as moving in a vertical direction 'from below upwards, and from above downwards' (1963: 680). This is very different from understanding how religion and culture interpenetrate one another, since a vertical interpretation refers to revelatory acts given from beyond the limits of the human religious quest. The vertical dimension to religion thus is not a phenomenon; it does not appear. It can be grasped only in experience. This means that the vertical dimension can be seen from no other perspective than that of the believer, since that which does not constitute a phenomenon cannot be studied scientifically. Only the believer's feeling for the transcendent, the sense of the infinite, or what van der Leeuw calls Otto's idea of the holy, the numinous, can become a phenomenon and thus open to a phenomenological investigation (1963: 681). This, of course, describes the area of religious experience, not the numinous itself, which remains in Otto's terms, as van der Leeuw notes, 'Wholly Other' (1963: 681). Taken together, the horizontal and vertical directions demonstrate that religion in its essence concerns itself with one goal: salvation. This can be interpreted in numerous cultural contexts to mean such things as 'the enhancing of life, improvement, beautifying, widening, deepening', but everywhere power is translated finally into a quest for salvation. Life as it is given never defines the proper end of existence; there is always something more, something which religious people seek (1963: 681). Van der Leeuw concludes: 'In this respect all religion, with no exception, is the religion of deliverance' (1963: 682).

After having defined phenomenology and then religion, van der Leeuw moves in chapter 109 of *Religion in Essence and Manifestation* to discuss the two together, not by applying phenomenological principles explicitly to religion, something in effect he has already done, but by carving out a distinctive space for the phenomenology of religion as a discipline in its own right. He claims to have already made it clear that the aim of phenomenology is to attain understanding, defined in terms of *Verstehen*, which John Daniels (1995: 45) describes as 'literally a transposition ... of subject into object for the purpose of re-experiencing it'. Such understanding presupposes *epoché*, but in van der Leeuw's words, 'this is never the attitude of the cold-blooded spectator' (1963: 684). As Daniels suggests, this method implicates the observer personally with that which is observed, a point made absolutely clear by van der Leeuw when he likens phenomenological understanding to 'the loving gaze of the lover on the beloved object' (1963: 684). This approach to understanding is not restricted to the phenomena of religion since, as van der Leeuw explains, 'to him who does not love, nothing whatever is manifested' (1963: 684). In support of this, he cites both the Platonic and Christian epistemological traditions.

If van der Leeuw's contention is correct, that genuine understanding entails intense degrees of sympathy irrespective of the content for which understanding is sought, his method cannot be called theological, but as he says, quoting the Roman Catholic theologian Przywara, can serve only as a 'preparedness for revelation' (1963: 685). Nor can the phenomenology of religion be equated with the 'poetry of religion', despite the fact that by identifying structures within 'the chaos of the given' the phenomenologist performs acts of creative imagination. This is not poetry because the phenomenologist is 'bound up with the object' and must always submit 'again and again to correction by the facts' (1963: 685–6). The phenomenology of religion must also be distinguished from the history of religion, although, van der Leeuw admits, the historian and the phenomenologist 'work in the closest possible association', so much so that 'in the majority of cases they are combined in the person of a *single* investigator' (emphasis his) (1963: 686). The historian, however, can proceed without understanding the data by describing and cataloguing events. If the phenomenologist fails to understand, by contrast, 'he can have no more to say' (1963: 686). In other words, for van der Leeuw, phenomenology always requires understanding, since the task of the phenomenologist is to interpret the data. The historian, of course, also needs to understand, but the historian can carry on simply by describing 'what has actually happened' (1963: 686).

Since so much of the study of religion depends on an analysis of experience, phenomenology could be considered as closely allied to the psychology of religion. Van der Leeuw admits this, but argues that religion is constituted by so much more than psychic phenomena that it might be said 'that the phenomenologist of religion strides backwards and forwards over the whole field of religious life, but the psychologist of religion over only a part of this' (1963: 687). It is clear that van der Leeuw has drawn heavily on philosophy for his methodology in the study of religion, but philosophy in its metaphysical sense does not define the concerns of the phenomenologist. The philosopher, in the sense of Hegel, deals with the 'dialectical motion of Spirit' and may even be thought of as having something of God within him. The phenomenologist provides a stark contrast to the philosopher precisely at this point: 'The phenomenologist should not become merely frightened by the idea of any similarity to God: he must shun it as the sin against the very spirit of his science' (1963: 687).

This leads to van der Leeuw's explicit denial that the phenomenology of religion is theology. Like philosophy, theology searches for truth, whereas phenomenology, because it employs the method of *epoché*, suspends any comment on truth claims. The distinction between phenomenology and theology goes even deeper than this. Although phenomenology considers the horizontal and vertical dimensions of religion, and utilizes a sympathetic interpolation based on the premise that human sentiments are universally accessible to the researcher, theology speaks not about how the religious person experiences God, but 'about God Himself' (1963: 688). As he asserted in the very first paragraph of his book, van der Leeuw underscores in this section the fundamental principle that, for phenomenology, God can be neither the object nor the subject of religion, since in order to be either of these, God would need to 'appear', that is become a phenomenon. If God can be said to 'appear', the meaning of the term is entirely different from what is meant by a phenomenological appearance. Theology, in stark contrast to phenomenology, is concerned not with 'intelligible utterance' but with 'proclamation' (1963: 688).

For van der Leeuw, therefore, the phenomenology of religion can be said to constitute an academic discipline in its own right, one that is entirely distinguishable from all other disciplines concerned with the study of religion. It should not be confused with methods peculiar to history, philosophy, aesthetics, psychology or theology. On this account, phenomenology can be described best as fully *sui generis*, utterly irreducible to any other scholarly methodology.

Van der Leeuw: Anticipating the religion–theology debate

Jacques Waardenburg (1978: 222) claims that van der Leeuw's phenomenology of religion arose out of his earlier 'ethical theology' and was closely connected with his 'later theological and philosophical anthropology and his liturgical and sacramental theology' (1978: 222). Timothy Fitzgerald (2000: 36) adds that van der Leeuw, as a Christian theologian, was concerned primarily with 'European philosophical and theological concerns more than anything else', and thus when he distinguished phenomenology from theology, he gave 'a theological account of their differences'! One aspect of the theological charge against van der Leeuw results from his use of 'insider' perspectives, what he likened to 'the loving gaze of the lover upon the beloved object', which he believed was entailed in the stage of sympathetic interpolation. He thought that it was necessary methodologically for the phenomenologist to employ an active and imaginative intuitive involvement in the perspective of believers, an act which could easily be confused with endorsing the believer's point of view. This approach can be related also to van der Leeuw's personal Christian convictions and to his own theological background, as Waardenburg does in his analysis. George James (1995: 224) notes, 'For some, the role of theological categories in van der Leeuw's interpretation, in the very appearance of these phenomena, would be sufficient to discredit the claim that his phenomenology of religion is something other than theology'. Fitzgerald's point is somewhat different, in that he accuses van der Leeuw of an inherent contradiction by distinguishing phenomenology from theology on the basis of categories that operate exclusively within theological discourse.

These discussions anticipate the subsequent debate over the relationship between theology and religious studies, which I outline in Chapter 7. Nevertheless, that they persist in many current writings about van der Leeuw indicates that his argument provides a critical source for the contemporary controversies surrounding the role of theology in the study of religion (see also Wiebe, 1999: 176–81). In this context, it is important to underscore that van der Leeuw himself recognized the problem when he outlined differing roles for the scholar, even when such roles were employed by the same person. He consciously and deliberately separated his work as a phenomenologist of religion from his other tasks while operating as a theologian, an ethicist or even as a philosopher of aesthetics. As James notes, for van der Leeuw, the phenomenologist first and foremost operates as a scientist who studies 'a person who practices religion, who, sacrifices, who prays' (1995: 225). Van der Leeuw insisted that phenomenological research can never be reduced to an artificial objectivity or a cold empiricism, since the aim of the phenomenologist is to consider believers on their own terms, as those who perform religious acts in the firm conviction, in the words of James, that 'these acts are evoked by a power' outside of themselves (1995: 225). Nevertheless, van der Leeuw repeatedly stressed that the role of the phenomenologist is to describe, interpret and understand the acts of religious people and not the 'power' to which believers attribute the source of their religious experience. In the final analysis, for van der Leeuw's phenomenology of religion, the use of the twin techniques of *epoché* and sympathy led to a unique understanding (*Verstehen*) of religion, a result which he contended established phenomenology as a scientific discipline rather than a theological enterprise.

C. Jouco Bleeker: The scope of the phenomenology of religion

The third key figure in the Dutch school of phenomenology, after Kristensen and van der Leeuw, is C. Jouco Bleeker (1898–1983), who was Professor of the History of Religion in the University of Amsterdam and General Secretary of the International Association for the History of Religions from 1950 to 1970. Bleeker, like van der Leeuw, had studied under Kristensen at Leiden, and, like his mentor, specialized in Egyptian religions. Nevertheless, he seemed to have been more influenced by van der Leeuw than Kristensen. The Swedish scholar of religions, Geo Widengren, writing in 1969, observed: 'When I first met Bleeker – more than twenty years ago – it was obvious that the influence of G. van der Leeuw had outweighed the influence of his own teacher in Leyden, W. Brede Kristensen. He was more attracted by the phenomenological study of religion in general than by the historical investigation of some special religion' (cited by Molendijk, 2000: 34). Bleeker retired from his post at the University of Amsterdam in 1969, but continued to be involved in discussions concerning the academic study of religions well into the 1970s. One of his most important contributions after his retirement occurred at a conference on studies in the methodology of religions held in Turku, Finland in August 1973. The volume which resulted from the conference, edited by Lauri Honko under the title, *Science of Religion: Studies in Methodology*, contains one of Bleeker's most succinct summaries of his understanding of the phenomenology of religion and at the same time anticipated the debates that later in the century would surround the phenomenological method.

In response to papers delivered by Haralds Biezais (1979: 143–61) and Carsten Colpe (1979: 161–73) at the Turku consultation, Bleeker (1979: 173–7) outlined initially what the phenomenology of religion is not. First, he argued, the phenomenology and history of religions (what he called collectively a religio-historical science) do not aim to foster world peace and social harmony, despite the widespread popular belief that the study of religions creates 'mutual understanding among the adherents of the different religions' (Bleeker, 1979: 174). Although this may be a side-effect of academic studies, 'theoretically these

126

studies are a purely scholarly affair' (Bleeker, 1979: 174). Secondly, he noted that many people believe that the study of religions should act as a force to influence the future direction of religion. This would give to the study of religions a somewhat theological task by replacing traditional theology with what in popular circles would be seen as promoting a new, inclusive universal religious understanding. Again, Bleeker argued, this confuses the proper role of the student of religion with that of a religious activist (a point to which I will return in Chapter 7). What unites scholars of religion is not some kind of inter-religious cooperation, but 'the unbiased study of religion as a phenomenon of historical and current interest' (Bleeker, 1979: 175). These disclaimers staked out quite important territory for Bleeker's understanding of the phenomenology of religion as a non-theological methodology with little interest in promoting religious causes, no matter how liberal and humanitarian such causes might appear.

In his same response to Biezais and Colpe, Bleeker affirmed the traditional role for the history and phenomenology of religions as it had been outlined by Kristensen and van der Leeuw before him: 'The aim of the two disciplines at stake can be no other than the descriptions and the understanding of religion as a human phenomenon' (Bleeker, 1979: 175). He added that this study is fundamentally distinct from other disciplinary approaches, such as anthropology, psychology and sociology, which frequently reduce 'religion to non-religious factors' (Bleeker, 1979: 175). This point needs to be stressed because religion 'is not in the first place man's attempt to integrate himself into society, but is the effort to maintain his stand over against the overwhelming cosmic forces which influence human life, both individually and collectively' (Bleeker, 1979: 175). To reduce this apprehension of life to social or psychological factors would be to miss the distinctively 'religious' element within religion.

Bleeker acknowledged that the phenomenology of religion had come under severe criticism in recent years, but he contended that it should persist both in its name and in its primary methodology. In a separate article published in 1972 under the title 'The Contribution of the Phenomenology of Religion to the Study of the History of Religions', which he delivered at a Study Conference organized in 1969 by the Italian Society for the History of Religions, Bleeker (1972: 35–54) noted that since van der Leeuw's major publication on the phenomenology of religion first appeared in 1933, numerous interpretations of the purpose of phenomenology have been offered. These fall into three main categories:

1) the descriptive school which is content with the systematisation of religious phenomena, 2) the typological school, which aims at the research of the different types of religion, 3) the phenomenological school in the specific sense of the word, which makes inquiries into the essence, the sense and the structure of the religious phenomena. (Bleeker, 1972: 39)

In his contribution to the Turku Conference, Bleeker contended that each of these interpretations is justified, but that the third category most fully describes the aim of a genuine phenomenological approach. Of course, he argued, phenomenology produces typologies in the sense that 'it can clarify the sense and the structure of the constitutive elements of religion, such as sacrifice, prayer and magic'. Phenomenology, however, does far more than classify historical data; it provides, as the third category suggests, 'the heuristic principles which help historians of religions to find the clue to difficult problems' (1979: 176).

Bleeker concluded his contribution to the Turku Conference by reiterating the main tenets of what by 1970 had come to define the classical phenomenological method: 'one should study the religious phenomena both critically, unbiasedly, in a scholarly manner, and at the same time with empathy' (1979: 176). It is significant that Bleeker, near the end of his academic career, defended these basic phenomenological principles, derived from Kristensen and van der Leeuw, so vigorously against attacks from two opposing camps, one from liberal theologians who saw no difference between the study of religions and encouraging religious values and the other from social scientists who showed disregard for the fundamental religious sentiment in humanity by explaining it away in terms of non-religious social, psychological, or even economic or political forces. Bleeker, however, cannot be understood properly as one who simply restated phenomenological principles as they were developed by Kristensen and van der Leeuw. Over his long career, he made substantial contributions that refined phenomenological methods and in some ways made them more sophisticated. He did this chiefly by clarifying the relationship of the phenomenology of religion to Husserl's form of philosophical phenomenology and by demarcating the positive principles of phenomenology as a genuine, and hence irreducible, science of religion.

Bleeker's phenomenology as structural and dynamic

Bleeker accepted that for the phenomenology of religion to refrain from imposing pre-formulated biases into the study of religion, it would need to follow van der Leeuw's use of three philosophical concepts: the *epoché*, the eidos of religion and intuitive insight. In his discussion of this subject at the Italian study conference of 1969, he defined *epoché* as the 'suspension of judgement, in this case of the decision in regard to the question of the truth of religious phenomena' (1972: 40). He added that 'the concept indicates the attitude of impartiality, the attentive listening which is the absolute condition for a right understanding of the import of the religious phenomena' (1972: 40). He also affirmed with van der Leeuw that the aim of phenomenology was to discover the eidos, or the essence of religious phenomena: 'The eidetic vision is the search for the eidos, i.e. the essence and the structure of the religious facts' (1972: 40). He admitted that the terms *epoché* and eidos were obtained from the school of philosophical phenomenology founded by Edmund Husserl, but he insisted that phenomenologists of religion do not use these terms in any other than a 'figurative' way in order to provide historical data with a systematic structure. To attempt to apply these terms in the technical sense used by philosophers, as van der Leeuw had done, took the phenomenology of religion away from its empirical aims towards an oftentimes obscure analysis of ways of knowing. He explained:

> In my opinion a vulnerable side of Van der Leeuw's phenomenology is that too many elements of the philosophical phenomenology have been incorporated into it, in the form of speculations about the deeper meaning of the concept: phenomenon. Thereby the phenomenology of religion transgresses the boundary of its competence. Any one who has had a serious talk with the followers of the school of Husserl or Heidegger, knows that much expert knowledge and even penetrating thinking is required to solve such questions. The student of the history of religions is a layman in this matter and he should refrain from meddling in these difficult affairs. Phenomenology of religion is no philosophy of religion, but a systematization of historical facts in order to grasp their religious value. (Bleeker, 1972: 41)

Although he distanced himself in this way from Husserl, Bleeker borrowed from van der Leeuw the concept 'intuition', which he closely connected to the essence or eidos of religion (see Cox, 1998: 244–62). In an article entitled 'The Key Word of Religion', which appeared in his book *The Sacred Bridge* (1963), Bleeker argued that detecting the structure of religion would be aided by identifying 'a key word of religion, by which the heart of religion would be touched, so to say in one shot' (1963: 40). His 'key word' is intuited: 'The student of the history of religions and the phenomenology of religion starts his study with an intuitive, hardly formulated, axiomatic notion of what religion is'. Empirical studies follow on from the intuition in order to produce 'an inclusive formulation of the essence of religion', which, when constructed, comprises 'the crowning of the whole work' (1963: 36). The key word provides the connection between the initial intuition and the completion of empirical studies through the identification of the fundamental structure of religion.

After rejecting a number of possibilities for the key word, such as power (van der Leeuw) because it is too vague (1963: 40–2), 'personal god' (Geo Widengren) because it is too limited (1963: 42–3) and 'the holy' (Rudolf Otto) because it is too individualistic (1963: 44–5), Bleeker offered his own suggestion for the key word of religion: 'the divine'. The selection of 'the divine' as the key word of religion assigns historians and phenomenologists of religion the task of enquiring 'into the original significance of the terms by which the deity is indicated in the different religions' (1963: 46). After first intuiting 'the divine' as the key word of religion, the scholar then can study how the concept has been understood and practised in various religions. Bleeker admitted that to produce conclusive results his method would have to involve a vast empirical study, one that would extend beyond the range of any one scholar. It would be impossible for a researcher to catalogue all the ways the divine has been understood throughout the history of religions. An alternative would be to study a limited number of the notions of God in various religions which could then serve as a paradigm for all religions. The conclusions of such a limited study would be open to debate and modification, but this process 'would at any rate shed new light on the question of the key word of religion' (1963: 46).

e notion of intuitive insight into the eidos of religion, which could then be empirically in the history of religions, led Bleeker (1972: 41) to distinguish the work of the phenomenologist of religion not just from philosophy but also from anthropology. 'Some phenomenologists', he argued, 'think that religious phenomena can best be understood from the anthropological angle', but religion can never be understood simply as a human activity. If the key word of religion is 'the divine', the phenomenologist must conclude that, although the human factor in shaping religion is important to understand, it does not exhaust the meaning of religion. Since religion 'is always a relation to God or to the Holy', in order to understand its structure, the scholar must analyse 'a certain notion of God and not the mentality of the people who are religious' (1972: 41). Bleeker was not arguing theologically at this point, that behind all religion can be detected a divine reality, but was contending that religion contains its own inherent structure and dynamic operation which the phenomenologist of religion needs to uncover and disclose as part of an overall science of religion. In order to accomplish this, three processes are involved: 1) discerning the *theōria* of the phenomena; 2) identifying the *logos* of the phenomena; 3) examining the *entelecheia* of the phenomena (1972: 42).

For Bleeker, the term '*theōria*' referred to the shades of significance found within various types of religious belief and practice, such as the idea of the deity, the meaning of sacrifice, the use of magic and the place of symbol in ritual and art. The impartial manner of phenomenologists and their direct study of the phenomena, taken together, allowed them to discover the *theōria* and thus to disclose the essential meaning within religious typologies. Bleeker (1972: 42) explained: 'The theoria of the phenomena discloses the essence and the significance of the facts, [for instance] the religious meaning of sacrifice, of magic, of anthropology'. He added that the *theōria* of the phenomena 'leads to an understanding of the religious implications of various aspects of religion which occur all over the world' (1963: 16). By the term *logos*, Bleeker was referring to the work of the phenomenologist in identifying the hidden structures within the different religions by showing that they are built up according to strict inner laws. Bleeker understood a structure as an objective entity, which, when uncovered by the phenomenologist, demonstrates that religion possesses a logic which is just as detectable and capable of description as are the laws of the natural sciences. He explained: 'Religions always have an inner logic, that dominates the structure' (1972: 45). For example, the phenomenologist describes what the religion under study teaches about the divine and how such beliefs produce cultic or religious practices. In turn, quite logically, the ideas about the ultimate and the ensuing cultic practices lead to a sense of what it means to be saved within the particular religion under study. By discerning these connections, the phenomenologist is able to describe in a purely logical manner how ideas about the divine result in religious practices and produce ideas about the nature of what it means to be human. 'The logos of the phenomena', Bleeker concluded, demonstrates that 'religion is never an arbitrary conglomerate of conceptions and rites, but always possesses a certain structure with an inner logic' (1972: 42).

Bleeker's use of the *theōria* and the *logos* of the phenomena enlarged on van der Leeuw's theories of ideal types and the structure disclosed by religious phenomena. By introducing the notion of *entelecheia* into an overall science of religion, Bleeker made his own contribution to the phenomenology of religion by applying scientific principles to an analysis of the dynamic nature of religious change. In this way, he regarded the *theōria*, the *logos* and the *entelecheia* of religion as essential components in an overall science of religion. The *theōria* revealed the essential meaning within religious types; the *logos* disclosed the inner logic within religious beliefs and practices; the *entelecheia* demonstrated how religions change following a purposeful direction. Because he considered *entelecheia* so important for understanding religion, including its persistence in the modern world, he devoted a good deal of attention to explaining this concept.

Entelecheia: The logic of religious development

One of the central criticisms of the phenomenology of religion, particularly when it was depicted primarily as classifying religious beliefs and practices into types in order to identify the essence of each category, was that it adopted an entirely ahistorical approach to the study of religion. Bleeker (1963: 16) admitted that the 'task of the phenomenology of religion generally is taken as a static one' since to discern the significance of any phenomenon of religion requires examining the subject as if it were presented to the scholar 'as arrested pictures'. This certainly is not the case, since religions are dynamic and respond to numerous internal and external influences. They are better understood therefore as 'moving pictures'. In order to counter this central critique of phenomenology and to account for the dynamic nature of religion, Bleeker introduced the term '*entelecheia*', which he understood in an Aristotelian sense, 'namely the course of the events in which the essence is realized by its manifestations' (1963: 17). The classical Aristotelian example of this is the acorn, which possesses the potential to manifest its essence as a towering oak tree (see Popkin and Stroll, 1986: 137). Bleeker applied this teleological interpretation to religious change under four categories: 1) the question concerning the origin of religion; 2) the search for a historical logic in religious change; 3) the problem of 'impure religion'; 4) the question regarding the persistence of the religious impulse in the modern world. He dealt with these questions in depth in his essay, 'Some Remarks on the "Entelecheia" of Religious Phenomena', which comprised chapter two of *The Sacred Bridge* (1963: 16–24). At the outset of this chapter, he notes that 'by asking for a possible entelecheia of the religious phenomena the student of the history of religions approaches the object of his study in an unbiased way', fully aware 'that history is a narrative of continuous changes'. A search for the *entelecheia* of religious phenomena therefore asks 'whether there is any sense in this endless flux of happenings' (1963: 18).

One way scholars have sought to make sense of religion is to search for its historical origin. Bleeker asserts that although history provides vague clues as to the origin of religion, no conclusive evidence exists on this subject. If the scholar examines ancient practices, such as worship of the dead or references to female deities or to veneration of the sun, these cannot explain how religion began among humans. Some scholars have suggested that by studying the 'primitive' religions of Africa or Australia, we could obtain insight into the beginning of religion in human societies. But, Bleeker argues, this is based on the fallacy that primitive religion as it exists today represents the oldest form of religion. In fact, primitive religion has its own history; it has just not been written down. 'This means that at the time when European explorers met primitive religion it has long ago passed the initial religious stage where religion is supposed to have come into existence' (1963: 18–19). Thus, neither by studying ancient religions nor by analysing ethnographic accounts of primitive religions do we learn how religion originated as a human activity. This represents no loss to human knowledge, since the question itself is based on a fundamental misunderstanding of the purpose of religion. People seek to determine how religion began, because 'they hope to discover a point at which they can observe how religion arises from non-religious factors' (1963: 19). This discloses another fallacy, since 'religion always starts by itself and cannot be deduced from non-religious elements' (1963: 19). This does not mean that the scholar should not try to understand how religions appear, what circumstances prompt them or how they reinvent themselves today. What it does mean is that the assumption that religion has been produced from wholly non-religious forces should be challenged on the basis that religion provides its own clues about its purpose and meaning. Although it is not possible to detect scientifically how religion began initially in history, by studying its development throughout history, the scholar is led to conclude that 'religion appears spontaneously'. This must have been the case in primeval times just as it is today (1963: 19).

This leads Bleeker to consider if there can be identified some logic in the way religions develop throughout history. The main theories which have sought to analyse the role of religion in historical change, in Bleeker's view, are the hypotheses of a 'primordial monotheism', 'historical evolutionism' and a 'catastrophic end of history'. The notion of a primitive or primordial monotheism, as we have seen, was advocated by Max Müller, but was promoted most forcefully by the Austrian priest and ethnologist, Wilhelm Schmidt, who argued that originally humans were all believers in a single high God, but over time developed beliefs in many gods and spirits, which gradually displaced the importance of the high God in cultic practice. Bleeker argues that historical facts are too scant to make a judgement on this theory, but it is possible for the phenomenologist to uncover the prior assumption which underlies it. Those who endorse this theory presuppose that divine revelation explains the universal human belief in God. The science of religion is incapable of testing this claim since it 'is built on presuppositions which pure historical research can neither prove nor reject' (1963: 17).

The theory that religion has evolved over time is predicated on an application of Darwinian principles to social and cultural change whereby religion is thought to have 'evolved from a simple initial stage into the highest forms known to humanity' (1963: 17). Bleeker dismisses this theory out of hand as entirely 'obsolete' when seen in the light of the tragic occurrences of two world wars in the twentieth century. The idea that there could be discerned a gradual progress within cultures has been shown by recent history to be entirely unfounded. This, of course, could lend support to the idea that recent events tend to confirm that history is leading not to a gradual development based on progress but to a catastrophic end. This view has been promulgated within Christian theology, particularly by those emphasizing an apocalyptic interpretation of the biblical tradition, or by those employing similar formulations of an Islamic eschatology. It has also been promoted by secularists who have adopted a pessimistic view of modern history by concluding that European civilization is 'worn out'. Whether it has been formulated by Christian or Islamic theologians or contemporary secularists, Bleeker asserts, 'there is no convincing proof that religion after having flourished in former times is on its decline since the beginning of modern history so that a total collapse of the spiritual forces must be feared in the near future' (1963: 18).

Despite the failure of previous theorists to identify correctly consistent patterns operating in the history of religions, Bleeker argues that it is possible for scholars to discern a logic in religious change by applying the principle of *entelecheia*, which shows that religions are transformed according to certain inbuilt processes very much like life cycles. Since the religions of antiquity are now dead religions, a process of birth, maturity and eventually death would seem to apply to each of them. Religions which for centuries flourished, such as the religions of ancient Egypt and Greece, or Manicheism, have died, but they have been replaced by new ones. The historian analyses such cycles of birth, death and rebirth by noting that the new religions are never entirely new, since they incorporate old ways of believing and practising into processes of religious innovation. Nowhere can this be seen better than in the religions which have had founders, such as Buddhism, Judaism, Zoroastrianism, Christianity and Islam. Each founder of these new religions utilized the traditions that had preceded them, but each also made original contributions that produced something quite distinctive. Bleeker explains:

> It is a well known fact that Zarathustra made use of several ancient Iranian numina for his theology, that Moses went back to the faith of the Patriarchs, that the background of the teaching of Buddha is formed by certain currents in Brahmanism, that many sayings of Jesus are to be found in the rabbinical writings and that Mohammed borrowed his main ideas from Judaism and Christianity. (Bleeker, 1963: 20)

Yet each founder also vitalized the truth in the teachings of those who preceded them and each laid down patterns which would bind all those who followed after them. Religious founders thus show quite explicitly how *entelecheia* works as a logic within religious history.

Bleeker was quick to assert that *entelecheia*, although a non-evolutionistic process, nevertheless reflects a discernible pattern in which 'creative forces are at play'. Much like Aristotle's acorn, *entelecheia* reveals religion as fulfilling its inherent potential through history. This offers a plausible explanation for why a flourishing of religious activity occurred between the eighth and fourth centuries BCE, when founders of religions emerged who 'by their work religion suddenly reached a higher level of self-realisation' (1963: 20). Bleeker cites Heinrich Frick as having shown that events in Israel, Greece, China and India during this same period displayed a remarkable 'series of parallel religious events'. The first prophets of Israel appeared in the eighth and seventh centuries BCE; Greek religion moved towards a higher plane in the sixth century BCE. During the same century 'Brahmanism comes about in India' and 'history starts in China'. Between the sixth and the fourth centuries BCE, 'Ezekiel and the Second Isaiah proclaim a universal monotheism, Plato and Aristotle teach their philosophies, Lau-tse and Confucius spread their wisdom, Buddha preaches his doctrine of salvation' (1963: 20). Frick's analysis builds on the work of other scholars, like Karl Jaspers (1953), who called this period of history 'the axial age', and, although Bleeker regards these events as 'striking', he admits that they do not provide enough evidence on which to build a comprehensive theory of religion.

It is better to illustrate how *entelecheia* works through a study of religious founders, about which there exist far more data than was available to Frick and around whom a much more coherent logic can be identified. Founders develop followers, some of whom become theologians and teachers whose main task is to interpret the teachings of the founder, often by defending the truth that the founder revealed. Usually, a class of priests is created, for whom following the founder has become a profession. Eventually, an authoritative tradition is established that is associated with orthodoxy in the religion. Through time, the authoritative tradition can become so stifling that it distorts the original revelation of the founder, which in turn gives rise to reformers who try to return the religion to its original purity. Sometimes, after the founder's death, divisions among adherents develop concerning the nature of the founder (the founder may be deified) or the founder's teachings are interpreted differently, or disputes concerning succession occur. Bleeker concludes that the events which transpire following the original founding of a religion demonstrate that cycles of life and death occur within religions, but in the end 'religion never totally loses its original creative force' and that it contains an inherent 'power of regeneration' that is 'part of the entelecheia of the religious phenomena' (1963: 21).

If religion possesses its own potential that is fulfilled in history, this raises for Bleeker the problem of what he called 'impure religion'. By this he was referring to movements in history which have many of the characteristics of religion, but can only be labelled properly as 'pseudoreligion'. He insisted that this did not involve a judgement on the truth of such movements, which would have violated the principle of *epoché*. Rather, impure religion indicates entirely human ideas or values, which have been magnified to the level of divinity. Recent examples of these include Nazi ideology or communism, both of which have demanded total allegiance from followers, but which lack the purity of a genuine religion. These cases are rather obvious, but for the scholar of religion it becomes more difficult to distinguish pure from impure religion in the instances of modern religious sects. Bleeker asks: 'Do these forms of belief represent superstition, a substitute for religion or religious truth?'. This is not an easy question to answer since 'the phenomenology of religion is lacking unequivocal criteria to test the religious validity of these phenomena'. The case of Satanism is easier to treat, since the dark side of beliefs has always included the idea that the devil or witches have powers greater than those of God. These clearly represent pseudo-religions because 'Satanism is a spiritual aberration' and 'the belief in witches is mainly part of mass-psychology' (1963: 22).

Instances exist where religion has been used as a tool of sexual exploitation, a fact which has led some theorists, primarily Sigmund Freud, to posit that the sexual urge is at the core of the religious impulse in humanity. Bleeker counters that such scholars 'refute themselves by their exaggerations' (1963: 22). In cases where the cult of the phallus or temple prostitution exist, as in the religions of antiquity, the ethical sentiments consistent with contemporary religion are offended. The phenomenologist of religion, however, employs the principle of empathy to achieve understanding of such practices based on the awareness that 'in Antiquity people had another evaluation of sexuality than modern man' whereby sexuality was regarded 'as a manifestation of the divine' and not as 'deliberate debauchery' (1963: 22). Bleeker concludes that the problem of impure religion is an important and complex one and that scholars of religion need to pay much more attention to it than they have done previously, not least because it poses serious challenges to the theory of *entelecheia*. It would appear that Bleeker attempted to resolve the issue by suggesting that 'pure' religion focuses on the 'key word' of religion, the divine properly understood rather than the human elevated to the position of the divine, and those who practise 'pure' religion do so from sincere and honourable motives unlike those who deliberately pervert the nature of religion by confusing it with purely human interests, some of which are entirely selfish and prurient.

Finally, Bleeker turned his attention to the issue of the persistence of religion in the contemporary world by asking 'whether the phenomenology of religion produces evidence of a gradual rising of the religious level' (1963: 23). Bleeker did not mean by this to suggest that the forces of secularization were undermining traditional religions, although he admitted that 'a large part of our generation is spiritually uprooted' (1963: 21). He believed that the history of religions demonstrates that the contemporary person no longer thinks the way people did in ancient times, when the religious and the secular were confused and intermingled. The contemporary person has a clearer idea of what religion entails and thus 'makes higher demands as to the quality of religion'. This largely can be attributed to the superior level of spirituality that has been introduced into the world religions by their founders, who gave humanity 'a new and purer conception of God'. The power of religion to renew itself is seen clearly in the capacity of the world religions to initiate reform movements, which often ice new or at least more advanced degrees of religious awareness: 'Thereby ring about a rising of the religious level'. It is clear from these comments that Bleeker understood religion in the contemporary period as having become more complete in a teleological sense and thus as nearer to achieving the maturity that has been inherent in the religious impulse throughout history. Although this is not an evolutionary idea, for Bleeker, a study of history confirms that each seeming relapse of religion has produced a response that has made religion stronger and purer. This awareness has come from the phenomenological approach to the study of religion which 'teaches us that religion is man's inseparable companion' and is 'an invincible, creative and self-regenerating force' (1963: 24). The power of religion in the present day, although different from prior periods in history, thus bears evidence of the principle of *entelecheia*, which Bleeker regarded as his most substantial contribution to the phenomenology of religion. Not only do particular types of religious behaviour manifest their essence as van der Leeuw had demonstrated, but religion itself shows that it is an entity in its own right with its own inner logic that drives it towards a purposeful end.

Dutch phenomenology in relation to philosophy, theology and the social sciences

The influential Dutch school in the study of religions, by 1970, had clearly delineated a science of religion that included phenomenology (synchronic, typological and structural) and history (diachronic, descriptive and evidential), both of which could be distinguished from philosophy, theology and the social sciences. Kristensen, van der Leeuw and Bleeker, when considered together, defined what came to be regarded as axiomatic for the academic study of religions, the main characteristics of which can be summarized in the following points:

1. Religion constitutes a subject matter in its own right, and thus possesses its own methodology.
2. The methodology peculiar to religion clarifies the relationship between historical data and the interpretation of the data by the researcher and distinguishes both history and phenomenology from all other academic approaches.
3. Testing interpretations from phenomenological research always takes place by recourse to the data of religions themselves.
4. Phenomenological interpretations privilege the perspectives of believers.
5. Evaluating data is to be distinguished from interpreting meanings and delineating structures; evaluation does not belong within a scientific approach to the study of religions, whereas interpretations leading to understanding define its fundamental purpose.

With respect to its philosophical heritage, Bleeker argued that the *epoché* and the eidetic vision are employed figuratively in the phenomenology of religion and thus should be distinguished from a specialized philosophical analysis of knowledge as found particularly in the writings of Edmund Husserl. This differed from the perspective of van der Leeuw, whom, as we have seen, deliberately borrowed these terms from Husserl by applying them to the subject–object distinction in the study of religions. Kristensen, as the mentor for both van der Leeuw and Bleeker, held that the phenomenology of religion was not a philosophical discipline, and that the search for essences belonged properly to the philosophy of religion rather than to phenomenology. Kristensen nonetheless connected the specific functions of history, phenomenology and philosophy into a general science of religion, since he regarded each as distinct, but complementary. The seriousness of Bleeker's objection to van der Leeuw depends on how he meant the term 'figurative' when applied to Husserl's use of the *epoché* and the eidetic vision. It would seem that Bleeker intended that these terms should be applied in order to limit prior assumptions in the study of religions and to indicate how typologies can be constructed. He opposed efforts to engage in epistemological discussions, and thus wanted to take Husserl's terminology at face value. Certainly, there is a difference of emphasis between van der Leeuw and Bleeker on this point, but too much can be made of this. Van der Leeuw's aim, like Bleeker's, was to eliminate prejudice in the study of religion and to understand how it is possible for the study of religious data through history to uncover the hidden or underlying structures. This seems not far removed from Bleeker's use of *theōria* and *logos*, and even *entelecheia*, as categories for understanding religions. Van der Leeuw, although clearly more inclined to draw on Husserl's philosophical method than either Kristensen or Bleeker, nevertheless did so in Bleeker's 'figurative' sense. Van der Leeuw, in other words, was not concerned to apply a technical theory of knowledge to religion, but was intent on preserving religion as a category in its own right with methods specific to its investigation. Philosophical phenomenology in the Dutch school thus remains a significant formative influence on the phenomenology of religion, but no more than that.

Theological influences may have been more influential on the Dutch school than philosophical ideas. This is clear historically when, as we have noted, the theological roots to Dutch phenomenology are traced directly through the theologians, Tiele and Chantepie de la Saussaye. If we look beyond the situation in the Netherlands, moreover, we can see that Kristensen, van der Leeuw and Bleeker each regarded religion in a highly sympathetic way. This did not mean that they naively or uncritically regarded religion as a force for good in history, but each maintained in similar ways that the religious response in humanity is best described as a relationship with the divine, the numinous, the holy or a transcendent power. This relationship produced in humans a sense of deep respect and awe, so much so that the core element in religion is described by each as reflecting this elemental human experience. Each was careful not to comment on the ontological reality of the object of the religious response, but the influence of theologians, most specifically Rudolf Otto, is evident in the way the sense of the numinous is depicted as a universal and elemental religious category. As we have seen, this idea has its roots in earlier theologians in the line of Schleiermacher, but perhaps most specifically in the school of Allbrecht Ritschl, where the religious judgement of value is interpreted largely as a sense of emotional satisfaction with respect to the highest and best that humans know. Although the Ritschlians are not explicitly cited by phenomenologists in the Dutch school, the judgement of value is reflected in Kristensen's insistence that the believers are always right, in van der Leeuw's likening the adherent's faith to a loving gaze on the beloved, and in Bleeker's concept of *entelecheia*, where religion fulfils its inherently positive potential in history.

As I have argued earlier, it would be easy to describe Kristensen, van der Leeuw and Bleeker as simply reacting against evolutionary theories of religion that came out of earlier anthropological and sociological writings, which were rooted in Feuerbach's projectionist theory, and applied to primitive cultures through writers such as Tylor, Frazer and Spencer. We have seen in this chapter that indeed each of the Dutch phenomenologists regarded the use of *epoché* as a necessary safeguard against these preconceived prejudices in the study of religions. Nevertheless, we also find the positive use of social scientific categories, influenced by Weber's sociology, where the 'ideal type' not only is coherent and rational, but active in directing historical processes. This notion corresponds to the phenomenological emphasis on the dynamic nature of religion in history, most forcefully expressed by Bleeker's notion of *entelecheia*.

I will return to the issue of how the formative influences over key figures in the phenomenology of religion have produced subsequent debates in the academic study of religions in Chapter 7. I consider in the next chapter the second major grouping of phenomenologists in the study of religion, those I refer to as comprising the 'British school'. We will see that this is a far less coherent alignment of scholars than we have discovered in the Dutch school of phenomenology, a fact which at least in part reflects the African origins of departments of Religious Studies in British universities.

References

Biezais, H. (1979), 'Typology of religion and the phenomenological method', in L. Honko (ed.), *Science of Religion: Studies in Methodology. Proceedings of the Study Conference of the International Association for the History of Religions, held in Turku, Finland, August 27–31, 1973* (The Hague, Paris and New York: Mouton Publishers), pp. 143–61.

Bleeker, C. J. (1963), *The Sacred Bridge: Researches into the Nature and the Structure of Religion* (Leiden: E. J. Brill).

Bleeker, C. J. (1972), 'The contribution of the phenomenology of religion to the study of the history of religions', in U. Bianchi, C. J. Bleeker and A. Bausani (eds), *Problems and Methods of the History of Religions* (Leiden: E. J. Brill), pp. 35–54.

Bleeker, C. J. (1975), *The Rainbow. A Collection of Studies in the Science of Religion* (Leiden: E. J. Brill).

Bleeker, C. J. (1979), 'Evaluation of previous methods: Commentary', in L. Honko (ed.), *Science of Religion: Studies in Methodology. Proceedings of the Study Conference of the International Association for the History of Religions, held in Turku, Finland, August 27–31, 1973* (The Hague, Paris and New York: Mouton Publishers), pp. 173–7.

Capps, W. H. (1995), *Religious Studies: The Making of a Discipline* (Minneapolis: Fortress Press).

Carman, J. B. (2000), 'Modern understanding of ancient insight: Distinctive contributions of W. B. Kristensen's phenomenology of religion', in S. Hjelde (ed.), *Man, Meaning, and Mystery: Hundred Years of History of Religions in Norway. The Heritage of W. Brede Kristensen* (Leiden: E. J. Brill), pp. 157–72.

Chantepie de la Saussaye, P. D. (1891) *Manual of the Science of Religion* (London and New York: Longmans, Green, and Co.

Colpe, C. (1979), 'Symbol theory and copy theory as basic epistemological and conceptual alternatives in Religious Studies', in L. Honko (ed.), *Science of Religion: Studies in Methodology. Proceedings of the Study Conference of the International Association for the History of Religions, held in Turku, Finland, August 27–31, 1973* (The Hague, Paris and New York: Mouton Publishers), pp. 161–73.

Cox, J. L. (1998), 'Religious typologies and the postmodern critique', *Method and Theory in the Study of Religion*, 10, 244–62.

Daniels, J. (1995), 'How new is neo-phenomenology? A comparison of the methodologies of Gerardus van der Leeuw and Jacques Waardenburg', *Method and Theory in the Study of Religion*, 7, 43–55.

Fitzgerald, T. (2000), *The Ideology of Religious Studies* (New York and Oxford: Oxford University Press).

Flood, G. (1999), *Beyond Phenomenology: Rethinkng the Study of Religion* (London and New York: Cassell).

Hjelde, S. (2000), 'Introduction', in S. Hjelde (ed.), *Man, Meaning, and Mystery: Hundred Years of History of Religions in Norway. The Heritage of W. Brede Kristensen* (Leiden: E. J. Brill), pp. xii–xxii.

Hofstee, W. (2000), 'Phenomenology of religion versus anthropology of religion? The "Groningen School" 1920–1990', in S. Hjelde (ed.), *Man, Meaning, and Mystery: Hundred years of History of Religions in Norway. The Heritage of W. Brede Kristensen* (Leiden: E. J. Brill, pp. 173–90).

Honko, L. (ed.) (1979), *Science of Religion: Studies in Methodology* (The Hague and New York: Mouton Publishers).

James, G. A. (1995), *Interpreting Religion. The Phenomenological Approaches of Pierre Daniel Chantepie de la Saussaye, W. Brede Kristensen, and Gerardus van der Leeuw* (Washington, DC: The Catholic University of America Press).

Jaspers, K. (1953), *The Origin and Goal of History* (translated by Michael Bullock) (London: Routledge and Kegan Paul).

Kristensen, W. B. (1954), *Religionshistorisk Studium* (Oslo: Olaf Norlis Forlag).

Kristensen, W. B. (1960), *The Meaning of Religion* (translated by J. Carman) (The Hague: Martinus Nijhoff).

Leeuw, G. van der (1963), *Religion in Essence and Manifestation. Volumes I and II* (translated by J. E. Turner with Appendices incorporating the additions of the second German edition by H. H. Penner) (New York and Evanston: Harper and Row Publishers). (Original publication in one volume, 1938. London: George Allen and Unwin Ltd.)

Molendijk, A. L. (2000), 'At the cross-roads: Early Dutch science of religion in international perspective', in S. Hjelde (ed.), *Man, Meaning, and Mystery: Hundred Years of History of Religion in Norway. The Heritage of W. Brede Kristensen* (Leiden: E. J. Brill), pp. 19–56.

Naguib, S.-A. (2000), 'Lieblein, Kristensen and Schencke and the quest for Egyptian monotheism', in S. Hjelde (ed.), *Man, Meaning, and Mystery: Hundred Years of History of Religion in Norway. The Heritage of W. Brede Kristensen* (Leiden: E. J. Brill, pp. 101–13).

Pals, D. L. (1996), *Seven Theories of Religion* (Oxford: Oxford University Press).

Pettersson, O. and Åkerberg, H. (1981), *Interpreting Religious Phenomena. Studies with Reference to the Phenomenology of Religion* (Stockholm: Almqvist and Wiksell International).

Platvoet, J. G. (1998), 'Close harmonies: The science of religion in Dutch *duplex ordo* theology, 1860–1960', *Numen*, 45, 115–61.

Platvoet, J. G. (2002), 'Pillars, pluralism and secularisation: A social history of Dutch sciences of religions', in G. Wiegers (ed.), *Modern Societies and the Science of Religions. Studies in Honour of Lammert Leertouwer* (Leiden: E. J. Brill), pp. 82–148.

Sharpe, E. J. (1986), *Comparative Religion: A History* (London: Duckworth, 2nd edn).

Thrower, J. (1999), *Religion: The Classical Theories* (Edinburgh: Edinburgh University Press).

Tiele, C. P. (1973) [1896 and 1898], 'Extracts from *Elements of the Science of Religion*', in J. Waardenburg, *Classical Approaches to the Study of Religion: Aims, Methods and Theories of Research. I. Introduction and Anthology* (The Hague and Paris: Mouton), pp. 96–104.

Waardenburg, J. (1978), *Reflections on the Study of Religion* (The Hague, Paris and New York: Mouton Publishers).

Waardenburg, J. (2000), 'Progress in research on meanings in religion (1898–1998)', in S. Hjelde (ed.), *Man, Meaning, and Mystery: Hundred Years of History of Religion in Norway. The Heritage of W. Brede Kristensen* (Leiden: E. J. Brill), pp. 255–85.

Wiebe, D. (1999), *The Politics of Religious Studies: The Continuing Conflict with Theology in the Academy* (London: Macmillan).

Chapter 5

From Africa to Lancaster: The British School of Phenomenology

In this chapter I identify four key figures in the phenomenology of religion who have influenced the academic study of religion in British contexts: Edwin W. Smith (1876–1957), E. Geoffrey Parrinder (1910–2005), Andrew Walls (b. 1928) and Ninian Smart (1927–2001). By selecting this list of scholars, I am pointing towards a somewhat unexpected source for phenomenology in Britain. Each scholar, with the exception of Ninian Smart, began his work in Africa. As I will note, the first Department of Religious Studies as we now know it in Britain was founded by Parrinder at Ibadan, Nigeria in 1949. I shall argue that Parrinder was influenced in part by the thought of Edwin W. Smith, who worked in central Africa during the early part of the twentieth century and who much later wrote the foreword to Parrinder's published doctoral thesis on West African religion. The first Department of Religious Studies in Scotland was established by Andrew Walls in 1970, shortly after he had returned from many years of teaching in Sierra Leone and Nigeria. He soon enlisted numerous members on his staff at Aberdeen who had worked in Africa, including Harold Turner, who was best known for his phenomenological studies of new religious movements. In European and North American contexts, Ninian Smart has been associated almost exclusively with religious studies in Britain, particularly since 1967, when he founded the first entirely non-theological Department of Religious Studies in the United Kingdom at the University of Lancaster. In his contribution to a recent book outlining the history of theological studies in Britain, Keith Ward (2003: 271) calls Smart 'the doyen' of religious studies in Britain 'until his death in 2001'. In this chapter, I will stress that Smart was preceded by scholars of religion who had worked in Africa, each of whom contributed significantly to the British emphasis within the phenomenological movement as a whole.

The influence of Edwin W. Smith on British phenomenology

Edwin W. Smith was born in 1876 in the Cape Province of South Africa to missionary parents of the English Primitive Methodist Church. His parents arrived back in England in 1888 and sent Smith to Elmfield College, a Primitive Methodist school near York. He returned to Africa ten years later, first going to

Basutoland (now Lesotho) to work with French Protestant missionaries before arriving in 1899 at Aliwal North in South Africa, where he was born and had spent his early youth. In 1902, he was assigned by the Primitive Methodist Church as a missionary to what is now the southern part of Zambia. He completed his missionary work in Zambia in 1915 when he was offered a post in England with the British and Foreign Bible Society, where he stayed until 1939, eventually taking charge of the Society's translation work. During his long career with the British and Foreign Bible Society, Smith made numerous contributions to the field of African studies as a whole. He was a founding member of the International Institute of African Languages and Cultures, whose other founding members included such notable figures as F. D. Lugard, Lucien Lévy-Bruhl, J. H. Oldham (Secretary of the International Missionary Council), Wilhelm Schmidt and C. S. Seligman. In addition, Smith served as President of the Royal Anthropological Institute from 1933 to 1935. After leaving the British and Foreign Bible Society in 1939, he held visiting professorships in the United States, first at Hartford Theological Seminary and then at Fisk University in Tennessee. He was editor of the journal *Africa* from 1945 to 1948 and concluded his career in 1950 with what might be regarded as his most lasting achievement, the edited volume *African Ideas of God* (Smith, 1950; see also van Rinsum, 2003).

Smith is described in a recent biography written by W. John Young in glowing terms:

> Was there, indeed, any other person so expert in so many parts of African studies as well as the field as a whole? Anthropology in British Central Africa – he was the founder. Research interest in Africa – he set it going in earnest. African traditional religion – he outlined its chief characteristics. African Christian theology – he was its *'fons et origo'*. (Young, 2002: 216)

Young makes only vague references to Smith as a phenomenologist, probably because Smith did not identify himself explicitly with the phenomenology of religion, and because he was motivated by missiological concerns. Nevertheless, the way he approached the study of African religions reflected the phenomenological emphasis on employing a sympathetic attitude towards any religion in order to identify its overarching core concern. Smith did this with African religions, and found that, despite earlier descriptions of them by anthropologists and ethnologists as primitive and void of any idea of a divine being, Africans have universally expressed their religious ideas under three carefully refined categories: belief in a Supreme Being, a sense of dynamic power and a localized, kinship focus on ancestor spirits. These three main typological classifications enabled Smith to make generalizations about African religions, very much in the same way that phenomenologists of religion in the Dutch school had sought to identify the unifying characteristics within religious beliefs and practices. The way he organized and classified African religious beliefs and practices led Young to conclude that Smith's contribution to the academic study of religions should not be underestimated:

Smith's headings ... along with his later awareness of ritual symbolism have helped to elucidate a great deal of religion in Africa and other places. With good reason, therefore, it can be claimed that he founded the study of religion in Africa as a serious academic discipline. (Young, 2002: 217)

Smith's phenomenology and the universal belief in God throughout Africa

We see most clearly how Smith employed a phenomenological approach in his edited volume, *African Ideas of God* (1950), which is comprised of 12 chapters describing the belief in God throughout vast areas of sub-Saharan Africa, including Anglophone and Francophone countries in West Africa, and incorporating what is now Malawi, Zambia, Namibia and South Africa in the central and southern regions. Each contributor, with the exception of Harold Beken Thomas, who was a British colonial officer working in Uganda, was a British missionary associated with the Church of England, the Methodist Church or the Baptist Church. In his introductory remarks, Smith, as editor, suggests that the book is intended to provide a 'factual' answer to two questions: 1) 'Among Africans who have not come under the influence of Islam, Judaism or Christianity, is there any awareness of God?' 2) 'If so, what idea of Him have they formed?' The answers must be based on 'evidence' which is examined and interpreted 'with such open-mindedness as we can command' (1950: 1). He describes the contributors to the book as having lived for many years in Africa, as knowing Africans in 'intimate fellowship' and as speaking their languages. They have, in Smith's words, 'sought to get at the back of the black man's mind' (1950: 2). Simpler peoples, 'such as Bushmen and Hottentots are represented as well as those peoples of the higher culture, Yoruba and Akan' (1950: 2). Clearly, in these introductory remarks, Smith affirms the principle behind *epoché* (not the word itself), and urges an empathetic approach to those who comprise the objects of the study. At the same time, he rejects the distorting biases of evolutionary thinkers who suggest that peoples with basic technologies and subsistence patterns possess a lower concept of God than those with more sophisticated technologies and complex social arrangements.

At the outset of *African Ideas of God*, Smith raises methodological issues and discusses the primary sources on which the conclusions of the book are based. He acknowledges that interpretations of the data, even by those who know the African mind intimately, could be subject to bias and distortion. In one sense, the possibility of bias could even result from the close relationship between those who have written the ethnographic data and the Africans who are being described in the reports. Since missionaries have been working in many of the regions included in the survey for nearly one hundred years, Smith draws attention to the 'twin dangers' of 'reading in' what is not in fact there and of 'reading out' what is not in fact indigenous (1950: 3). In order to 'arrive at the truth' on the questions posed, certain methods have been followed carefully. One method is etymological, although one must be extremely careful when drawing conclusions, particularly about names assigned from the vernacular languages for God. The praise names for God, however, are instructive since in Africa 'names are not mere labels, but often express qualities for which the owners are conspicuous' (1950: 4). Another point for analysis must include proverbs, which 'enshrine the gathered wisdom of the past'. These frequently bear testimony to the Africans' conception of God. Related to proverbs are African myths, which have been recorded by ethnologists, whose findings offer insight into 'African cosmology and theology'. Myths often tell of beginnings of human life, particularly referring to a time when 'God and man lived together on earth and talked one to another' (1950: 7). Another source of understanding the African mind can be found in the everyday speech of the people, where the name of God is typically invoked (1950: 9). Finally, an excellent source for understanding African ideas of God comes from prayers. 'These are addressed more commonly to the ancestral spirits than to the Supreme Being, but the general belief is that the spirits are intermediaries who relay the prayers to Him' (1950: 10).

Smith notes that the Supreme Being in Africa needs to be understood in terms of a hierarchy of ubiquitous living forces (including ancestor spirits), some of which even inhabit objects that Westerners generally regard as inanimate, such as stones, mountains or rivers. Under the influence of Father Placide Tempels' idea of vital force as the core element within Bantu philosophy (Tempels, 1959) and the earlier theory of dynamism as advanced by the British anthropologist R. R. Marett (van Rinsum, 2003: 50), Smith divides the levels of spiritual development within African ideas from the most basic, dynamism, through spiritism to the highest form, which is theism (1950: 16). Smith acknowledges that Africans identify power and force throughout nature and that their world is comprised of many spirits, but this does not mean that Africans have no idea of a high God nor that Africans think of the supreme power as an 'It'. The evidence of the contributors to *African Ideas of God*, Smith claims, clearly shows that Africans think of God as a person.

To distinguish what he means by God, as opposed to a vital force, Smith then sets out seven criteria which mark the high God apart from some sort of dynamic force or what he calls 'cosmic Mana'. These are:

1. God has personality and a personal name.
2. God has a life and consciousness 'analogous to that of man'.
3. God is a being who is not and never has been human.
4. God is creator of, if not all things, at least of some things.
5. God is the ultimate power and authority.
6. God is worshipped through prayers and sacrifices, even if it occurs rarely.
7. God is a judge, or at least has an ethical relationship to humanity. (Smith, 1950: 21–2)

Smith adds that these characteristics do not exclude a belief in other or many deities. Belief in God, as opposed to a force or power, therefore, need not be construed as a 'strict monotheism'. Smith concludes that the authors of his symposium have proved conclusively that throughout the regions studied, which, although not exhaustive, cover the main areas of sub-Saharan Africa, 'we have to do with a High God and not with "an abstract Power or natural potency"' (1950: 22).

Even the casual reader, perhaps aided by the historical distance of over 50 years, can recognize within Smith's criteria a liberal theology and a Christian missionary agenda. Indeed, Smith confirms in his introductory chapter that he is engaged in a theological debate, particularly against the school of Karl Barth, who denied, in Smith's words, 'that a general revelation is an integral part of Christian doctrine' (1950: 32). Smith finds quite the opposite to have been the experience of pioneer missionaries: 'they have discovered some belief in the existence of God among the Africans' (1950: 33). This means, theologically, that God has revealed himself to the Africans through their own apprehensions of nature and the world, and through their power of thought. They already have some knowledge of God before they ever hear one word about Jesus Christ. 'When the Christian missionary comes with the Good News of God revealed in Jesus Christ as a loving Father – whatever else in his teaching they find it hard to accept, this [belief in God] at least they readily take to their hearts' (1950: 34).

Smith's theological assumptions were made even more explicit in a book he wrote in 1936, largely for teaching purposes in African theological colleges, under the title, *African Beliefs and Christian Faith*. In his introductory chapter (1936: 15–26), Smith likens the African to a pupil in 'God's school', in which gradually he is instructed in the higher knowledge. The African has a rudimentary understanding of God. In the full light of Christian teaching, he is given a more complete knowledge, but the knowledge Africans have had from the beginning, in so far as it has come from God, is true and should be respected. Smith puts it this way:

> They, like all other men, were learners in God's school. We may put their beliefs and acts to the test, saying: What is the opinion of the Lord Jesus about this and that? And when we come upon things which are good and true we will give praise to God from whom all the good and the true and the beautiful comes. (Smith, 1936: 26)

Elsewhere in this chapter, Smith likens the African to a child, but he does not regard this as a disparaging or patronizing comparison. An adult, he argues, is not a different person from a child: 'men are children made complete'. The stage of childhood is necessary and 'for children it is the best thing'. Of course, 'the condition of being a man or a woman, which takes its place, is a better thing' (1936: 24). It is possible to understand that Africans, before they had knowledge of Jesus Christ, were like children, not yet complete, but still possessing 'true knowledge about a number of things'.

After this introduction, Smith then devotes the first part of his book to the 'Belief in God among the Africans' and asserts that Africans, like all humans, are able to recognize God in the natural world of sun and rain, thunder and lightning, and through the power of human reason (1936: 29). This, of course, constitutes the Christian doctrine of natural or general revelation. Smith then divides his analysis of the African belief in God into categories focusing particularly on God's personality, his creative activity and the ways Africans worship him. Clearly, when he wrote *African Beliefs and Christian Faith*, 14 years prior to his symposium on God in Africa, Smith was committed fully to the idea that Africans believe in God, not just in a power, and that this has resulted from God's universal revelation in nature and in the human mind. When the Christian Gospel was presented, therefore, the African readily recognized God in the message and embraced enthusiastically the personality of God made known in Jesus Christ.

In light of this earlier writing with its unmistakable statement of a missionary theology, *African Ideas of God*, composed as it was by missionaries and constructed by Smith to address the issue of a Christian conception of God, cannot be treated, as Smith wanted it to be, as firm evidence of the universal belief in God throughout Africa. Nor can it be used uncritically as source material for a hierarchical phenomenological typology of religious beliefs held throughout Africa. What we find in Smith that is consistent with phenomenological themes is his effort methodologically to reverse the order for reaching his conclusions from one based on theological presuppositions to one resulting from the evidence of field research. Smith's missionary fieldworkers are depicted as acknowledging potentially distorting biases, as adopting a sympathetic attitude towards the beliefs and practices of adherents and as drawing conclusions derived exclusively from the data themselves. The distinctive contribution Smith made to British phenomenology of religion thus was his emphasis on field data derived from those competent in vernacular languages, and who in many ways had become 'insiders' within the cultures they were seeking to describe, understand and interpret. The motivation beneath this approach, which resulted from the missionary engagement with African cultures, set British phenomenology squarely within colonial and missionary contexts. This can be seen clearly in the development of the new field of religious studies pioneered in Africa by E. G. Parrinder.

E. Geoffrey Parrinder as a phenomenologist of religion

Geoffrey Parrinder was appointed as Reader in the Comparative Study of Religions in King's College, London in 1958. Later, he was promoted to a Professor in the same institution, from which he exercised an extensive influence over students of religion both in Britain and more widely in continental Europe and North America. He was a longstanding member of the British Association for the History of Religions (later the British Association for the Study of Religions), serving as its Secretary for 12 years when E. O. James was its President, and then on James' death in 1972, succeeded him as President until 1977.[1] Although Parrinder described himself primarily as a scholar of comparative religions, his many writings demonstrate that he employed methods consistent with phenomenological principles set forth within the Dutch school, particularly by Kristensen and Bleeker.

Initially, as we have noted, Parrinder was an Africanist, beginning his work as a missionary of the Methodist Missionary Society in Francophone West Africa where he taught in Protestant theological seminaries in Dahomey (now Benin) and in the Ivory Coast, and worked as Superintendent of a Methodist Circuit in the Ivory Coast. For three years, he served as Principal of the Protestant Theological Seminary in Porto Novo, Dahomey, after which he returned to England, where in 1938 he completed his BA in French, English and Ethics in the University of London, which he followed in 1940 by earning a first class Bachelor of Divinity Honours degree from London. During the early years of the Second World War, due to political pressures exerted by the Vichy government, Parrinder was unable to travel to French-speaking West Africa. It was not until 1943 that he was allowed to return as Principal of the Theological Seminary in Porto Novo, a post he held until 1946. During this time, according to Martin Forward's extensive work on Parrinder (1998: 18), he explored 'in detail West African indigenous religion'.

It was partly from material he obtained during the period from 1943 to 1946 that he based his doctoral thesis submitted in 1947 to the University of London, and published in 1949 as *West African Religion*. In the same year of this publication, he accepted a post at the University College Ibadan in Nigeria, as a lecturer, in an entirely new subject called 'Religious Studies', a department he served until 1958 when he took up his post at King's College, London. During his years at Ibadan, Parrinder wrote extensively about African religion. He was one of the contributors to Smith's symposium on *African Ideas of God*, and was read widely by students of African religion throughout the English-speaking world, chiefly through his book, *African Traditional Religion* (first published in 1954). Andrew Walls (1980: 146) comments about Parrinder that in Ibadan 'a European became the teacher of Africans on some aspects of their cultural and religious inheritance that some had learned to despise or be ashamed of'. After he moved from Ibadan to London, Parrinder remained interested in African religions but largely as they fitted into a larger comparison of the world's religious traditions. His book, *Religion in Africa* (1969),[2] compares the main features, which he sorted into broad categories, of what he called Africa's three principal religions: Christianity, Islam and Traditional Religion.

In his approach to the comparative study of religions, the influence of Edwin Smith over Parrinder's thinking can be traced, particularly on the question of the African view of God. In *African Traditional Religion* (1974), the first edition of which was published four years after Smith's *African Ideas of God* (1950), Parrinder refers the reader to Smith's edited volume as a solid resource to correct 'the wild assertions sometimes made about African belief or lack of belief in God' (1974: 37), and cites the many names for God recorded by the contributors to the book as evidence 'that there is much more general belief in him than has been thought in the past' (1974: 24, 39–40). Parrinder's later works on this topic frequently refer to Smith, even when they deal with matters much wider than Africa. For example, in Parrinder's 1965 publication, *Sexual Morality in the World's Religions* (1980), the section on 'Traditional Africa' (1980: 127–31) begins with a lengthy quotation from Smith, which is then followed in the next few pages with a discussion based on Smith's work in Northern Rhodesia. As we have noted, Smith wrote the foreword to *West African Religion* (Parrinder, 1949: ix–xii), and he served as one of the doctoral examiners of the thesis on which the book was based. Despite this close connection, Smith's influence over Parrinder should not be exaggerated. According to Adrian Hastings (2001: 356) 'it was the books on Ashanti religion by R. S. Rattray, together with Margaret Field's *Religion and Medicine of the Ga People* (1937), which most influenced his early Africanist development'. Certainly, *West African Religion*, although receiving Smith's full endorsement, carried the marks of Parrinder's own approach to the study of religions.

Parrinder begins *West African Religion* (1949: 2) with a statement outlining the aims of the book and the theoretical perspective he has adopted. He describes the book as having two main objectives: 1) to provide an overview for comparative purposes of the main religious beliefs of the Yoruba, Ewe, Akan and what Parrinder calls 'kindred peoples'; and 2) to correct prior descriptions of African religions that classify them as the religions of 'savages' which are characterized primarily by the term 'fetishism'. On the first point, Parrinder explains that most tribes of West Africa share 'common elements' which 'need drawing out and collating, so that the European student may have presented to him the nature of the chief beliefs and practices of these deeply religious peoples' (1949: 2). On the second point, Parrinder argues that the tribes of his study should not be confounded with the idea of a 'primitive religion', since each group has 'a religion which deserves consideration as a distinct entity and that it is well advanced beyond the first dawn of the religious sense' (1949: 12). In particular, the term 'fetish', along with its association in English with 'juju', is confusing and inaccurate, and thus should be 'relegated to the museum of the writings of early explorers' (1949: 14). From the outset Parrinder thus describes his project as identifying typological classifications for understanding West African religions, which can be derived only after the distorting and biased categories found in earlier popular literature on West Africa have been exposed as incorrect and replaced with fair and objective descriptions.

In the early sections of the book, Parrinder considers various theories about African religions, including that of Wilhelm Schmidt, whom, as we have seen, posited an original primal monotheism which had degenerated into polytheism, and that of E. B. Tylor, who traced the development from a primitive animism to more developed forms of polytheism. Parrinder's aim, however, was not to engage in theoretical discussions concerning the origin of the religious impulse in humanity: 'It is not our purpose to enter into these speculations here' (1949: 15). Rather, he was concerned to classify West African religious beliefs according to a four-fold scheme: a supreme God; the chief divinities (generally non-human and associated with natural forces); the cult of ancestors; and powers associated with charms and amulets. He calls this a 'workable classification' based on distinctions already occurring within the 'African mind' (1949: 16–17). Much later, Andrew Walls (1980: 145) called Parrinder's four-fold classificatory scheme 'axiomatic', adding that 'one finds it right across the plethora of works that now exist on African religions'.

In the chapter devoted to the 'Supreme God', Parrinder poses two main questions: 'whether the conception of a "Supreme Being", a "high God", is native to West Africa' and 'whether such a God is worshipped, and to what extent' (1949: 18). Parrinder concludes that 'most tribes in our area believe in a high God', although variations exist regarding 'his worship' (1949: 29). In support of this conclusion, he cites Rattray's research as having shown conclusively that the supreme God is believed in by the Akan of Ghana, a fact further corroborated in Twi language designations for the High God. Rattray found evidence that God is worshipped among the Akan through the numerous altars dedicated to him (on Rattray, see Platvoet, 1996: 108). Other groups in Parrinder's study of West Africa, however, seem not to worship the Supreme Being directly. Parrinder observes: 'In other parts of our field, a supreme God is recognized by everyone, but his cult is not practised by all who take his name on their lips' (1949: 29).

Parrinder then asks if the designation 'Supreme Being' corresponds to the Christian usage of the same term. He notes, 'The word "supreme" is easily comprehensible in European theology, but it may be fatally misleading to transfer our ideas into the African hierarchy. God the Father is 'supreme in Christian theology' (1949: 30). In West Africa this cannot be regarded as the case, since generally 'worship is scant enough'. Parrinder concludes: 'In short, while to us belief in God is the "highest" article of religion, and practised as such, it is not in the forefront of practised West African religion' (1949: 31). Parrinder asserts that his investigations confirm the same conclusion reached earlier by the German missionary and linguist Diedrich Westermann that 'the African's God is a *deus incertus* and a *deus remotus*', an unknown being who remains far away and uninvolved with day to day human details' (Parrinder, 1949: 32; Westermann, 1937: 74; see also Stoecker, 2004). In his book *African Traditional Religion*, first published five years later, Parrinder insisted that the experience of the Supreme Being as remote from everyday life in many parts of Africa should not be interpreted by outsiders as diminishing for Africans the significance of God, since belief in God defines a central feature within the African worldview as a whole.

> The belief in, prayers to, names of and myths about God show clearly that nearly all Africans, 'untutored' though some may be, do conceive of God. For most of them God was the creator of all things, but he has withdrawn to that remoteness which is part of his greatness. Like the most mighty of kings, he is only rarely approached, and the intervening gods and ancestors receive much more attention from the common man. In time of need, however, when all else has failed, God can be resorted to directly without priest or temple. (Parrinder, 1974: 42)

Parrinder later enlarged his discussion of African religions to include the religions of other oral societies in order to compare them as a unit with the so-called world religions. In his widely read book *The World's Living Religions*, first published in 1964, Parrinder devotes a chapter to 'Africans, Australians, and American Indians', whom he describes as among the 'many people of the world who do not follow one of the great historical religions, but have religious beliefs and practices that derive from ancient ideas and traditions' (1964: 124). They are similar, despite their geographical diversity, because 'they have no scriptures, since they live in countries where writing had not been invented' (1964: 124). In this same chapter, Parrinder again follows his four-fold system of classification beginning with the sub-heading, 'God and Divine Power', although in this case after discussing 'Ancestors and God' as the second classification, he employs the phrase 'Totem and Taboo' to refer to a broader category than might be implied by West African divinities. He explains: 'The word "totem" comes from the American Ojibway Indians where it meant "brother-sister kin", and was used of kinship of men and animals' (1964: 133). Power is discussed under the heading 'Magic, Sorcery and Witchcraft', but he means by it the same category he employed in his earlier writings, since he discusses in this section 'the attempt to control the course of nature by special powers or actions' (1964: 136) and 'dynamism or vital force' (1964: 137).

A similar approach is employed in his later volume *Religion in Africa* (1969), particularly in chapter three which is entitled, 'Unity, the Supreme Being'. After first establishing the traditional view of God, he then introduces his other three categories used for comparison, in this case with Christianity and Islam: ancestors (coupled with a further discussion of God), totems and taboos and magic and witchcraft. In this way, beginning with *West African Religion*, Parrinder gradually built up a standard system for comparing religions, from the traditions of West Africa through African Traditional Religion in general, then through a discussion broadly of non-literate traditions and finally to a description of the world's literate and historically documented religious traditions. Throughout this process, he followed a phenomenological method whereby the religions of the world, including non-literate traditions, were compared fairly and empathetically by organizing them typologically into categories of religious belief and practice. This same point has been underscored recently by Andrew Walls, who refers to Parrinder's 'principle of structure' as:

> a framework for explaining both congruence between African religious systems and regional and ethnic difference. It also provides the beginning of typological classification. Parrinder had, even in his earliest work, proceeded by *comparison.* (emphasis his) (Walls, 2004: 214)

A 'religious' approach to the study of religion: Parrinder's methodology

One of Parrinder's most explicit statements of his methodological approach to the study of religions is found in the concluding chapter of his book *Comparative Religion* (1962) in which he outlines what he calls the 'tasks' of his discipline. He notes that comparative religion has been given a small space in English universities, and where it does exist, it is located in departments of theology where 'religions are studied ... from the Christian point of view only' (1962: 119). One way to overcome this is for those teaching a particular religion to be an adherent of that religion already, so that a Buddhist would teach Buddhism, a Hindu, Hinduism and so on. This is not always practical, but at any rate, he argues, the scholar of religion should already possess a feeling for what religion is, and indeed should be in this sense 'religious'. He explains: 'One who has no religion is likely to oppose religion, to seek to debunk it, or to be sceptical of its claims. So today it is more common to claim that only those who have a religious faith can understand the faith of others' (1962: 120). This means that it is far better, for example, for a Christian to teach Islam than for an agnostic or atheist to do so, at least if the Christian adopts a proper methodology that not only empathizes with the Muslim out of a common religious understanding, but also presents Islam fairly and accurately.

Parrinder then reviews the main branches of the academic study of religion as they relate to the comparative study of religions. He argues that comparative religion should not be equated with the history of religion since this runs the risk of implying that the study of religion deals only with what is 'dead'. On another level, it also introduces a bias against those religions with ancient oral traditions, such as the religions of India and China, in favour of more recent scriptural traditions, such as Christianity and Islam. The emphasis on history also encourages a search for the origins of religion, which means that the study of religion devolves to those with specializations in prehistory or archaeology. It also leads easily to the assumption that 'modern illiterates', such as those living in Africa or Australia are ' "primitive", in the sense that they retained ancient beliefs and customs almost unchanged and so could tell us what our own remote ancestors were like' (1962: 122). Finally, a disadvantage of the 'history of religions' approach is that it deals 'with the externals of religion almost exclusively' (1962: 122). This has resulted directly from its methodological focus on the past and its neglect of what it is to be religious today. Parrinder insists that his views are not intended to disparage historical studies of religion, but it is right to acknowledge that historical work in the study of religion is largely completed: 'The history and the externals of religion have been well studied' (1962: 123). The overriding task for the current scholar of religion 'is to go on to study what are now called significantly "living religions"' (1962: 123).

Which disciplinary approach then is most appropriate for the study of the world's living religious traditions? In Africa and among other so-called 'primitive' societies, this task has been taken up by anthropologists who employ fieldwork methods to learn about their subjects. In a way similar to historians, anthropologists have described the externals of religion, but in their case it was done in terms of outward forms, such as portraying rituals and assigning religion a social function. This has led to the opposite extreme whereby fieldworkers frequently 'have been unconcerned with history and the development of ideas' (1962: 123). Parrinder goes on to complain that despite the very careful notes compiled by anthropologists, 'they never believed in the religions they studied' (1962: 123). Most of them, he argues, were atheists or agnostics who consigned their observations and interpretations inevitably to 'a study of externals, or of delusions at best' (1962: 124). Their great mistake is to 'have taken the observable signs of the religion as if they were the religion itself' (1962: 124). Philosophy is in a similar situation to anthropology. Although it could be used as a source for promoting understanding between religions by 'helping to bridge the gap between East and West', over the past 50 years, in Europe at least, philosophy has come under the sway of positivism, which seems 'to deny all possibility of expressing metaphysics in a meaningful way' (1962: 125).

Parrinder then spells out the approach he thinks is best suited for a comparative study of the world's living religions. For this, he draws on the thought of Wilfred Cantwell Smith, whom we will consider in the next chapter, but also on the contributions of C. J. Bleeker and the Dutch theologian and missiologist, Hendrik Kraemer. Parrinder asserts as a first principle that 'when statements are made about another religion they must be recognized as fair within the context of that religion' (1962: 125). This applies to all scholars, regardless of their religious convictions, but it must be heeded particularly by agnostics who tend to dismiss the value of religion or reduce it to non-religious factors. His second principle dictates that any statement made by a scholar about a religion must be acceptable within three traditions: the academic, the Christian and the other religion under study (1962: 126). Parrinder does not mean to privilege the Christian position over others by this formula, but he is suggesting that the study of religions in the West begins, at least implicitly, from within a Christian framework. Westerners understand the inter-relationships between an academic perspective and a Christian one, and many will be able to affirm both simultaneously. It is much more difficult for a Westerner to affirm an academic perspective and a non-Christian world view simultaneously, but it is even more challenging to embrace the academic, the Christian and the non-Christian perspectives at the same time.

By the academic perspective, Parrinder is referring to what he calls dissecting and labelling in an objective manner the religious beliefs and practices under study. This, he repeats, is not enough; it is 'a kind of dry-as-dust approach' which misses 'the living organisms' (1962: 126). The scholar of religion must be involved in religion, since those 'who try to understand another religion must both know what it means to have faith, and also be prepared to meet men of different faith on equal terms' (1962: 127). In support of this view, he cites Bleeker, whom Parrinder commends for showing those living in the West that Indian and Chinese scholars study religion in order to gain 'a deeper insight into the value of religion, in which intuition as well as ratiocination is needed' (1962: 127). Involvement in the study of religions thus implies dialogue, a point Parrinder notes has been made by Hendrik Kraemer. Although he registers disagreement with Kraemer's Barthian theological stance, he concurs with Kraemer's challenge to take other religions seriously and thus to make the comparative study of religions a fundamental part of theological training, and not just 'a queer option' (1962: 127).

In Parrinder, therefore, we meet a scholar of religions in the British tradition, who began in Africa, was moulded in his thinking by the African experience and by his personal and academic contact with Edwin W. Smith regarding Africa, and who gradually expanded the principles he applied to African religions to include the broader, more comprehensive comparative study of religion. He discovered from Africa how earlier popular and academic biases had distorted the genuinely religious nature of African peoples by labelling them as either 'fetishistic' or 'primitive'. Moreover, he gradually and increasingly emphasized that at the heart of any adequate methodology is demanded a 'sympathetic spirit', on the part of the scholar, 'because comparative religion deals with the faith of persons, which men hold as dear as life itself' (1962: 19). The academic method of classifying for comparative purposes thus must be done by acknowledging and limiting the power of prior assumptions and by cultivating a feeling for the religious sentiment in humanity. Finally, for Parrinder, the academic study of religions, although not ignoring the historical work that has been undertaken and largely completed, must consist primarily of synchronic studies to enhance understanding of living religious communities and, further, to promote dialogue and active engagement between the scholar, who already is religious or at least deeply sympathetic to religion, and the adherents within the religions the scholar is studying.

Andrew Walls: Missionary theory applied to the study of religions

During the latter part of his academic career, Andrew Walls has been recognized primarily for his work with the Centre for the Study of Christianity in the Non-Western World, which he founded in 1982 in the University of Aberdeen and transferred to Edinburgh University in 1987, serving as its Director until his retirement in 1995. Because of this, he normally is associated with mission studies, but in the context of this chapter, it is important to stress that he began his work as a church historian in Africa and became known after his return to Britain, within religious studies circles at least, as a scholar of religions with a specialization on the recent history of Christianity in sub-Saharan Africa. Walls succeeded Parrinder as President of the British Association for the History of Religions in 1977, and very soon afterwards made his impact on the Association. The theme of the BAHR Annual Conference for 1978 bears Walls' distinctive influence: 'The Understanding of Mission in the Study of Religion'. As we shall see, Walls employed the same phenomenological emphases we found in Parrinder, and increasingly stressed, again like Parrinder, the study of living religions, in Walls' case, by focusing on the exponential growth of Christianity in Africa.

In the late 1950s Walls, as a missionary of the Methodist Church, was appointed to a lectureship in church history in the Department of Theology in the University of Sierra Leone. In the early 1960s, he transferred to the Department of Religious Studies at the University of Nigeria, Nsukka, and returned to Britain in 1966 to take up a post in ecclesiastical history in the University of Aberdeen. Four years later, he established the first Department of

Religious Studies in Scotland at Aberdeen in the Faculty of Arts and Social Studies rather than the Faculty of Divinity, where it might have been expected to be located. He described the purpose of the Department as 'the study of religion, in its own terms and in its social, phenomenological and historical aspects' (Walls, 1990: 42). This approach underscored a theme, which was repeated throughout many of his later writings, that religion is a subject in its own right, distinct from either theology or the social sciences, one which is best understood through an empathetic approach that includes the perspectives of believers in any eventual interpretation of its meaning. He was also convinced that a department of religious studies, as entirely distinct from divinity programmes or the social sciences, would enable his students to attain broad, global perspectives on religion generally.

A special edition of the journal *Religion*, published in 1980, was dedicated to Geoffrey Parrinder on the occasion of his seventieth birthday. As one of the contributors to the journal, Walls chose to explain his own theoretical understanding of the study of religions. He argued that 'the study of religion is a field in its own right', irreducible to the many disciplines which contribute to its understanding. He acknowledged that the disciplines 'can be, and must be applied' to the study of religion, but he insisted that religion is 'a well-nigh universal dimension of human life' and that 'its manifestations in one place may illuminate its manifestations elsewhere' (1980: 148). His non-reductionist position is confirmed when he writes: 'Religion can best understand religion'. He explained further that 'religious commitment' provides the best 'entrance gate' for understanding religion because 'it at least presupposes the reality of the subject matter' (1980: 143). He regarded these sentiments as entirely consistent with Parrinder's own approach and as a necessary prelude to his interpretation of the relationship between what he called 'primal' religions and the great religions of the world (Walls, 1987: 250–78).

If religion is 'a universal dimension of human life', Walls reasoned that there must be a basic, primary form that underlies its concrete manifestations in diverse cultures everywhere throughout history. He called this the 'primal' component, which is found at the base of every religion and provides for all religions their elemental and common understandings. This explains why many societies around the world share foundational ideas and why the universal religions cannot be understood apart from the way they incorporate primal concepts within their world views. As the universal religions interact with primal religions, new religious movements emerge. This can be seen particularly in Africa where, Walls notes, 'the old religions' form 'the sub-structure of African Christianity'. The scholar cannot ignore this vital relationship: 'Neither in life nor in study can the two now be separated' (2004: 215).

The role of primal religions in contemporary contexts was addressed by Walls in his contribution to Frank Whaling's edited volume, *Religion in Today's World* (1987). Walls argued that the primal religions have been exposed to numerous threats created by the forces of globalization. These include the expansion of world economic systems, the Western model of democratic nationalism, the widespread influence of Western education into primal societies and, of course, the spread of proselytizing religions, particularly Christianity and Islam. Globalization endangers traditional values of worth, obligation and rules of permission and prohibition by upsetting customary patterns of authority, which, in turn, produce changes in social attitudes towards ancestors, gender relations, political structures and community relationships. Walls calls these 'disturbances of focus' created by the new world order, which coerces primal societies into engaging with forces well beyond their localized kinship contexts (1987: 266–7).

Primal religions have responded to the threats posed by globalization in a number of ways. In many cases, they have receded and have been replaced by universal religions, which are better equipped to respond to global challenges. The recession of primal religions is shown by the very large number of people who have converted throughout primal societies to Christianity and Islam. Walls explains why this is the case:

> Christianity and Islam, with their capacity to link into a wider universe, their provision of alternative codes of behaviour and their demand for symbolic change requiring some sort of act of decision, continue to provide keys to meaning and a means of adjustment to new conditions when a people's traditional lore is no longer able to do so. (Walls, 1987: 269)

This might suggest that primal religions are simply dying away. Wall argues that this is not the case. Rather, they are absorbed into the universal religions, which take over many of the features that originally characterized primal religions. In this sense, primal religions live in and through the universal religions, but their local concerns are transformed and magnified by universal beliefs, primarily about God, thereby enabling primal societies to deal with global issues (see also Cox, 1998: 23–7).

The translation principle

Throughout his analysis, Walls maintained that the world religions are built on a primal sub-stratum. If this is the case, the world religions not only will have adopted elements of primal world views into their own belief systems; they will already have been in some senses prefigured in the primal religions before they ever appeared on the scene. To demonstrate this, Walls introduced the concept of 'cultural translation', whereby, just like a language, the ideas, thoughts and beliefs of the universal religion are translated into the primal world view, which turns them into appropriate conveyers of meaning within their own cultural contexts. Walls argued that if the universal religions had not been built on the primal perspective, this process of translation could not have occurred. The universal religious ideas would have been entirely alien to the primal consciousness, rendering impossible any translation of meaning from one cultural context to another.

Walls defined the way universal religions become incorporated into local contexts as the 'translation principle', which he applied specifically to the

Christian notion of the incarnation. In an article he contributed to a book exploring the relationship between Bible translation and the spread of Christianity (1990: 24–39), he outlined precisely how the process of 'translation' works: 'When God in Christ became man, Divinity was translated into humanity, as though humanity were a receptor language'. Yet, language is specific to a particular culture and people. 'No-one speaks generalized "language"' (1990: 25). When God became a human being, therefore, he necessarily incarnated himself into a particular culture, during a specific historical period and within a localized ethnic group. 'The translation of God into humanity, whereby the sense and meaning of God was transferred, was effected under very culture-specific conditions' (1990: 25).

The translation principle bears directly on the process of conversion. Walls explains that in translating from one language to another there is an attempt 'to express the meaning of the source from the resources of, and within the working system of the receptor language' (1990: 25). Although the source comes from the outside and thus introduces something new into the local language, 'that new element can only be comprehended by means of and in terms of the pre-existing language and its conventions' (1990: 25). This means that the source is expressed in terms of the receptor but that the receptor must adapt to that which is introduced into it. Both the source and the receptor are altered by this exchange. The new element can make no sense or bear any relevance unless it can be comprehended by the original. When the new is thus incorporated into the old, it is indigenized, but it is also elevated 'to realms it never touched in the source language' (1990: 25). This process finds a direct parallel in the conversion of adherents within primal religions to Christianity. The Christian message represents the source from the outside, which, when translated, becomes comprehensible only in terms of the local culture. Walls argues that this is not a case of substituting the primal with the Christian, but implies a 'transformation' whereby the 'already existing' is turned 'to new account' (1990: 26).

In an earlier article published in 1982, Walls described what he would later call the 'translation principle' as the 'indigenising principle' (1982: 97–8). African independent churches provide an excellent example of how this principle operates, precisely because they typify how the local and the universal interact. Walls notes that within many African independent churches strict rules are enforced that prohibit the participation of women in worship during their menstrual periods. He remarks that this may seem to Westerners a rather odd thing to worry about, but it is a major topic 'for certain African Christians' (1982: 101). He adds: 'There often turns out to be a sort of coherence in the way in which these churches deal with it, linking Scripture, old traditions and the Church as the new Levitical community – and giving an answer to something that had been worrying people' (1982: 101–2). In this way, the traditional preoccupation with purity is incorporated into African Christian worship but, at the same time, because purity is now couched in Christian terms, believers must reformulate their understanding of it in reaction to globalizing processes.

Walls explains that the meeting of the universal with the local has prompted

African Christian theologians, like E. B. Idowu and John S. Mbiti to address the relationship between African forms of Christianity and the traditional cultural practices associated with Africa's past. Both Idowu (1994 [1962]) and Mbiti (1969, 1970) draw attention to the misrepresentations of African traditions by Westerners. In Idowu's case (1994: 7–10), the notion that Africans have no concept of a Supreme Being is carefully refuted by showing that the *orisas* (divinities) of Yoruba religion are manifestations of the one God, 'Olódùmarè', and not evidence of polytheistic or animistic religion. Walls points out that this issue would not have been raised at all had there been no contact between Christianity and Islam as world religions and Yoruba traditions as indigenous religious expressions. In the new, global, situation, however, the reinterpretation provided by Idowu, and other African Christian theologians, defines a vital component for understanding the contemporary practice of African primal religions in relation to universal religions, and to their permutations within the African independent churches (1982: 103).[3]

It will be clear from this analysis that Walls is not simply restating the old missionary theory of *preparatio evangelica*, which upholds the theological dogma that God had been at work everywhere in the world preparing non-Christians for the Christian message when it was brought to them. Walls, of course, was arguing from a Christian theological starting point, but his primary aim was to explain how cultural interaction produces conversion, in this case, in the ways universalizing forces stimulate cultural change from the outside and how, once the changes are effected, they are incorporated into and owned within specific cultural contexts. This is why Walls argues that every translation involves revision: the universal becomes the local; the local transforms itself into global applications. This is particularly evident in recent history, where we have witnessed 'within the last century ... a massive southward shift of the centre of gravity of the Christian world, so that the representative Christian lands now appear to be in Latin America, Sub-Saharan Africa, and other parts of the southern continents' (1982: 100).

Walls as a key figure in the phenomenology of religion

Walls' influence within the academic study of religions must be emphasized at this point, since, as I noted earlier, he has come in the past 20 years to be associated more with a theology of mission than with phenomenology proper. Nevertheless, it is clear that at Aberdeen Walls steered the Department of Religious Studies towards what I have called elsewhere a programme of 'global religionism' that was rooted firmly in his African experience and which applied non-reductive and empathetic approaches to the study of religions (Cox, 2004: 255–61). His stress on the primal sub-stratum for the universal religions resulted in his inaugurating in 1976 at Aberdeen a one-year taught master's programme, 'The M.Litt. in Religion in Primal Societies'. The aim of the course was described in an appendix to John B. Taylor's edited volume, *Primal World Views*, as:

providing instruments for the study of the 'primal' (or 'ethnic' or 'traditional') religions characteristic of many societies in Africa, the Americas, Asia and Oceania, the effects on belief systems, practices and religious institutions of the meeting of these religions with 'universal' religions (notably Christianity and Islam), and the new religious movements arising after contact with Western influences. (Taylor, 1976: 128)

It was further claimed in Taylor's volume that:

there are considerable documentary resources on this field in Aberdeen, including what is probably the largest specialized collection in Britain of literature from and about the new religious movements; and the Department holds a large corpus of bibliographical and classificatory information on these movements, on primal religions generally, and on non-Western manifestations of Christianity. (Taylor, 1976: 128).

As I noted above, in 1982, Walls established within the Department of Religious Studies at Aberdeen a 'Centre for the Study of Christianity in the Non-Western World (CSCNWW)' which carried forward his emphasis on studying the global impact of Christianity within primal societies. In an announcement appearing in the *Bulletin* of the British Association for the History of Religions shortly after the Centre had been established, Walls (1983: 10–11) described it as having been founded on the conviction 'that the churches of Africa, Asia, Latin America and the Pacific are now central to the Christian faith, and lie at the heart of most questions about the present and future of Christianity'. He added: 'To provide materials for understanding their life, teaching, history and setting is ... a prime task of scholarship'. In order to fulfil this programme, the CSCNWW would 'hold seminars, publish aids to study, maintain links with all continents, and aim to make information available and accessible'. This programme attracted a strong following, particularly among Africans, some of whom have become highly influential and internationally recognized theologians, like Kwame Bediako, a graduate from Aberdeen, and Lamin Sanneh, a former lecturer under Walls. Both Bediako and Sanneh have carried forward into their new expressions of African Christianity many of the ideas Walls advanced originally as principles within a globally transformed phenomenology of religion.[4]

By calling Walls a phenomenologist in this discussion as opposed to a missiologist, I am emphasizing in his writings the analytical tools he employed that are consistent within the broad phenomenological tradition. He clearly adopted a sympathetic approach to the study of religions that privileged the perspectives of believers. He also sought to establish the study of religions within a clear academic context that was distinguishable from theology and that employed a combination of historical and typological methods for understanding religions. He underscored the relevance of the social sciences and theology to his approach, but safeguarded a place for the study of religions that was neither purely social scientific nor theological. None of this was particularly innovative, but it is the way that Walls analysed the impact of cultural change on indigenous societies that marks his primary contribution within a phenomenology of religion. Through the idea of the translation or indigenizing principle, Walls set out a method for understanding the dynamic processes of religious change, in

terms of universal and local situations, without losing the balance betwee[
In this sense, he echoed a thesis consistent with Weber's dynamic typol(
change, but applied this within the new southern configuration of worla
Christianity.

In British phenomenology of religion, Walls has been at least partially eclipsed
by Ninian Smart and the new programme in religious studies Smart initiated in
Lancaster in 1967. I recall a remark made to me by Andrew Walls during one of
my many conversations I enjoyed with him during the mid-1970s, which I
paraphrase as follows: 'There are two centres of excellence for Religious Studies
in Britain today, Aberdeen and Lancaster. The difference is this: one believes in
God, the other doesn't'. Walls was probably referring to Smart's theory of
'methodological agnosticism', which Smart employed even when recruiting
lecturers to his newly established department. Robert Morgan, who was
appointed to the academic staff at the very beginning of the Lancaster
programme, recalls, 'The impending Department, planned to expand to five, had
already raised eyebrows by its advert for a professor "of any religion or none"'
(Morgan, 2001: 349). This was entirely consistent with Smart's approach to the
study of religions and shows clearly that Walls took his phenomenology of
religion in a very different direction from Smart, by studying the variations of
Christian faith and life among primal societies. Although he wrote widely on the
world religions and India in particular, Smart remained throughout his career
very much a theoretician of religion who analysed much more explicitly than
Walls had done how religion should be studied non-theologically without
thereby becoming a mere epiphenomenon of the social sciences. It is important
nonetheless in this chapter on the key figures in British phenomenology to
ensure that neither Walls, nor the tradition that he represented with its deep
African roots from Smith through Parrinder, are overshadowed by the
international reputation Smart achieved as a scholar of religion, particularly in
Europe and North America.

Ninian Smart and British phenomenology

Ninian Smart was educated at Glasgow Academy and Queens College, Oxford.
His first academic post was at the University College of Wales, Aberystwyth from
1952–1955. In 1961 he was appointed as the H. G. Wood Professor of Theology
at the University of Birmingham and in 1967, as we have just noted, he
established the first Department of Religious Studies in Britain at the University
of Lancaster. Later, he shared posts between Lancaster and the University of
California, Santa Barbara and finally became the J. F. Rowny Professor of
Comparative Religions in the University of California, Santa Barbara. He died in
2001. Like Parrinder and Walls, Smart was a lifelong member of the British
Association for the History of Religions, and became its President in 1981, a post
he held until 1985. This coincided with a period when many departments of
religious studies in Britain were under threat due to government funding cuts. As
President, Smart became an advocate for religious studies by calling for 'urgent

action' on the part of university administrators to proceed responsibly so that the academic study of religions could continue to 'broaden people's knowledge and understanding of the multicultural and multireligious world which is already upon us' (1981: 2; see also, U. King, 1982). It was also under Smart's presidency that discussions were initiated to change the name of the association to the British Association for the Study of Religions, a change which was inserted finally into its Constitution in 1988.

The extent to which Ninian Smart is regarded as a key figure in the phenomenology of religion, and more broadly as a scholar of religions, is demonstrated by the wide praise he received in a special volume of the journal *Religion*, which was dedicated to him in 2001 after his death earlier that year. In his contribution to the volume, Adrian Cunningham, who worked with Smart at Lancaster, made the bold claim: 'In thirty-three years as a Professor of Religious Studies, Ninian Smart was the single most important figure in the development of the subject in British education, and a strong influence more widely in Australia, North America and New Zealand' (Cunningham, 2001: 325). Smart contributed to the development of the phenomenology of religion in four main ways: 1) he made popular and accessible the philosophical concept *epoché* through his theory of 'methodological agnosticism'; 2) he organized classical phenomenological typologies into the 'dimensions of religion'; 3) he addressed problems of defining religion by employing Wittgenstein's 'family resemblances' analogy; 4) he included 'secular worldviews' within the academic study of religions, while maintaining a firm line of demarcation between religion and secularism. In none of these areas was Smart a thoroughly innovative thinker, but, as Donald Wiebe (2001: 379) observes, he did 'provide a framework within which religion and religions constituted appropriate objects for detached, scientific analysis' (Wiebe, 2001: 379).

Epoché as 'methodological agnosticism'

Ever since the publication of Smart's *The Science of Religion and the Sociology of Knowledge* (1973a), the academic study of religions has been dominated by the concept that the scholar of religions, for methodological purposes, not only refuses to comment on the realities postulated by religious communities, but meticulously avoids posing any question about truth or value (Cox, 2004b: 259). What earlier phenomenologists referred to in Husserlian terms as performing the *epoché*, Smart called 'methodological agnosticism', by which he meant simply that 'we neither affirm nor deny the existence of the gods' (Smart, 1973a: 54). He contrasted this with the sociologist Peter Berger's notion of 'methodological atheism', which Berger defined as a sociological theory that strictly brackets 'the ultimate status of religious definitions of reality' in order to remain 'value free' (1969: 180). Smart argued that, although Berger did not assert literally an atheistic position, in practice the theory of 'methodological atheism' entails a comment on the 'true state of affairs' and thus becomes 'effectively indistinguishable from atheism *tout court*' (Smart, 1973b: 59). Smart intended by

Bergen becomes → atheism = Smart = Agnostic.

methodological agnosticism that the scholar of religion would remain agnostic about the ontological status of the transcendental focus of religion, a position he contended would be difficult to maintain using Berger's approach.

In his book, *The Phenomenon of Religion* (1973b), Smart painstakingly explained how phenomenological bracketing works. He begins by arguing that the *epoché*, which seems simple at first glance, requires careful analysis. To illustrate this, he examines in depth how a phenomenologist might approach a study of the ritual of Holy Communion as it is practised in the Church of England (1973b: 53–8). One of the first things to be observed is that the people pray that Christ might be made present in the bread and the wine. What, asks Smart, are we to bracket in this instance? Do we suspend judgements as to whether or not Christ is actually present in the bread and wine? On first glance, that seems correct, but this is not what is put into brackets, precisely because this is just what the Christian would say. Christians pray that Christ will become present in the bread and the wine and thus in some sense they actually bracket out the reality of Christ's presence in the bread and the wine. Rather than saying, 'Christ might or might not exist in the bread and the wine', the phenomenologist is better advised not to raise the question of his existence in the first place. This principle applies generally to all phenomenological studies of religion. The scholar simply does not ask the question about the reality of the object of faith, but maintains, for methodological purposes, a position of agnosticism. Like Berger, Smart argued that this does not force the scholar literally to adopt an agnostic position (or in Berger's case, an atheistic one), since it remains purely a methodological position. The question of truth, Smart explains, 'is a question not asked, not a question left undecided' (1973b: 62).

Simply refusing to ask a question, however, does not resolve the problem of what should be bracketed out in phenomenological descriptions. This problem can be clarified by returning to the case of the Anglican Eucharist (1973b: 62–8). Smart argues that the scholar cannot bracket out the fact that the people are praying for Christ to be present in the bread and the wine, since this merely describes what the people are doing or saying. The act of prayer, nonetheless, points to something characteristic of what the people believe, that is to the doctrinal dimension of religion. It may also infer myths concerning the creation and the relation of Christ to the creation, and thus represent the mythic dimension. In so far as the prayer implies that the people become united with Christ and with one another, an ethical dimension can be interpreted. It would seem, therefore, that the phenomenologist should bracket out the doctrinal, mythic and ethical dimensions implicit in the act of praying. Yet, even this needs qualification, because the doctrinal, mythic and ethical dimensions are means through which the people express their faith in what Smart calls the 'Focus' of the religious tradition. The Focus in this sense transcends the doctrines, myths and ethical practices of a religion, since the Focus defines that to which the dimensions of religion point or refer. The dimensions of religion, moreover, express adherents' faith in the Focus, but at the same time the expressions affirm that the Focus has manifested itself in various ways within human experience.

The expressions of faith in the Focus for the phenomenologist are organized into dimensions or typologies; the manifestations of the Focus indicate how believers themselves understand how the Focus becomes known. For Smart, the distinctions between expressions, manifestations and the Focus resolve the problem as to what the phenomenologist is supposed to put into brackets. Both the manifestations and expressions of the Focus, he explains, 'are to be bracketed so long as they are seen under the categories of manifestation and expression' (1973b: 63).

These distinctions may seem unduly complicated until Smart adds a further clarification. A manifestation is an appearance of what a tradition regards as the Focus of faith. Thus, in the Anglican Eucharist, the presence of Christ in the bread and wine corresponds to a manifestation of the Focus. An expression represents a statement of faith in the Focus by the community. For example, in the service of Holy Communion, the believer will utter the words of the Lord's Prayer: 'Our Father, who art in heaven'. For Smart, this is an expression of faith in the Focus and not a manifestation (1973b: 63–4). Both manifestations and expressions are bracketed in the sense that questions about their reality or truth, either as real manifestations of or as true expressions about the Focus, are not posed. In addition, Smart recognized that expressions of faith in the Focus, for believers, evoked feelings, sentiments and values. He argued that these also need to be included within the idea of bracketing (1973b: 32). This detailed analysis clearly represents Smart's interpretation of the phenomenological method whereby every theory, idea, doctrine, ethical assumption or story, which is observed as manifesting or expressing the Focus of faith, including the feelings they induce among believers, is placed in brackets, making it possible for the scholar to attend solely to the phenomena themselves without distinguishing between what is real or apparent.

The dimensions of religion

As we have just seen, even when he sought to clarify the meaning of methodological agnosticism, Smart required the concept of the 'dimensions' of religion. In the case of the Anglican Eucharist, for example, he referred to the doctrinal, mythic and ethical dimensions, which helped him clarify the difference between an expression and a manifestation of the Focus of faith. Smart first introduced the idea of 'dimensions' in his influential book on the world's religions, *The Religious Experience of Mankind* (1977: 15–25), in which he listed the dimensions, without great explanation, as the ritual, the mythological, the doctrinal, the ethical, the social and the experiential. He restated and clarified his use of the dimensions in many of his subsequent books, particularly with respect to secular world views. It is clear that Smart employed the dimensions in support of his lifelong contention that religion should be understood beyond its doctrinal aspect, usually interpreted as a belief in God or gods, which had so dominated the traditional Western approach to the study of religions. Instead, he preferred to describe religion as multidimensional and organic, a view he summarized in the introductory chapter to *The Religious Experience of Mankind*: '[Religion] is a six-dimensional organism, typically containing doctrines, myths, ethical teachings, rituals and social institutions and animated by religious experiences of various kinds' (1977: 31).

In 1989, Smart produced a much enlarged study which he entitled *The World's Religions: Old Traditions and Modern Transformations* (first paperback edition, 1992). Just as he did in *The Religious Experience of Mankind*, he begins with a discussion of the dimensions of religion, but expands them in number from six to seven and gives each a 'double-barrelled' name or dual classification. What formerly he called the ritual dimension, he now labels 'the practical and ritual dimension', explaining that these refer not only to 'formal or explicit rites of religion' but also to 'practices' that develop 'spiritual awareness or ethical insight ... such as yoga in the Buddhist and Hindu traditions' (1992: 12–13). The 'experiential and emotional dimension' comprise his second category, and point to 'the *emotions* and *experiences* of men and women' (emphasis his), which outsiders must try to enter into for purposes of understanding (1992: 13). He identifies a third classification as the 'narrative or mythic dimension', which he calls 'the story side of religion', including both historical and quasi-historical narratives, 'some about great founders, such as Moses, the Buddha, Jesus, and Muhammad', and myths about 'that mysterious primordial time when the world was in its timeless dawn' (1992: 15). The 'doctrinal and philosophical' dimension constitutes Smart's fourth category, and it is this which underpins the narrative dimension. Doctrines refer to beliefs, as exemplified in the Christian notion of the Trinity, which resulted from the meeting of early Christianity with the great philosophical traditions of the Graeco-Roman world (1992: 16–17). Smart combines the ethical and the legal dimensions into one classification to construct his fifth dimension of religion. Some religions, like Islam and Judaism, he notes, stress that ethics are expressed in legal formulations prescribing how believers are to live. Other traditions place less importance on law, but 'still display an ethic which is influenced and indeed controlled by the myth and doctrine of the faith' (1992: 17).

Smart then moves to the final two dimensions, the 'social and institutional' and the 'material', both of which, he explains, 'have to do with the incarnation of religion'. Because every religion 'is embodied in a group of people', students of religion need to understand 'how it works among people' (1992: 18). This entails various levels of organization, such as hierarchical systems or forms of democratic governance, but all descriptions of institutions require an analysis as to how they relate to the wider society of which they form a part. The material dimension, which Smart later called 'the material or artistic dimension' (1997: 12), refers to the expressions of religion in forms such as buildings, works of art or even natural features of the world, like rivers, mountains, trees or pools (1992: 20). Both Islam and Protestant Christianity shun elaborate material representations, but even these sometimes iconoclastic traditions express themselves artistically, in the case of Islam by the use of calligraphy in the Qur'ānic script or in Protestant Christianity by symbols engraved on pulpits or a communion table (1992: 327).

For Smart, each of the seven dimensions not only has a dual name to demonstrate the breadth of the concept, but each is related to every other dimension on the list. For example, he explains, in the case of Buddhism, the doctrine of impermanence is a central belief, but it is also expressed within the ritual or practical dimension since philosophical reflection aids meditation, and meditation in turn helps the individual experience personally the force of the doctrine. The point of the seven dimensions, Smart concludes, is to provide 'a balanced description of the movements which have animated the human spirit and taken place in the shaping of society, without neglecting either ideas or practices' (1992: 21). Throughout *The World's Religions*, Smart applies the dimensional analysis to each religious tradition or sub-tradition he considers, which he then interweaves within historical accounts.

Religion, world views and family resemblances

Not only did he outline the historical development and dimensions of religious traditions in *The World's Religions*, but Smart also included what he called 'world views' in general, and quite specifically, 'secular' world views (1992: 21–5). In one sense, he argued, all religions constitute a world view and contribute to wider world views, and thus should not be defined too narrowly. If religious world views are broad and interact with wider perspectives, Smart then asks if the dimensions of religion can be applied to secular world views, and, if so, what implications follow from this. Smart includes among secular world views, as primary examples, nationalism, Marxism, scientific humanism and existentialism. In the case of nationalism, he argues that the modern world has been shaped by the nation state, particularly since the era of colonial expansion. The nation state possesses the same dimensions as those which appear in what otherwise would be called religions. It performs rituals, such as in the United States where memorials to the history of the nation are surrounded by the faithful on special days and respected through recitation of creeds ('I pledge allegiance to the flag of the United States of America') or in songs, such as 'America the Beautiful'. It follows from this that nation states induce strong emotional attachments: they produce common narratives; they promulgate doctrines to which loyal citizens must adhere; they inculcate values which are enshrined in law; they are highly organized social institutions; and they symbolize all of the above dimensions in material and artistic forms. Smart concludes: 'It is, then, reasonable to treat modern nationalism in the same terms as religion' (1992: 24).

To treat them in the same terms as religion, however, is not the same as calling secular world views religions as such. At this point, Smart seems to opt out of engaging in a strictly methodological debate, or even being drawn into complicated problems entailed in defining religion. His dimensions are pragmatic tools aimed at promoting understanding of 'the various ways in which human beings conceive of themselves and act in the world' (1992: 25). It is out of respect for the 'secularity' of secular world views that Smart refrains from calling them religions, or even 'quasi-religions'. He explains: 'Though to a greater or lesser extent our seven-dimensional model may apply to secular worldviews, it is not really appropriate to try to call them religions' (1992: 25). To do so would be to demote them 'below the status of "real" religions'. It would enforce a standard for understanding human responses to the world in a graduated way, with real religion at the top and religion-like world views somewhat inauthentically positioned below them. In fact, secular world views, Smart argues, often see themselves as opponents of religion. It is here that the dimensional approach becomes useful: 'The various systems of ideas and practices, whether religious or not, are competitors and mutual blenders, and can thus be said to play in the same league' (1992: 25).

Despite his unwillingness to discount the importance of secular world views as part of overall human responses to common existential situations, Smart persisted with a distinction between a secular world view and religion. His language, as evidenced in the citations I noted in the previous paragraph, continued to draw distinctions between religion and non-religion, although he argued convincingly that the seven dimensions could apply equally to both. It is important to underscore at this point that Smart did not introduce the dimensions of religion as a method for defining religion, but, as he explained towards the beginning of *The World's Religions*, they were intended to provide 'a scheme of ideas which will help us to think about and to appreciate the nature of the religions' (1992: 11). By refusing to define religion in a narrow way, Smart could be accused of using terms quite indistinctly and indeed in an ambiguous and incoherent manner. It is not quite correct to conclude from this, however, that Smart never defined religion. Instead, his dimensions of religion should be understood as a definition in practice, or as a 'working' definition.

For this, he found Wittgenstein's model of 'family resemblances' useful as a means for creating categories generally, but primarily for constructing the category 'religion'. The dimensions of religion had demonstrated for Smart that different religions stress some of the dimensions so greatly that other dimensions seem hardly present at all within them. Moreover, the variations within what we call the same religion also are frequently quite pronounced. In Christianity, for example, an outsider might regard the practices of one of its forms as an entirely different religion from the practices of another of its forms. 'A Baptist chapel in Georgia is a very different structure from an Eastern Orthodox church in Romania, with its blazing candles and rich ikons' (1992: 11). We can understand the relationship between the Baptist chapel and the Orthodox church better if we think of them as somewhat distant cousins, part of the same family but far removed from one another. Certain core narratives are held in common, but the ritual dimension, and even the doctrinal one, may display quite sharp differences. Smart suggests that it may even seem in some situations 'that a person within one family of subtraditions may be drawn closer to some subtradition of another family than to one or two subtraditions in her own family' (1992: 11). He admits, on a personal level, 'I happen to have had a lot to do with Buddhists in Sri Lanka and in some ways feel much closer to them than I do to some groups within my own family of Christianity' (1992: 12–13). Nevertheless, if one describes family resemblances according to the seven-fold dimensions of religion, it will be seen that families of religions do exist.

When this is put within the context of religions in general, although the relationships become even more distant and at times appear unrecognizable, there is retained a certain common content with respect to the dimensions among religions that is sharply different from the content of secular world views. The dimensions are structural, formal, and thus descriptive of how humans respond to common life situations. When the religious is placed alongside the secular, content is filled into the forms or dimensions, making the differences much clearer than when the dimensions are considered strictly as formal categories. Although it is not possible to designate strict lines of demarcation between religion and secularism, following the analogy of families, it becomes clear that the two are unrelated except in the most general sense.

To understand how the content of religion differs from secular world views, and thus makes them into different species rather than part of the same family, we need to return to Smart's idea of the Focus of faith. For Smart, the Focus transcends the dimensions of religion. It is 'what these ideas and practices refer to or are directed at' (1973b: 62–3). For this reason, the Focus is expressed, that is explained and understood cognitively, 'most sophisticatedly in Theology'. As a transcendent referent, over and above theology, it operates as what Smart calls the 'real' Focus, which is ineffable, beyond description, at the level of faith or human aspiration. Of course, both the transcendent Focus and its appearances are bracketed for the phenomenologist, but that a transcendent Focus manifests itself for religious people remains for Smart the core defining element of religion, that which distinguishes religious from secular world views. He explains:

> It [phenomenology] is not just concerned with how a faith manifests itself or appears ... but rather with how it actually is. Its aim is to give insight and understanding of the substance of men's faiths, the way they actually operate, impinge on institutions, exist in human consciousness and so on. (Smart, 1973b: 69)

In his summary of Smart's contribution to the academic study of religion, Walter Capps (1995: 310) calls Smart's phenomenology 'straightforward' because 'it is specific, detailed, and, most importantly, self-possessed'. Capps explains that Smart avoided 'complex theoretical analyses' and refused to 'treat the religious traditions as testing ground for more extensive methodological or cultural issues'. Capps did not mean that Smart was theoretically unsophisticated, but that he articulated the principles of phenomenology without undue complication. Indeed, we have seen how he falls fully within a phenomenological tradition by adopting the principle of bracketing or *epoché* through the theory of methodological agnosticism, by employing typologies in a renamed form as dimensions, by distinguishing the religious from the non-religious in terms of a core transcendental focus and by clarifying the key phenomenological concepts of expression and manifestation in relation to each other component within his method. We will return to Smart's legacy in fostering subsequent debates in the study of religions in Chapter 7, but there can be little doubt that his influence as a British phenomenologist has shaped quite fundamentally to this day what came to be adopted internationally as the mainstream approach within the academic study of religions.

Conclusion

In this chapter, I have argued that British phenomenology owes its inspiration to the study of religions in Africa, first through the widespread influence of Edwin W. Smith and thereafter through Geoffrey Parrinder and Andrew Walls. British departments of religious studies, certainly in part, were created from an approach to the study of religions that adopted a sympathetic attitude towards African religions by seeking to redress abuses perpetrated by early missionaries, anthropologists, ethnologists and colonial administrators. This encouraged an emphasis on the central place of God in the study of African religions, as demonstrated quite early on in the writings of Edwin W. Smith which, although consistent with the Christian theological assumptions of other European phenomenologists, developed against the backdrop of British colonialism. This helps to explain why Andrew Walls founded a Department of Religious Studies in Aberdeen with an emphasis on the 'non-Western world' and on new religious movements in 'primal societies'. Geoffrey Parrinder developed a comparative approach at King's College, London, that moved beyond his African experience, but remained rooted in the principles he first applied when he wrote *West African Religion* in the 1940s. Ninian Smart, much more than his counterparts with African experience, reflects a scholar who, inspired by Dutch phenomenology, applied its main principles to stamp his own brand of the phenomenological tradition into religious studies.

By the 1970s two main centres of British phenomenology had developed as a result of these factors, one with its non-Western focus in Aberdeen under Andrew Walls and the other with its world religions, comparative and theoretical perspectives in Lancaster. Although these centres were related, they were entirely distinct in emphasis, each fostering different approaches to the study of religions in British universities that persist to this day. One, with African roots in Aberdeen and now in Edinburgh, has maintained a close connection with missionary theology and has fostered links with academic institutions in the third world, and in recent years with similarly minded programmes in North America, such as the Overseas Mission Study Center near Yale University and the Yale Divinity School itself. The other at Lancaster has retained its emphasis on (and may even have moved further towards giving prominence to) theory and method in the study of religion, but at the same time has encouraged an interest in the application of theory to the contemporary British religious scene. If there is a thread connecting the two, it must be seen in the phenomenology of religion, where, as we have noted, the same themes recur in the writings of Smith,

Parrinder, Walls and Smart: the use of sympathetic bracketing, raising awareness of prior distorting assumptions in the study, particularly, of non-Christian religions, the attempt to create classifications for understanding religion, and maintaining a firm commitment to the irreducible core of religion and hence to the inherent 'religiousness' of the scholarly enterprise.

References

Berger, P. L. (1969), *The Sacred Canopy. Elements of a Sociological Theory of Religion* (Garden City, New York: Doubleday Anchor Books).

Cox, J. L. (1998), *Rational Ancestors. Scientific Rationality and African Indigenous Religions* (Cardiff: Cardiff Academic Press).

Cox, J. L. (2004a), 'From Africa to Africa: The significance of approaches to the study of African religions at Aberdeen and Edinburgh Universities from 1970 to 1998', in F. Ludwig and A. Adogame (eds), *European Traditions in the Study of Religion in Africa* (Wiesbaden: Harrassowitz Verlag), pp. 255–64.

Cox, J. L. (2004b), 'Afterword. Separating religion from the "sacred": Methodological agnosticism and the future of religious studies', in S. Sutcliffe (ed.), *Religion: Empirical Studies* (Aldershot: Ashgate), pp. 259–64.

Cunningham, A. (2001), 'Obituary: Ninian Smart', *Religion*, 31 (4), 325–6.

Forward, M. (1998), *A Bag of Needments. Geoffrey Parrinder and the Study of Religion* (Bern: Peter Lang).

Hastings, A. (2001), 'Review Article: Geoffrey Parrinder', *Journal of Religion in Africa*, XXXI (3), 354–9.

Idowu, E. B. (1962), *Olódùmarè. God in Yoruba Belief* (London: Longmans).

Idowu, E. B. (1995), *Olódùmarè. God in Yoruba Belief* (revised and expanded edition) (Old Bethpage, New York: Original Publications).

King, U. (1982), 'Current state of the study of religions in British universities. Report on the response of the July questionnaire', *British Association for the History of Religions Bulletin*, 38 (December), 1–5.

Mbiti, J. S. (1969), *African Religions and Philosophy* (London: Heinemann Educational Books Ltd).

Mbiti, J. S. (1970), *Concepts of God in Africa* (London: SPCK).

Morgan, R. (2001), 'Religious studies in Britain: Lancaster in the sixties', *Religion*, 31 (4), 349–52.

Parrinder, G. (1949), *West African Religion: Illustrated from the Beliefs and Practices of the Yoruba, Ewe, Akan and Kindred Peoples* (London: Epworth Press).

Parrinder, G. (1962), *Comparative Religion* (London: George Allen and Unwin Ltd).

Parrinder, G. (1964), *The World's Living Religions* (London: Pan Books Ltd).

Parrinder, G. (1969), *Religion in Africa* (London: Pall Mall Press and Harmondsworth: Penguin). (Reprinted, 1976: *Africa's Three Religions*, London: Sheldon Press).

Parrinder, G. (1974), *African Traditional Religion* (London: Sheldon Press, 3rd edn).

Parrinder, G. (1980), *Sexual Morality in the World's Religions* (Oxford: Oneworld).

Platvoet, J. G. (1996), 'From object to subject: A history of the study of religions of Africa', in J. Platvoet, J. Cox and J. Olupona (eds), *The Study of Religions in Africa: Past, Present and Prospects* (Cambridge: Roots and Branches), pp. 105–38.

Rinsum, H. J. van (2003), '"Knowing the African": Edwin W. Smith and the invention of African Traditional Religion', in J. L. Cox and G. ter Haar (eds), *Uniquely African? African Christian Identity from Cultural and Historical Perspectives* (Trenton, New Jersey: Africa World Press), pp. 39–66.

Smart, N. (1973a), *The Science of Religion and the Sociology of Knowledge* (Princeton: Princeton University Press).

Smart, N. (1973b), *The Phenomenon of Religion* (New York: The Seabury Press).

Smart, N. (1977), *The Religious Experience of Mankind* (Glasgow: Collins Fount Paperbacks). (First published, 1969, New York: Charles Scribner's Sons).

Smart, N. and Pye, M. (1981), 'Press Release: Religion under attack on campuses', *Bulletin of the British Association for the History of Religions*, 35 (October), 2.

Smart, N. (1992), *The World's Religions. Old Traditions and Modern Transformations* (first paperback edition) (Cambridge: Cambridge University Press).

Smart, N. (1997), *Dimensions of the Sacred. An Anatomy of the World's Beliefs* (London: Fontana Press).

Smith, E. W. (1936), *African Beliefs and Christian Faith: An Introduction to Theology for African Students, Evangelists and Pastors* (London: The United Society for Christian Literature).

Smith, E. W. (ed.) (1950), *African Ideas of God: A Symposium* (London: Edinburgh House Press).

Stoecker, H. (2004), '"The gods are dying": Diedrich Westermann (1875–1956) and some aspects of his studies of African religions', in F. Ludwig and A. Adogame (eds), *European Traditions in the Study of Religion in Africa* (Wiesbaden: Harrassowitz Verlag), pp. 169–74.

Taylor, J. B. (ed.) (1976), *Primal World Views: Christian Dialogue with Traditional Thought Forms* (Ibadan, Nigeria: Daystar Press).

Tempels, P. (1959), *Bantu Philosophy* (translated by C. King) (Paris: Presence Africaine).

Walls, A. F. (1980), 'A bag of needments for the road: Geoffrey Parrinder and the study of religion in Britain', *Religion*, 10 (Autumn), 141–50.

Walls, A. F. (1982), 'The Gospel as the prisoner and liberator of culture', *Missionalia*, 10 (3), 93–105.

Walls, A. F. (1983), 'Centre for the Study of Christianity in the Non-Western World, University of Aberdeen, Scotland', *British Association for the History of Religions Bulletin*, 39 (April), 10–11.

Walls, A. F. (1987), 'Primal religious traditions in today's world', in F. Whaling (ed.), *Religion in Today's World* (Edinburgh: T. and T. Clark), pp. 250–78.

Walls, A. F. (1990), 'The translation principle in Christian history', in P. C. Stine (ed.), *Bible Translation and the Spread of the Church: The Last 200 Years* (Leiden: E. J. Brill), pp. 24–39.

Walls, A. F. (2000), 'Eusebius tries again: Reconceiving the study of Christian history', *International Bulletin of Missionary Research*, 24 (July), 105–11.

Walls, A. F. (2004), 'Geoffrey Parrinder (1910) and the study of religion in West Africa', in F. Ludwig and A. Adogame (eds), *European Traditions in the Study of Religion in Africa* (Wiesbaden: Harrassowitz Verlag), pp. 207–15.

Ward, K. (2002), 'The study of religions', in E. Nicholson (ed.), *A Century of Theological and Religious Studies in Britain* (Oxford: Oxford University Press), pp. 271–94.

Westermann, D. (1937), *Africa and Christianity* (London: Oxford University Press).

Wiebe, D. (2001), 'Ninian Smart: A tribute,' *Religion*, 31, 379–83.

Young, W. J. (2002), *The Quiet Wise Spirit: Edwin W. Smith 1876–1957 and Africa* (Peterborough: Epworth Press).

Notes

[1] Parrinder sent the following message to the Annual General Meeting of the BAHR held on 14 September 1977: 'On 14 September I am reading a paper to the 36th Annual Convention of the Japanese Association for Religious Studies, at Nagoya, on "Recent Trends in the B.A.H.R.". I will give them greetings from the B.A.H.R. as the last act of my presidency. Please thank all members for their support in the last five years, and previous twelve as secretary' (*BAHR Bulletin*, 21, October 1977, 3).

[2] The second edition of this book, published in 1976 by Sheldon Press in London carried the title, *Africa's Three Religions*.

[3] Walls refers, among others, to works dealing with God in African religion, particularly to Idowu's *Olódùmarè. God in Yoruba Belief* (1962) and Mbiti's *Concepts of God in Africa* (1970).

[4] Kwame Bediako directs the Akrofi-Christaller Memorial Centre for Mission Research and Applied Theology in Akropong-Akuapem, Ghana. Lamin Sanneh is the D. Willis James Professor of Missions and World Christianity and Professor of History at Yale Divinity School.

Chapter 6

Interpreting the Sacred: North American Phenomenology at Chicago and in the Thought of W. C. Smith

In his discussion of Mircea Eliade, Daniel Pals (1996: 161) notes that when Eliade came to the University of Chicago in 1958 to take up a post in the history of religions, there were three significant professorships in the academic study of religions in the United States. Twenty years later, 'there were thirty, half of which were occupied by his students'. Eliade succeeded Joachim Wach at Chicago, who had assumed the Chair of History of Religions in 1945 at a time when, as Russell McCutcheon (2003: 60) observes, the religion programme 'was considerably weakened' and had 'lost its distinct disciplinary location as a humanistic study' by becoming absorbed into theology. Wach turned this around and became a leading international figure in what became known as the 'hermeneutical approach' in the study of religions, a tradition that was carried forward and expanded by Eliade. The so-called Chicago school emerged from the widespread influence of Wach and Eliade, producing important figures at Chicago, including Wach's close colleague, Joseph Kitagawa, and Charles Long, Jonathan Z. Smith and Bruce Lincoln. In the context of key figures in the development of the phenomenology of religion, I will focus in this chapter on the consistent 'hermeneutical' approach that can be traced from Wach through Eliade to J. Z. Smith. Many other important figures in the study of religions in North America could be included, but it would not be possible to omit the Canadian Islamicist, Wilfred Cantwell Smith, whose bold book *The Meaning and End of Religion*, first published in 1962, called into question the category 'religion' itself and thereby forced religious studies scholars very early on into a form of self-reflexivity, a mode of knowing Smith had introduced into some of his earlier works and that he continued to develop throughout his later writings.

Each of the scholars I will consider in this chapter, Wach, Eliade, J. Z. Smith and W. C. Smith, aligned themselves with disciplinary approaches in the study of religions that can be distinguished from the phenomenology of religion. Wach became most widely known as a sociologist of comparative religions; Eliade referred to himself as a historian of religions; J. Z. Smith, as a specialist primarily in ancient Near Eastern religions, retained Eliade's emphasis on history; and W. C. Smith thought of himself chiefly as a scholar of comparative religions. My contention in this chapter is that, despite their stated or preferred disciplinary identities, each figure quite deliberately employed principles derived from the phenomenology of religion, including the attitudinal, descriptive and interpretative methods I have already identified as prominent themes in the Dutch and British versions of phenomenology.

Joachim Wach: A phenomenological interpretation of sociology

Joachim Wach, who was born in Germany in 1898, undertook his university studies in Leipzig, Munich and Berlin, where he specialized in the history and philosophy of religion, and in Oriental languages. He earned his Ph.D from Leipzig with a dissertation entitled, 'Basic Elements of a Phenomenology of the Idea of Salvation', which had been influenced by attending lectures delivered by Husserl at Freiburg in 1922 (Waardenburg, 1973: 487). Eric Sharpe (1986: 238) adds that at Leipzig, Wach was 'introduced to the comparative study of religion by Friederich Heiler, and counted among his mentors Troeltsch, Harnack, Söderblom, Weber and Otto'. In 1924, he wrote his *Habilitation* thesis on the disciplinary autonomy of the science of religion, in which he considered, according to McCutcheon (2003: 61), the ' "specifically religious" aspect of life'. Later, Wach was awarded a doctor of theology degree from the University of Heidelberg. It is evident from his early years of study and teaching in Germany that Wach's thought emerged out of the twin formative influences on phenomenology I described in Chapters 1 and 2: the philosophy of Husserl and theological studies in the broad German tradition of Albrecht Ritschl, both of which influenced his theory that the study of religion constitutes an irreducible subject in its own right.

Wach left Germany in 1935 for the United States when his position at Leipzig was eliminated under the Nazi regime (Sharpe, 1986: 238). He was appointed to teach religion in Brown University in Providence, Rhode Island, a post he held for ten years before being awarded the Chair in the University of Chicago. His most important books written for an English audience after coming to the United States include, *Sociology of Religion* (1944), *Types of Religious Experience: Christian and Non-Christian* (1951) and his posthumous publication, *The Comparative Study of Religions* (1958). In addition, several of his essays were translated into English after his death, including his important introduction to Joseph Kitagawa's edited volume *The History of Religions: Essays on the Problem of Understanding*, published in 1967. Wach's contribution to this book, which was written while he was still in Germany, first appeared in 1935 in the journal *Zeitschrift für Missionskunde und Religionswissenschaft*. The English translators, Karl W. Luckert and Alan L. Miller entitled the article 'The Meaning and Task of the History of Religions', but have helpfully put in brackets at the end the title the German word *Religionswissenschaft*, which more properly conveys Wach's meaning than does the phrase 'History of Religions'. In his later English publications, according to Pettersson and Åkerberg (1981: 40), Wach revived the ideas he had advanced earlier in Germany, with the specific aim of developing hermeneutics as 'a method for the study of comparative religion'. Walter Capps (1995: 173) argues that Wach was concerned primarily with the concept *Verstehen*, particularly in the cross-cultural and interdisciplinary studies of religion. In this section, I will argue that Wach extended these emphases to produce a sympathetic sociological theory of religion based on an in-depth understanding of religions attained through a hermeneutical method that combined an intuitive feeling for the religious sentiment in humanity with a scholarly and informed theoretical approach. I conclude by asserting that Wach's sociology of religion was phenomenological through and through.

Wach begins his *Sociology of Religion*, which was based on lectures delivered to undergraduates at Brown University, with what he calls a 'methodological prolegomena'. He distinguishes initially theology from a general science of religion (*Religionswissenschaft*), calling theology a 'normative discipline' that 'is concerned with the analysis, interpretation, and exposition of one particular faith' (1962: 1). Within the general science of religion, he includes phenomenology, history, psychology and the sociology of religion, each of which 'is essentially descriptive, aiming to understand the nature of all religions' (1962: 1). He adds in a footnote that by phenomenology he does not mean a study in 'the sense of Husserl', but 'to indicate the systematic, not the historical, study of phenomena like prayer, priesthood, sect, etc.' (1962: 1). He thus defined phenomenology as a typological or classificatory system only, which, when accompanied by other branches of the science of religion, builds up a full picture in its many dimensions of the human religious experience. Wach went on to describe the philosophy of religion as a third major discipline, which is 'akin to theology in its normative interests', but shares with the science of religion 'its subject matter' (1962: 2).

Wach explains that many scholars have written about the history of religions, but argues that there have been few 'systematic and comparative studies of the varied forms of the expression of religious experience' (1962: 2). Most studies of religion that have been undertaken thus far have tended to emphasize 'theoretical forms', such as 'myth, doctrine or dogma' at the expense of 'the practical expression in cultus and forms of worship'. He explains that his own study aims at examining a 'third field of religious expression' beyond either doctrines and rituals, the sociology of religion, which he defines as 'religious grouping, religious fellowship and association' fashioned into an 'individual, typological and comparative study' (1962: 2). Just like his Dutch counterparts had done when outlining a place for the phenomenology of religion within academic studies, Wach underscores the need for the sociologist of religion to gain an in-depth understanding of the historical development of the various religions: 'Without the work of the historian of religion, the sociologist would be helpless' (1962: 2). Again, like phenomenologists had done in relating phenomenology to history, Wach argues that 'it is the sociologist's hope that his categories will prove fruitful for the organization of the historian's material' (1962: 2).

Wach's sociological approach is built on the pioneering work of Max Weber, whom he credits as 'having been the first to conceive of a systematic sociology of religion' (1962: 3). It is a mistake, according to Wach, to focus almost exclusively on Weber's analysis of the economic impact of Calvinism in the West, since Weber outlined in many other works a quite systematic description of 'the non-Christian religious world', although he overlooked so-called primitive religions 'as well as Mohammedanism'. According to Wach, Weber nonetheless exhibited an overly critical attitude towards religious adherents. His categories for classifying religious phenomena 'are not entirely satisfactory, because not enough attention is paid to their original meaning' (1962: 3). Some of the attitudinal weaknesses Wach attributes to Weber he sees as having been avoided by Ernst Troeltsch, but Troeltsch's failure is to have limited his studies 'exclusively to Christianity' (1962: 3–4). Wach emphasizes that his own approach aims at improving on the works of Weber and Troeltsch by exposing 'the manifold interrelations between religion and social phenomena' in order 'to understand better the various aspects of religious experience itself' (1962: 5). By stressing the socially embedded nature of religion, Wach concludes that 'the impartial observer' becomes 'strikingly aware of the intricacy and variety of the relations existing between society and religion', and in turn is 'impressed with the tremendous fomenting and integrating power possessed by religion' (1962: 6).

The sociology of religion that Wach then expounds in the remainder of the book is entirely typological in form and content. In the first part, he examines religious experience under the categories of theory, by which he means doctrine and practice (cultic and ritual), and sociological expression, through which he examines religion as an individual experience and a collective manifestation. In the second part of the book, Wach considers the broad topic 'religion and society', by examining what he calls 'natural groups', which include family, kinship, local, racial and national cults, and cults formed on the basis of sex and age. He then outlines how religion is organized specifically in societies according to categories such as secret societies, mystery religions of Greece and Rome, the syncretistic communities characteristic of religion in India and finally various forms of founded religions and social movements that such founded religions have fostered. The next section analyses how religions have been differentiated and stratified according to social organizations, particularly from lesser to more complex systems. Finally, Wach examines relationships between religion and the state in various contexts and concludes with a typology of religious authority, in many ways similar to that described by Weber, including the relationship between 'charisma' and leadership in religious organizations. In his conclusion to the book, Wach restates his typological approach, indicating that one of the key subjects of the book has entailed an enquiry into 'the variety of forms in which religious communion manifests itself' including a 'study of the typical development of specifically religious organization in society' (1962: 379).

Wach's criticisms of Weber's attitudinal approach to religion, accompanied by his emphasis on religious experience as a social phenomenon, indicate that he was interested in far more than describing religion socially, following typological categories, but was intent on interpreting it sympathetically. Wach's understanding of a sociology of religion, as it fitted into a broader science of religion, is what Joseph Kitagawa (1959: 21) labelled the analysis of religious data 'religio-scientifically'. By this, Kitagawa meant that scholars of religion not only describe the phenomena of religion, but they direct their investigations 'to the meaning of religious phenomena' (1959: 21). In this sense, Wach can be regarded as introducing a sympathetic attitude within the sociology of religion in order to avoid reducing religion to sociological functions. In his discussion of Wach's contribution to the study of religions, Kitagawa notes that Wach may have been too optimistic about avoiding reductionistic conclusions about religion, since sociology assumes that religion results from social organization and group life (1959: 21). Nevertheless, Wach emphasized in his introductory comments on methodology in his *Sociology of Religion* that scientific studies of religion do not extinguish 'the sense for the numinous' but, on the contrary, the study of religion 'is awakened, strengthened, shaped and enriched by it' (1962: 3).

Wach: Understanding through a hermeneutical method

In the article he wrote originally in 1935, which was published later in Kitagawa's edited volume on the problem of understanding in the history of religions, Wach defined the 'practical significance of *Religionswissenschaft*' as broadening and deepening 'the *sensus numinis*' (1967: 4). He explained that the study of religions actually reinforces the scholar's own faith because 'it allows a new and comprehensive experience of what religion is and means' (1967: 4). It does this particularly when religion is seen in a social context, where 'the effectiveness of the religious genius' finds expression in religious communities and organization, in the practice of rituals and cult activities. When the scholar describes and penetrates to the meaning of the social expressions of religion, the scholar's own religious impulse is fortified. Clearly, Wach was writing in this context as a Christian, who saw his own faith enhanced by the comparative study of religions because, he explains, 'it ought to lead to the examination and preservation of one's own faith' (1967: 4). The scholar who possesses a personal faith is thus able to understand in a way that those without religious experience cannot. This does not detract from the scientific purpose of the study; rather, it enhances it, precisely because it facilitates understanding (*Verstehen*). Understanding results from a combination of the scholar's own religious experience and the ability to describe accurately the rudiments of a religion, both of which reveal an underlying numinous power. Wach derived this idea from his studies with Heiler and Otto, a fact which prompted Walter Capps (1995: 176) to observe that Wach's 'entire schema was ordered to honor the conviction that varieties of religious expression testify to the universality of the religious reality'.

To achieve genuine understanding, the student of religion must overcome three major obstacles. The first Wach calls 'quantitative in nature', which refers to the problem of the distance in time and space between the scholar of religion and, for example, the dead religions of antiquity, or even the culturally remote religions of primitive peoples (1967: 8). A second obstacle entails 'qualitative difficulties', such as the complexity of attaining an understanding of religious beliefs, practices and experiences that are genuinely foreign to the researcher (1967: 9). The third major problem affecting understanding is 'the secret of plurality among religious experiences' (1967: 9). This is related to the second obstacle, in the sense that plurality fosters foreignness for the scholar who necessarily comes from a specific religious or cultural background. For example, Wach says that as a Protestant, the Roman Catholic mass appears strange to him, but stranger still are the rites of the Eastern Orthodox Church and further 'a Jewish, or an Islamic, or even a Buddhist worship service' (1967: 10). Each of these obstacles demonstrates how the major task of *Religionswissenschaft* is to attain an understanding of other religions. At this point, the scholar must adopt a hermeneutical approach.

By hermeneutics, Wach means a seeing into 'the specific and perhaps unique spirit (*Geist*) of a certain religious context' (1967: 10). This produces an understanding 'into the depths' (*Tiefenblicke*) (1967: 10). The process of hermeneutics involves an interpretation of the data emerging from this vision into the depths, but it does not occur simply by intuition. It requires a trained mind, using intuitive skills, to synthesize the data (*Kombination*). This is possible because at the core of religion, no matter how foreign it may appear to the trained and sympathetic observer, an astoundingly similar religious experience resonating with one's own can be found. 'In this it is a great help for the human understanding that in the structure of spiritual expressions (of such great and deep experiences as are the productively religious ones) there is inherent an amazing continuity (*Folgerichtigkeit*)' (1967: 11). The process that unfolds from this method engages the scholar in identifying the key or 'central intuition' of the religion one is seeking to understand. In Islam, for example, it is possible to grasp the sense of the 'experience of the deity' as the clue to understanding its 'doctrine, theology, cosmology, anthropology, soteriology, and cult' (1967: 11). Or in Buddhism, by sensing the deep experience of human suffering, the scholar can 'unlock an otherwise strange-appearing world of expressions' (1967: 11). The hermeneutical principle thus implies that 'he who wishes to understand other religions must have a sense (*Organ*) for religion and in addition the most extensive knowledge and training possible' (1967: 13). In this way, a hermeneutical approach provides the basis on which a genuine science of religion can be constructed. Wach concludes that after the data of religion have been described and studied, what he calls the 'desire' of *Religionswissenschaft*

> will always be to place the individual beliefs and ideas, the customs and communal modes, into that context in which alone they live; to connect them and to show them together with the spirit of the entire religion, with the basic intention that animates them, and with the creative religious intuition at their source. (Wach, 1967: 6)

On the basis of what we have seen thus far, and in the context of this book, it seems fully justified to include Wach among a list of key phenomenologists of religion. That he declared his interest in a general science of religion suggests that he regarded himself as much more than a phenomenologist, but there are many solid reasons to suggest that his methodology was firmly phenomenological and anti-reductionist. Certainly, his sociology is thoroughly typological, reflecting the influences of Troeltsch and Weber, but it is used in a way consistent with the methods of the Dutch phenomenologists, particularly van der Leeuw. His stress on religious intuition as the foundation for a genuine understanding through which accurate interpretations of religious data can be produced also thoroughly reflects a phenomenological approach. In addition, he voiced his opposition in many writings to a facile positivism by arguing for a numinous experience at the core of religion. In his sections on methodology in his book *Types of Religious Experience: Christian and Non-Christian* (1951), he makes a case forcefully for the *sui generis* nature of religion, which he interprets as referring to an essence which can be found nowhere else in human experience. Again, in a way reminiscent of van der Leeuw, he defines religious experience as 'the total response of man's total being' whereby 'he confronts a *power* greater than any power which he controls by his own wit or strength' (emphasis his) (1951: 33–5). The study of religion thus, in its individual and collective forms, describes the universal human response to a sense of the numinous, which, because it is expressed in diverse ways, the scholar of religion classifies and interprets systematically. It is only by acknowledging the compatibility between the scholar's own deepest religious sentiments and the religious responses of human beings to a sense of the numinous that a scientist of religion can organize and interpret religious data in such a way that leads to genuine understanding. This combination of attitudinal empathy with what might be called 'insider' hermeneutical principles places Wach firmly within the phenomenological tradition.

Mircea Eliade: History and phenomenology in the hermeneutic tradition

As I noted at the outset of this chapter, no figure has exercised such an extensive influence over the academic study of religions in North America, and arguably elsewhere, as Mircea Eliade. Even today, some 20 years after his death, his writings are read and debated among students of religion in numerous international settings. Eliade continued in the same hermeneutical line as Wach, but he did so in a much more profound manner by outlining a formal structure of religion, which, when applied to culturally specific situations, he believed would enable scholars to achieve an accurate interpretation and understanding of religious expressions everywhere. It was by exposing the structure of the religious consciousness that Eliade made the almost limitless number of religious symbols around the world comprehensible according to one fundamental pattern.

Eliade was born in 1907 in Bucharest, Romania. In 1925, he enrolled in the University of Bucharest, where he read philosophy. At the age of 21, he went to India, to study at the University of Calcutta. Although he remained in India for just three years, his experience there had a lasting effect on his eventual theory of religion, particularly the six months he spent living in an *ashram* in Hardwar, a place of pilgrimage in the Himalayas. In his autobiography (1981: 202–3), Eliade attributes his Indian experience with helping him discern 'the leading role played by symbols and images, the religious respect for the earth and life, the belief that the sacred is manifested directly through the mystery of fecundity and cosmic repetition'. He also credits the understanding he gained of 'aboriginal Indian spirituality' with helping him appreciate 'the structure' of his own Romanian culture with its rich symbolism and iconography. After the Second World War, Eliade moved to Paris where he completed manuscripts for three of his most influential books, *Patterns in Comparative Religion* (first published 1949), *The Myth of the Eternal Return* (first published 1949) and *Shamanism: Archaic Techniques of Ecstasy* (first published 1951, and translated from the French into English in 1964). Most of the central themes delineating his theory and method in the study of religions were expressed in these early publications. In 1956, Eliade delivered the Haskell Lectures in the University of Chicago which were published in 1958 as *Birth and Rebirth: The Religious Meaning of Initiation in Human Culture* and which in 1965 were issued under the title *Rites and Symbols of Initiation*. In 1958 he assumed the Chair of the History of Religions there, as Wach's replacement, a post he held until his death in 1986. After coming to Chicago, he continued to write widely, making further important contributions to the study of religion. Some of his most important later books, many of which underwent numerous reprintings, include *The Sacred and the Profane* (1957, first English publication 1959), *Myth and Reality* (1963), *The Quest: History and Meaning of Religion* (1969) and *Australian Religion: An Introduction* (1973). In addition, he made numerous contributions to collections on methodology, one of the most significant of which was 'Methodological Remarks on the Study of Religious Symbolism', which appeared in a book he edited with Joseph Kitagawa entitled *The History of Religions: Essays in Methodology* (1959).

Interpreting hierophanies

Eliade described the scholar of religions as a 'hermeneutist' (1959: 91), who primarily is concerned with religious orientation around symbols of meaning. For Eliade, the key word that helps the scholar unlock the meaning of religious symbols is the 'hierophany', the manifestation of the sacred, which locates for the religious person (*homo religiosus*) points of orientation around sacred centres. Based partly on Rudolf Otto's idea of 'the holy', Eliade contended that the 'sacred' is unknown and unknowable in itself, but is revealed through manifestations in profane space and time (1987: 9–10). These manifestations, the hierophanies, constitute the subject matter of the history of religions. In *Patterns in Comparative Religion*, Eliade (1996: 29) explains: 'This paradoxical coming together of sacred and profane, being and non-being, absolute and relative, the eternal and the becoming, is what every hierophany, even the most elementary, reveals'.

Eliade's theory of religion, that which is replicated in all cultural contexts, depicts the religious person as focused on a time when the world came into being through an initial creative act of sacred manifestation. The religious person imagines a primeval moment, before the foundation of the world, dominated by the terror of profane homogeneity, where there were no indications of sacred orientation. In a homogeneous universe, everything is the same; no points of demarcation can be located. This is equivalent to being lost, where a person cannot identify any familiar landmarks and experiences utter despair and hopelessness as a result. In a like manner, for the religious person, homogeneity, the inability to recognize sacred points of orientation, results in a sense of absolute meaninglessness and total chaos. In the beginning, *in illo tempore*, sacred intrusions broke into the homogeneity of space and time revealing what would otherwise remain unknown and unknowable, and providing life with meaningful points of cosmic orientation by 'founding' the world (1987: 20–2). As religions develop in history, these primordial hierophanies become expressed symbolically chiefly through cosmogonic myths and their ritual re-enactments (1996: 416). The history of religions thus becomes a study of sacred manifestations, uncovering how they have been enshrined in myths and how they are brought into the present through rituals. For the religious person, myth and ritual are replete with symbolic meaning and provide the scholar with the tools necessary for interpreting religious experience.

Because religion primarily is about orientation, certain symbols recur in various forms throughout the world and across history. These primarily have to do with cosmic centres, which connect the layers of the world, the upper levels reaching to the heavens and hence to the gods and the lower levels extending to the foundations of the earth, often inhabited by murky figures, devils and demons, what Eliade calls 'the infernal regions' (1987: 36–7). Such centres, as we have observed, result from hierophanies, but Eliade explains that these 'not only project a fixed point into the formless fluidity of profane space, a center into chaos', they also open 'communication between the cosmic planes (between earth and heaven)' (1987: 63). This enables the religious person to enter what Eliade calls an 'ontological passage from one mode of being to another' (1987: 63). This is why myths and symbols frequently refer to natural objects extending to the sky, such as mountains, trees, birds, sun and moon. It also explains why the shaman constitutes such a central and universal religious figure, since the shaman primarily travels to the upper and lower worlds, in some cultures, such as in the Altaic regions of inner Asia, by climbing the cosmic tree towards the highest level of the gods (1989: 190–200). Although some hierophanies do not convey meanings beyond their own cultural contexts, the universal pattern whereby the sacred discloses itself can be discerned everywhere.

Clearly, on this model, Eliade has constructed a dichotomy between the sacred and the profane, what he calls 'the dialectic of the hierophany' (1996: 13), or more broadly elsewhere, 'the dialectic of the sacred' (1989: 32). To understand this, we need to note that anything at all can become a hierophany, a conduit for manifesting the sacred, but not everything does. A particular entity becomes sacred precisely because it has manifested what otherwise would remain unknowable. A mountain may be selected because it is the highest in the region and hence nearest the abode of the gods, or a tree may be identified because of its unusual shape indicating the presence of a mysterious force.

Eliade explains: 'What matters is that a hierophany implies a *choice*, a clear-cut separation of this thing which manifests the sacred from everything else around it' (emphasis his) (1996: 13). The sacred object also possesses a certain ambivalence, since potentially it can be dangerous if it is not treated in a prescribed fashion, or if it becomes polluted by contact with profane objects. Its ambivalence is also enhanced by its mundane character; a stone or a tree remains what it is even while manifesting the 'wholly other'. By understanding the dialectic of the hierophany, the scholar gains insight into the way the religious person apprehends and experiences the world, and thus is able, as a hermeneutist, to disclose for academic understanding the structure of the religious consciousness.

Cosmogonic myths and ritual re-enactments in Eliade's hermeneutical method

As we have just noted, the two most important categories for understanding religion for Eliade are myth and ritual, including the ways each category is linked to the other. This relationship is seen vividly in archaic and primitive religions, but less so in modern times, since 'modern man's originality, his newness in comparison with traditional societies, lies precisely in his determination to regard himself as a purely historical being, in his wish to live in a basically desacralized cosmos' (1966: ix). Eliade defines myth as narrating a sacred, as opposed to a profane, history: 'It relates an event that took place in primordial Time, the fabled time of the "beginnings"' (1975: 5). Its main characters are supernatural beings, who are recounted in the myth as bringing reality into existence, 'be it the whole of reality, the Cosmos, or only a fragment of reality' (1975: 5). In this sense, every myth speaks of origins, either cosmogonic, telling of the creation of the world, or as an origin myth, relating how humans have become as they now are, often, as for example in the case of the Inuit peoples of the northern circumpolar regions (whose name is translated tellingly 'the real people') in specific cultural contexts (1987: 42–7; see also Cox, 1991). For Eliade, therefore, all myths relate 'how something was produced, began to *be*' (1975: 6), and for this reason, they primarily communicate events surrounding hierophanies; they 'describe the various and sometimes dramatic breakthroughs of the sacred ... into the World' (1975: 6). Myths become in this sense ontologies, defining what it means for a human to *be* in a world that has resulted from the sacred interventions described in the myths (1987: 63–4). Because they are ontological, myths construct paradigms for human behaviour, explaining not only how humans came to be as they are, 'mortal, sexed and cultural', but also serving as 'the exemplary model for all significant human activities' (1975: 6). Eliade's view of myth thus is entirely consistent with his hermeneutical task, which is to make clear the structure of the religious consciousness, since myth teaches the religious person 'the primordial "stories" that have constituted him existentially' (1975: 12).

If myths tell of origins and construct being, it follows that the most important rituals bring the religious person back to the beginnings; they reconstitute life, as it were, 'fresh from the creator's hands' (1987: 65). This becomes particularly clear in the way new year rituals are celebrated almost universally. In new year festivals, the cosmogonic myth is re-enacted, so that participants experience an entirely new world beginning. 'Each New Year', Eliade explains, 'begins the Creation over again' (1975: 41). Of course, new year rituals diverge dramatically according to the type of societies to which the myth relates, displaying marked differences, for example, between agricultural economies and nomadic forms of subsistence. They also vary according to the degrees of complexity within their social organization. Nonetheless, 'there is always a cycle, that is, a period of time that has a beginning and an end' (1975: 42). New year festivals also include a period in-between when the old has not totally passed away and the new has yet to come into being. The in-between period represents a return to Chaos, a time when the homogeneity of space and time reappears in the destruction of the old order before the creation of the new. Examples of this phenomenon include 'extinguishing fires, expelling "evil" and sins, reversal of habitual behavior, orgies, return of the dead' (1966: xiii). The re-creation of the world then follows, usually in the form of symbols of the cosmogony, such as 'lighting new fires, departure of the dead, repetition of the acts by which the Gods created the world, solemn prediction of the weather for the ensuing year' (1966: xiii). Eliade says the simplest example of a new year ritual is found among the Australian aborigines, where the origin myth literally is re-enacted each year: 'The rock paintings, which are believed to have been painted by the Ancestors, are repainted in order to reactivate their creative force, as it was first manifested in mythical times, that is, at the beginning of the World' (1975: 43).

Rites of initiation provide another type of repeated and formal social activity that demonstrates how myths of origin are re-enacted, in this case, to effect a change from one state of being into another. Eliade says that such rituals are 'equivalent to a basic change in existential condition'; the novice emerges from the rite 'with a totally different being from that which he possessed before his initiation' (1966: x). The most consistent and prominent example of this type of ritual in primitive societies is the transition from childhood to adulthood that is marked by 'a series of initiation ordeals', which ensure that a young person at the stage of puberty assumes his or her role as a responsible member of the society. During the stage of transition, when the rites of initiation are being performed, the initiate learns about societal rules, sexual obligations and compulsory adult behaviour. Yet, for Eliade, instruction in the rules and mores of the society comprises just one function of the rite, but certainly not its most important one. More fundamentally, the novice 'learns the mystical relations between the tribe and the Supernatural Beings as those relations were established at the beginning of Time' (1966: x). In other words, the rites of initiation teach the young person about the society's myths of origin and become thereby much more than instruction as it is understood in contemporary Western education. Knowledge is not about information in primitive and archaic societies, but about reexperiencing in the ritual the myths of origin, whereby Supernatural Beings have constituted the world '*in the beginning*, in the Time of the myths' (emphasis his) (1966: xi). In its paradigmatic function, the myth becomes embodied in the rite of initiation, so that the novice emerges from the rites, not only 'knowing' how the world came to be, but understanding how the myth has laid the foundation for 'all social and cultural institutions' (1966: xi).

By disclosing the structure of the religious consciousness through the dialectic of the sacred, and through its primary expressions in myth and ritual, Eliade attempted to overcome what the Dutch scholar, J. G. Platvoet, calls the 'hermeneutical predicament', the problem of conveying 'the correct perception, interpretation and translation of the social, cultural or historical data studied' (Platvoet 1988: 2). Eliade insisted that for a hermeneutical approach to generate understanding, scholars must recognize religions as 'spiritual universes'. Without this, they simply collect data which 'serve to augment the number, already terrifying, of documents classified in archives, awaiting electronic computers to take them in charge' (1969: 70–1). Such a recognition involves far more than acknowledging that religious people experience the world in a spiritual way; it entails the necessary insight for scientific interpretations of otherwise inscrutable symbols. Eliade calls this 'an immediate intuition' which enables the student of religion to understand religious symbols as 'ciphers' of the world (1959: 98). For example, the Cosmic Tree, which is found widely within primitive societies, cannot be understood unless it is seen in its totality as revealing the world as living, 'periodically regenerating itself and, because of this regeneration, continually fruitful, rich, and inexhaustible' (1959: 98). Without a sensitivity to this religious apprehension of the world, the scholar is unable to 'decipher' what would otherwise remain hidden, that for the religious person, human life comes 'from "another part", from far off; it is "divine" in the sense that it is the word of the gods or of supernatural beings' (1959: 98).

The structure of the religious consciousness can be made objective through the categories of sacred and profane, and in the typologies of myth and ritual, but the symbolism which infuses both cannot be grasped by the scholar nor interpreted to outsiders without the scholar adopting an entirely empathetic attitude towards the religious world view. To understand in the fullest sense is to articulate that, for the religious person, symbols convey reality, meaning and being. The religious person thus longs to be as near as possible to the sacred, to the moment when everything became new, and the only way to do this is to re-experience the creation by telling the story of beginnings and re-enacting it in powerful and symbolic ritual dramas. The scholar simply cannot communicate the potent strength of a religious symbol without understanding it religiously: 'it is for him to restore to it all the meanings it has had during the course of its history' (1959: 105).

Eliade's place in the study of religions

Thus far, we have seen Eliade operating very much as a phenomenologist of religion. He has outlined the main structure of religion, identified its key typologies, emphasized the hidden nature of its symbols and stressed that none of these can be understood without a fully empathetic approach to the religious apprehension of the world. It remains for us to consider two further points: 1) how Eliade, who always called himself a historian of religion, related to phenomenology and 2) how a 'religious' apprehension of the world, which he believed was most enshrined in primitive societies and archaic religions, exposed the profanity within the contemporary Western desacralization of human experience. Both points are important to clarify Eliade's place in the academic study of religions.

We have met consistently in our previous discussions, starting as far back as Chantepie de la Saussaye and C. P. Tiele, a scholarly preoccupation with delineating various branches within the academic study of religion. We have seen in the Dutch school how Kristensen, van der Leeuw and Bleeker each sought to carve out separate places for the phenomenology, history and philosophy of religion, and distinguished each from theology. In Eliade, we find the academic study of religion persistently couched in terms of history rather than phenomenology (Pettersson and Åkerberg, 1981: 41). Bryan Rennie, one of Eliade's foremost academic advocates, has argued that 'the actual texts of Eliade's work emphasize *meaning*, not phenomenology' (emphasis his) (1996: 3). To underscore Eliade's anti-phenomenological position, Rennie cites what he calls Eliade's 'approving citation of Raffaele Pettazzoni's statement that "the only way to escape the dangers" of a phenomenological interpretation "consists of constantly referring to history"' (1996: 3). Despite this, many scholars, such as Douglas Allen (1978), Joseph Bettis (1969) and António Barbosa da Silva (1982: 142), maintain that in all but name Eliade was a phenomenologist. This has prompted Rennie to charge that 'Allen overemphasizes phenomenology' (1996: 3). It may be that Rennie himself is overemphasizing the antagonism between history and phenomenology since, as David Cave (1993: 26) notes in his discussion of Eliade, 'religious data can only be understood as historical data'.

One of Eliade's clearest statements concerning the relationship of the history of religions to other approaches in the study of religions is found in the introductory comments to his extensive and highly influential volume on shamanism. Because the shaman can be studied from many angles, Eliade distinguishes his method as a historian of religion from sociology, psychology and ethnology, each of which, he acknowledges, is 'indispensable to understanding the various aspects of shamanism' (1989: xi). Yet, the historical approach, which Eliade contends had largely been absent from studies of shamanism up until the time of writing his volume, promises to add a critical and essential dimension to an overall understanding of shamanism as a widespread and fundamental religious phenomenon. By examining Eliade's declared approach to the study of shamanism, we see clearly how he used history in a way that was entirely consistent with a broad interpretation of the phenomenology of religion, confirming Joseph Bettis's conclusion that Eliade 'applied broad phenomenological methods to the study of the history of religions' (1969: 1).

In his introduction to *Shamanism*, Eliade constructs a sharp distinction between history and historiography, the latter referring simply to recording events chronologically. History, on the other hand, extends to 'the philosophical and general meaning' of the facts, which are understood religiously as hierophanies. Even in their simplest forms, hierophanies always refer to an 'eternal new beginning', which is 'atemporal' and hence entails 'a desire to abolish history' (1989: xvi–xvii). The specific task of the historian of religions thus is 'to decipher the properly religious meaning of one or another fact' even if this does not follow 'the chronological course of historiography' (1989: xvii). A hierophany that appears in the present moment is 'structurally-equivalent to a hierophany a thousand years earlier or later' (xvii). That this reflects a phenomenological rendition of history, hardly distinguishable from van der Leeuw's 'ideal types', is confirmed in Eliade's observation that 'the hierophanic process' repeats 'the same paradoxical sacralization of reality ad infinitum' (1989: xvii). It is precisely this structural and typical process that 'enables us to understand something of a religious phenomenon and to write its "history"' (1989: xvii).

Hierophanies, as manifestations of the sacred in history, refer beyond themselves. This is why any historical account of religion is not sufficient without acknowledging the significance of the sacred. To return to the case of the Altaic shaman, Eliade notes that historical influences, particularly on the shaman's ascent to the sky on the branches of the World Tree, can be traced to the spread of ideas into central Asia from the ancient Near East. The historian can document the dissemination of ideas and their adaptations into various cultural situations, but it is unwarranted to conclude that the source of the shamanic ascent has come from outside central Asia. Certainly, the actual Altaic practice as it was observed and recorded by ethnologists in the late nineteenth century incorporated modifications under the influence of outside cultures, but the celestial ascent proper 'appears to be a primordial phenomenon'. As such, it belongs not to one culture alone, but to humanity as a whole as witnessed by 'dreams, hallucinations, and images of ascent found everywhere in the world, apart from any historical or other "conditions"' (1989: xiv). For Eliade, therefore, the work of the scholar of religions is 'not solely "historical"' because documenting and organizing 'historico-religious facts' only makes sense when they are presented as revealing 'boundary-line situations of mankind' (1989: xiv). Scholars from other disciplines, such as psychology or sociology, interpret religious phenomena from within their own specialized approaches, as they have done in the case of shamanism, but only the historian of religion makes 'the greatest number of valid statements on a religious phenomenon *as a religious phenomenon*' (emphasis his) (1989: xv).

Eliade claims further that the history of religions differs markedly from the phenomenology of religion because the phenomenologist 'rejects any work of comparison' by 'divining' the meaning of 'one religious phenomenon' (1989: xv). The historian, by contrast, does not arrive at an interpretation of meaning until 'after he has compared it with thousands of similar or dissimilar phenomena, until he has situated it among them'. Only 'by making use of all the historical manifestations of a religious phenomenon' can the historian 'discover what such a phenomenon "has to say"' (1989: xv). This is an entirely idiosyncratic interpretation of phenomenology since, at the very least, for the phenomenologists of religion we have studied thus far, the creation of typologies emerges from comparative studies and facilitates comparison in just the way Eliade ascribes to the history of religion. We have just seen that Eliade's interpretation of history, as opposed to historiography, is ahistorical, in the sense that the interpretation of meaning requires the scholar to discern the fundamental structure of religion according to the dialectic of the sacred. In so far as Eliade intuits the formal structure of religion from historical data, in every way he is operating as a phenomenologist. It was precisely this principle he affirmed when he declared: 'The historian of religion ... attempts to decipher whatever transhistorical content a religious datum reveals through history' (1989: xv).

A second important consideration relevant to understanding Eliade's place in the study of religions results from his analysis of what he called the 'archaic man' as the prototypical religious person. António Barbosa da Silva (1982: 196) argues that Eliade equates the term *homo religiosus* in an ideal sense with the 'archaic man' to designate the 'purest form of experience of the Sacred'. This is entirely consistent with Eliade's idea that in the religious imagination at some primordial time the sacred broke into the homogeneity of space, providing it with points of orientation and meaning. This is also consistent with Eliade's theory that by re-enacting the myth of origin through rituals of renewal, the religious person is brought back persistently to the moment of creation. To understand the archaic mind, Eliade argued, the scholar of religion must 'place oneself inside it, at its very center' (1987: 165), gaining thereby insight not only into the structure of the religious consciousness, but equally into the paucity of spirituality now present within contemporary Western society.

The fundamental character of the 'religious man of the archaic societies' is formed under the conviction that 'the world exists because it was created by the gods'. This awareness transforms human experience because it means that the world 'is neither mute nor opaque, that it is not an inert thing without purpose or significance'. The religious person lives in a sanctified cosmos, one, Eliade says, that 'lives' and 'speaks' (1987: 165). He adds: 'Whatever the historical context in which he is placed, *homo religiosus* always believes that there is an absolute reality, *the sacred*, which transcends this world but manifests itself in this world, thereby sanctifying it and making it real' (emphasis his) (1987: 202). This can be contrasted to the experience of 'a non-religious man', who 'refuses transcendence, accepts the relativity of "reality", and may even come to doubt the meaning of existence' (1987: 202–3). It is possible, Eliade admits, that such a person existed in archaic societies, 'but it is only in the modern societies of the West that nonreligious man has developed fully' (1987: 203).

The process under which the modern non-religious person evolves is one Eliade calls 'desacralization', which suggests an historical relationship between archaic man and the contemporary profane man. The primal religious sentiment persists even in the most secularized situations. Eliade says, 'The majority of the "irreligious" still behave religiously' and they retain 'a large stock of camouflaged myths and degenerated rituals' (1987: 204–5). This can be seen in the ways non-religious people continue to observe new year festivities or mark the movement to a new home, or celebrate marriages, births or attaining an advanced position in society. Eliade asserts: 'Strictly speaking, the great majority of the irreligious are not liberated from religious behavior, from theologies and mythologies' (1987: 205–6). Eliade thus sees the current beliefs and practices characteristic of the modern human as degenerated forms of religion. This can be called an historical process since 'the profane man ... cannot utterly abolish his past' (1987: 204). This process can even be discerned by studying the paradigmatic religious figure, the shaman, who by means of ecstatic trances travels to other worlds and communes directly with the gods and objects of sacred power (1989: 6–7). Shamans are paradigmatic human beings precisely because they are in the fullest sense hierophanies, the elect who are ' "lived by" the religious form that has chosen them (gods, spirits, ancestors, etc.)' (1989: 32). Yet, the shaman, as a mediator who incarnates the sacred, only becomes necessary when other human beings lose their power to do precisely what the shaman does and thus need this commanding figure to intervene for them to secure their well being. This notion, the 'degeneration of the shaman' thesis (1989: 505–6), corresponds with Eliade's general theory of religion, that at the moment of creation, the human being stood, mythologically at least, face to face in uninterrupted communion with the sacred. Although the religious person always resacralizes experience through myth and ritual, the process of desacralization has been ongoing from the beginning of time.

The historical connection to an archaic past is demonstrated by the symbolism that emerges out of the human unconscious. Eliade refers to 'impulses that come from the depths of ... being, from the zone that has been called the "unconscious"'. Images materialize from this zone that exhibit 'astonishing similarities' to the symbols found in religious myths (1987: 209). Eliade is careful not to reduce myth to psychological processes at this point, but he asserts that 'the contents and structures of the unconscious are the result of existential situations ... and this is why the unconscious has a religious aura' (1987: 210). It is here we find evidence that C. G. Jung's idea of the 'archetypal images', present universally in the human unconscious, strongly influenced Eliade. With Jung, Eliade contends that the existential crises of modern humanity can be overcome only by returning to what Eliade calls the 'archaic levels of culture' where '*being* and the *sacred* are one' (emphasis his) (1987: 210). In this way, religion becomes for Eliade, as it was for Jung, a therapeutic resolution to a sense of meaninglessness and despair. Since the unconscious is universal and results from 'countless existential experiences', Eliade concludes that 'religion is the paradigmatic solution for every existential crisis' (1987: 210). It is paradigmatic for life, just as the shaman provided the model for the ideal religious person, 'not only because it can be indefinitely repeated, but also because it is believed to have a transcendental origin and hence is valorized as a revelation from an *other*, transhuman world' (emphasis his) (1987: 210). In a manner that is not so clearly articulated in Jung's writings, for Eliade, the benefit of religion goes beyond therapy, precisely because it focuses on the transcendent. Eliade explains: 'The religious solution ... makes existence "open" to values that are no longer contingent or particular, thus enabling man to transcend personal situations' (1987: 210).

We see clearly from this analysis that Eliade's hermeneutics provides a thoroughly religious interpretation of the world. His investigations into 'religio-historical facts' were intended not only to enhance the understanding of religion academically, but in the end, to commend religion as the only way to achieve a sense of meaning in life and to attain what he called 'access to the world of spirit' (1987: 210). On the basis of this overwhelmingly 'religious' interpretation of the study of religions, Eliade's place as a scholar in the phenomenological tradition seems secure. How far this has cast him personally into the role of a pseudo-theologian raises issues not only for evaluating Eliade generally, but for considering more broadly how successful phenomenologists as a whole were in separating their methods from Christian theology. I will consider the significance of this question further in the next chapter, but additional light will be shed on this by considering the approach of Jonathan Z. Smith, one of Eliade's principal colleagues and chief critics in the University of Chicago.

Jonathan Z. Smith: Classifications and comparison in the study of religions

I am aware that by including a section on Jonathan Z. Smith in a book on key figures in the phenomenology of religion, I risk committing the kind of classificatory error Smith himself exposes, based either on vagueness (2000: 38) or, more seriously, making a wrong choice about 'generic matters' (1987: xi). Certainly, in a much more precise sense than we find in Eliade, Smith fits into the mould of a historian of religions. Nevertheless, the way he uses history as a hermeneutical device suggests that he follows broadly in the tradition of Eliade, and therefore justifies my decision to include him among the 'Chicago school' of phenomenologists. In one of his discussions of Eliade, Smith writes of his relationship to his colleague as 'the stance of the pygmy standing on the giant's shoulders'. He adds that 'the giant has taught all of us how and what to see; and, far more important how to understand what we have learned to see' (1978: 90). He suggests that his own theories have resulted from queries that have arisen 'for one who understands himself to be standing within Eliade's work' (1978: 90). Smith currently is the Robert O. Anderson Distinguished Service Professor of the Humanities in the University of Chicago. His website describes him as a 'historian of religions', but emphasizes that his interests have ranged widely to include 'ritual theory, Hellenistic religions, nineteenth-century Maori cults, and the notorious events of Jonestown, Guyana' (http://divinity.uchicago.edu/faculty/profile_jsmith.html). In this section, I begin by examining Smith's argument that the academic study of religions historically has suffered from an overwhelmingly Christian bias before analysing specifically his critique of Eliade. I will then outline what I consider to be Smith's main theoretical contributions to the study of religion: the emphasis on religion as location and the principle of incongruity in the comparative study of religions.

J. Z. Smith on the Christian bias in the study of religions

In many of his writings, Smith contends that religion has been studied, classified and interpreted historically in the West under a strongly Christian theological bias resulting in a largely unrecognized and unacknowledged Christian scheme having been imposed on religious studies. In lectures delivered in 1988 at the School of Oriental and African Studies in London, he asserted that 'the issue confronting the enterprise of religious comparison is not so much one of lack of data, as one of inadequate theory ranging from matters of classification to more complex matters of interpretation and explanation' (1990: 118). To illustrate this, he analysed the category 'soteriology', which he describes as having been classified by scholars in dualistic language, such as 'saved or not-saved' with its sub-classifications, 'this-worldly' salvation or 'other-worldly' salvation (1990: 118–19). He then asks in what sense it is legitimate to interpret religion in general as primarily soteriological and why it then becomes logical for scholars to create the next division based on salvation that occurs in this world or the next. The answer is found in the Christian origins to academic studies in religion. During the early centuries of the Christian Church, theologians debated which type of soteriology best applied within Christianity, the triumph over death in the form of resurrection ('this-worldly') or the attainment of eternal bliss in heaven ('other-worldly'). Although for scholars of religion, such questions should simply be 'taxonomic', the emphasis on soteriology and its principal types has resulted directly from Christian theological disputes (1990: 120, 142). Another example showing how a Christian bias has influenced scholars of religion, becomes clear when the hegemonic Protestant interpretation of Christian history as having been founded in a pristine state but as having been corrupted over time is seen as a backdrop for various theories of the degeneration of religion. Smith asserts that 'the variety of comparative endeavours, their theoretical goals and methodological entailments ... have not been recognised' within academic contexts (1990: 142).

Another signal of Christian bias in the study of religions can be seen in the way value has been assigned to categories used for comparative purposes. The academic task of classifying religions, although steeped in methodological difficulties, would appear on the surface to entail an analysis of which categories to select and which types of religious practices to include or exclude within each category. In theory it would be possible to discover a belief or action that did not correspond to any other, and thus could be called 'unique'. To locate such a religious phenomenon outside other categories carries with it no value judgement, since the production of a taxonomy of religious behaviours, undertaken correctly, would simply classify relationships between species according to variations of difference and similarity. To be unique thus is not a point for pride, but simply a way of noting how one member of a particular class relates to all others (1990: 37). The Christian bias in the study of religions has missed this point, attributing to uniqueness a sign of superiority based on an evolutionary grading of religions from lower to higher. This has resulted in 'the most fundamental classification of religions' as being constructed in terms of ' "ours" and "theirs", often correlated with the distinction between "true" and "false", "correct" and "incorrect" ' (2000: 39). This perspective has transformed the idea of uniqueness from a taxonomic qualifier to an ontological category. The 'unique' in this sense 'expresses that which is *sui generis* ... and, therefore, *incomparably* valuable' (emphasis his) (1990: 38).

This same criticism can be applied to the way the widespread category 'world religions' has been understood and used in academic contexts. Smith contends that a world religion is usually defined as a 'religion like ours' (1978: 295), presumably with written scriptures, an identifiable organization, religious practitioners and oftentimes a missionary outreach. Again, he implies, this is not just a taxonomic error, but entails clear power relations. Foremost, a world religion is 'a tradition which has achieved sufficient power and numbers to enter our history, either to form it, interact with it, or to thwart it' (1978: 295). Religions that do not resemble Christianity or those that have not engaged with the West in significant ways are disregarded under terms like 'primitive' or 'minor'. To support this view, he cites statistics he derived from a recent encyclopaedia providing details of the number of adherents to the world's religions. Included on the list are Christianity, Islam, Hinduism, Confucianism, Taoism, Shinto, Judaism, Primitive Religions, and Others or None. Numbers attributed to the categories 'primitive' and 'others or none' comprise more than any other religion on the list, apart from Christianity, prompting Smith to observe that 'more than one fifth of the world's population has just been informed that religiously they have no identity and might as well not exist' (1978: 296). Smith concludes that what is needed in the comparative study of religions is not so much a 'revised taxonomy' as a recognition that the history of comparative religions 'has been an enterprise undertaken in bad faith' where scholarly interests 'have rarely been cognitive, but rather almost always apologetic' (1990: 143).

Smith's critique of Eliade

As I indicated above, Smith regarded Mircea Eliade as one of the 'giants' in the academic study of religions, but many of the errors he attributes generally to scholars of religion he ascribes to Eliade specifically. He devotes most of the first chapter in his book *To Take Place* (1987) to criticizing Eliade's theory of religion as orientation, primarily the way Eliade depicts cosmic centres, symbolized by such things as trees, mountains or temples, as places of communication between the divine and human worlds. Smith selects Eliade's case of the Australian aboriginal people, the Tjilpa, as recorded in *The Sacred and the Profane*, for detailed analysis. Eliade's choice of the Tjilpa is intended to demonstrate how a 'primitive' people organized their lives around a mythic cosmic centre, symbolized by the 'world pole'. According to Eliade, the Tjilpa trace their ancestry to a divine being called Numbakulla who created the first Tjilpa ancestor, established their social institutions and 'cosmicized their future territory' (Eliade, 1987: 33). After completing these acts of creation, Numbakulla selected a gum tree, changed it into a 'sacred pole', anointed it with his blood and climbed on it up to the sky, where he disappeared. For Eliade, the pole 'represents a cosmic axis', since, as a nomadic people, the Tjilpa carry it with them wherever they go in order to bestow a territory they enter with meaning and thereby turn it 'into a world'. The ritual significance of the pole from these actions becomes apparent: 'During their wandering the Achilpa [Tjilpa] always carry it with them and choose the direction they are to take by the direction toward which it bends' (Eliade, 1987: 33). Not only does the pole provide orientation for the Tjilpa as they move across the landscape, it also maintains constant mythic connection 'with the sky into which Numbakula [Numbakulla] vanished' (Eliade, 1987: 33). If the pole is broken, it signifies a cosmic break and 'denotes catastrophe', like the 'end of the world' (Eliade, 1987: 33). Citing the nineteenth-century ethnographers, Spencer and Gillen, Eliade recounts how at one time the pole was broken causing confusion and consternation among the people. 'They wandered about aimlessly for a time, and finally lay down on the ground together and waited for death to overtake them' (Eliade, 1987: 33). Eliade concludes that this example supports his fundamental theory of religion: 'Life is not possible without an opening toward the transcendent; in other words, human beings cannot live in chaos' (Eliade, 1987: 34).

Smith subjects this example to a scrupulous analysis and a biting criticism on many counts, including the very loose way Eliade uses his sources and the misleading manner in which he presents the material. When reading Eliade's account, Smith notes, we are led to believe that the Tjilpa continue up to this day to orientate themselves around the cosmic pole, whereas the facts indicate that at the very most this is a mythic rendering of events that occurred during the famous aboriginal 'dreamtime'. 'Numbakulla's creation, his withdrawal, the Tjilpa's wanderings, the breaking of the pole, the corporate death – all occurred within the myths of the "Dreaming"' (Smith, 1987: 3). In other words, the evidence cited by Eliade is, in Smith's words, 'story, not history'. Smith then notes that Eliade summarizes the Tjilpa myth in just one paragraph, whereas the accounts from which it was derived, a 1927 'hybird' of an original 1896 version based on a summary by Spencer and Gillen, covered a full 34 pages. Smith (1987: 3–4) calls this 'a remarkable act of compression'. It also becomes clear, when comparing the 1927 publication (on which Eliade relied) with the original, that in the earlier text the name Numbakulla did not refer to a single deity but, in

Smith's words, 'to the autochthonous class of "totemic ancestors" ' (Smith, 1987: 4). In the later publication, this generic title is transformed into a creator deity and 'High God'. This has important ramifications for Eliade's theory, since in the earlier version reference is made to ancestors whose normal way of disappearing was not into the sky but into the earth. Smith concludes that the pole could not have symbolized a cosmic centre connecting the Tjilpa with the sky, since no such cosmology was implied until the later versions, which by then had been highly Christianized. Poles were used widely among Australian aboriginals, but not as symbols connecting the people to a high god (Smith, 1987: 5).

Smith has gone into such detail in this case to demonstrate how Eliade misconstrued the Tjilpa accounts in order to support his prior assumption that the Tjilpa, like all religious people of primitive and archaic traditions, operate 'within a celestial and transcendental context' (Smith, 1987: 10). Smith's analysis of this case conveys at least three implications for any reading of Eliade. First, if Eliade could use the material relating to the Tjilpa so loosely, we might question his many other works which rely almost exclusively on secondary accounts drawn widely from historical, ethnographic and anthropological literature. Second, Smith implies that Eliade's prior commitment to a world view, where religions everywhere direct attention to cosmic centres, motivated him to fit the data to conform to his theory of religion rather than testing his theory against the data. Third, Smith corrects Eliade's practical and theoretical errors by insisting on accurately presenting source material by studying it in its context, both culturally and in light of the potentially distorting biases employed by those, such as early anthropologists or ethnographers, who have written the accounts. This stress on cultural specificity and factual accuracy are evident when he concludes that symbols of cosmic centres, which Eliade repeatedly emphasizes, admittedly may be found in some cultures, but 'the horizon of the Tjilpa is not celestial, it is relentlessly terrestrial and chthonic' (Smith, 1987: 10).

Smith's theories of location and incongruity in religion

Despite his rejection of Eliade's claim that cosmic centres are found universally in religions, in Smith's own theories of religion the concept of location features prominently. In this sense, he follows an Eliadean tradition by interpreting religion in terms of space, but he gives space not a transcendental, but a political, significance. In support of this, he addresses Eliade's strongest case, that of Babylonian religion and in particular the myth of *Enuma elish*. In *The Sacred and the Profane*, Eliade used this myth as a primary example of a new year festival that re-enacted a combat 'that took place *ab origine* and put an end to chaos by the final victory of the god' (Eliade, 1987: 77). Smith argues that Eliade has borrowed from the 'Pan-Babylonian School', which based its interpretation of a central cosmic mountain on philological evidence that 'has all but evaporated' (Smith, 1987: 16). Smith does not deny that central geographical focal points are found in the Babylonian texts, but he insists that 'the language of "center" is preeminantly political and only secondarily cosmological' (Smith, 1987: 17). In the case of the *Enuma elish* myth, Smith contends that it is improper to call this cosmogonic, since it is primarily about building, 'the creation of the holy city of Babylon and the construction of its central shrine, Esagila' (Smith, 1987: 19). Again, we see the language of politics operating, since the myth talks about succeeding generations of gods constructing royal dwellings. After a detailed analysis of the texts and reference to the discovery in 1961 of missing sections of the myth, Smith concludes that the building that is described is not primarily concerned with constructing a cosmic centre, as Eliade thought, but with establishing kingship. Eliade's fabled time of the beginning, when the homogeneity of space was broken by sacred intrusions, crumbles under Smith's argument that place becomes central, precisely because it marks the location where 'a king or god happens to have decided to take up residence' (Smith, 1987: 22).

This critique of Eliade's interpretation of the *Enuma elish* myth confirms Smith's central objection to Eliade's method of selecting facts to support his preconceived notion of religion. Smith, however, does not reject Eliade's dualistic analysis, since he also relies on a dichotomy to explain how religions politicize space. Rather than pitting the sacred against the profane, Smith prefers to analyse two religious world views, the 'locative' and the 'utopian' (1990: 121) from which he deduces his own typologies of religion. The locative world view, as the word suggests, is concerned with 'keeping one's place' and with 'reinforcing boundaries'. He explains: 'The vision is one of stability and confidence with respect to an essentially fragile cosmos, one that has been reorganized, with effort, out of previous modes of order and one whose "appropriate order" must be maintained through acts of conscious labour' (1990: 121). Smith classifies locative world views as 'religions of sanctification', which, he says, emphasize 'emplacement as the norm', and, if transgressed, require social and ritual acts of rectification. Such societies tend to be stratified rigidly, as under imperial rule, but a clear 'democratic' demarcation of responsibilities is implied. For example, a 'king-god maintains distinctions between sky and earth', a 'divine king maintains the order of his cities' and others perform sometimes menial work according to their assigned tasks. A frequent myth in the locative world view is the hero who fails, since this figure steps outside accepted boundaries and 'must learn to keep place' (1990: 121–2).

A utopian world view values being 'in no place' and is marked by an open, dynamic society (1978: 101). Rather than conformity to the norm, such visions of the world stress 'rebellion and freedom'. People are encouraged to 'challenge their limits, break them, or create new possibilities'. Smith calls this a 'centripetal world which emphasizes the importance of periphery and transcendence' as opposed to a 'centrifugal view of the world' in which 'each being has its given place and role to fulfill' (1978: 100–1). By way of example, Smith (1978: 102) cites two anecdotes reported by Plutarch of a meeting of Alexander the Great with the 'naked sages of India'. The first story tells how Alexander 'threw down upon the ground a dry and shrivelled hide'. As Alexander placed his foot on the edge of the hide, air was forced to other parts causing the hide to rise up. He went around the entire edge of the hide pressing his foot down, only to have the hide pop up in other places. Finally, Alexander stood in the centre of the hide 'and lo! It was all held down firm and still'. The second anecdote relates how, on seeing Alexander, the Indian sages stamped their feet, which was interpreted to Alexander to mean that 'every man can possess only so much of this earth's surface as this we are standing on'. For Smith, these two stories illustrate the locative–utopian dichotomy: 'On the one hand the structures of the Center and conformity to place as represented by the Indian sages; on the other, Alexander the world conqueror, the utopian, relentlessly and restlessly testing the boundaries of the cosmos and seeking to transcend all limits'. Smith concludes that 'the alternation, the discoveries and choices of and between' the locative and utopian world views represent 'the history of man and the history of religions' (1978: 102–3).

The locative–utopian dichotomy leads to Smith's second important contribution to the study of religions, his emphasis on incongruity. Scholars of religion, he suggests, have been most interested in the locative world view, largely because it 'is a map of the world which guarantees meaning and value through structures of congruity and conformity' (1978: 292). It also results from the political domination within hierarchical cultures of those 'who had a deep vested interest in restricting mobility and valuing place' (1978: 293). For this reason, an emphasis on the centre predominated in scholarly accounts since students of religion often relied on texts that were produced by scribes working in a temple with its royal court. The ritual repetition of the texts ensured the stability of the priesthood, enshrined the power of the ruling elite and provided the propaganda necessary to maintain the role of the king, who was depicted as the 'guardian of cosmic and social order' (1978: 293). That such a conservative ideology had so overwhelmingly influenced approaches to the study of religion in its various forms – phenomenological, functional and structuralist – Smith says, led him to consider incongruity as a new way of interpreting myths and rituals.

By incongruity, Smith is referring to a method in the study of religions that looks for difference, rather than similarity, often expressed in ritualized contexts as a contrast between an ideal situation and an actual one. Although the incongruent contexts about which Smith is speaking frequently relate to life-sustaining or life-threatening situations, oftentimes they are manipulated through humour or in practical jokes. Smith cites many examples to illustrate what he means. One is of the Marind-anim people of South New Guinea as reported by the Dutch anthropologist, Paul Wirz, in the 1920s. One of the chief pastimes of the Marind-amin, according to Wirz, is to identify one's clan by reference to the shape of an individual's navel. A convex shape resembles a betel nut, and hence the totem assigned to that person designates the betel nut clan. A person with a bulging navel, again because of its appearance, is attributed to the coconut clan. Smith cites approvingly Wirz's conclusion that this is 'all mere play', but the subject itself is utterly serious since clan identification constructs a whole series of social obligations and prohibitions. Smith asserts: 'What is funny, what is interesting, what is provocative is the juxtaposition between the actual clan membership and the "theoretical" clan membership induced by the empirical science of navel-study' (1978: 298–9). A second example concerns the initiation of the Aranda people of Australia, where initiates are told that the sound of the bull-roarer (a piece of wood with a slit in it producing a whirring sound) is in fact the voice of Tuanjiraka, described by Smith as 'a monstrous being' who is the source of all suffering. The pain of circumcision during the initiation is attributed to Tuanjiraka, whose voice can be heard in the swinging of the bull-roarer. After the rituals of initiation are completed, the elders inform the initiates that they must 'abandon belief in Tuanjiraka and understand that Tuanjiraka is only this piece of wood' (1978: 301). According to Smith, scholars of religion traditionally would have regarded this admission as a degeneration of an original religious experience (as in Eliade) or as only appearing to be a deception to an outsider who misses the religious significance of the act (in a phenomenological way). Smith concludes that 'it is precisely the juxtaposition, the incongruity between the expectation and the actuality that serves as a vehicle of religious experience' (1978: 301).

The principle of incongruity aids scholars particularly as they interpret myths and rituals. Myths obtain their power in Smith's view precisely because they underscore the difference between idealized events described in myths and real life situations experienced in the present (1978: 300). Since a myth can cover any ground or contain any content, it is capable of universal application. Yet, its universality is made relative by the situation to which it is applied and the context from which it emerges. Hence, myth combines into one what Smith calls a 'fit' and a 'no fit', and it is just this that 'gives rise to thought' (1978: 300). To illustrate how ritual conveys incongruity, Smith refers to a 'set of traditions which are usually labelled "hunting magic" in which a discrepancy exists between what hunters say they do when they hunt and what they actually do, a discrepancy that is raised to thought in rituals that enact a perfect hunt' (1978: 302). An example of this is found in the bear festival of northern (Siberian, North American Arctic and Scandinavian) traditional hunters. During the actual pursuit of a bear, the hunter recalls rituals that idealize a reciprocal relationship with the animal. He knows that he is expected to invite the bear to offer itself for the kill by asking it to turn around, to offer it words of thanksgiving for agreeing to be killed, to avoid excessive spilling of blood and afterwards to treat the body with respect. What actually happens is quite different. A hunter faced with a life-threatening situation or with the prospect of losing the kill will not perform such acts. Nor will the bear respond to the hunter by turning around and sacrificing itself to the hunter's arrow. In the bear festival as performed in the village, however, the ideal situation is replicated. The bear is 'compelled to rejoice in its fate, to walk to its death rather than run away, to assume the correct posture for its slaughter, to have the proper words addressed to it before it is shot, and to be killed in the proper, all-but-bloodless manner (1980: 126). In this way, ritual, like myth, reveals incongruity. Ordinary life is marked by 'contingency, variability, and accidentality' but through myths and rituals it is granted 'a significance which the rules express but are powerless to effectuate' (1980: 127; see also Cox, 1998: 67–70).

From this brief summary of Smith's many contributions to the academic study of religions, we can draw some basic conclusions about his place in what I am calling the 'Chicago school' of phenomenology. Initially, he should be understood as a scholar who, in the tradition of phenomenology, has criticized prior methods in the study of religions, in Smith's case, by stressing the need for clarity in the use of classificatory language and by exposing the Christian biases that have informed the choice of such language. He also urges careful attention to the sources for religious analysis, particularly historical, textual and anthropological, by placing both the sources and the subject matter they treat under close scrutiny in the light of contextualized socially and culturally constructed facts. His emphasis on sacred place stresses not the cosmological elements of religions, but their political aspects, and he argues in the same vein that the comparative study of religions should proceed not by searching for similarities but for patterns of incongruity emerging from and reinforcing either a locative or utopian world view. In these ways, Smith reformulated the cosmological and theological assumptions hidden within the Chicago herme-neutical tradition to fit his own reading of religion based on careful analyses of a series of powerful, but culturally relative, political choices.

The phenomenology of Wilfred Cantwell Smith

The final thinker I consider in the broader North American phenomenological tradition is the Canadian scholar, Wilfred Cantwell Smith (1916–2000), who has figured centrally in discussions on theory and method in the study of religion, as I noted at the beginning of this chapter, since the publication in 1962 of his landmark book, *The Meaning and End of Religion*. After completing his undergraduate studies in the University of Toronto with a degree in classical and Semitic languages and after studying theology in Cambridge University in England, Smith went in 1940 for six years to Lahore, India (now Pakistan), where he taught Indian and Islamic history at the Forman Christian College. He completed his Ph.D in Princeton University in 1948 in Islamic studies and the following year was appointed to McGill University in Montreal as the W. M. Birks Professor of Comparative Religion. Three years later, he founded the Institute for Islamic Studies at McGill, which continues as an important centre for scholarly research on Islam to this day. In 1964, Smith moved to Harvard University under an appointment as Director of the newly founded Center for the Study of World Religions, a post he held until 1973, when he returned to Canada to establish the Department of Religious Studies in Dalhousie University in Halifax. Smith's early works, such as *Modern Islam in India* (1943) and *Islam in Modern History* (1957), applied his specialization in Islam to contemporary issues of his day, but increasingly his interests shifted more to methodological considerations in the study of religions. This is evident in a series of publications he produced during his latter years in McGill, including an important contribution to the 1959 Eliade and Kitagawa publication on methodology in the history of religions under the title, 'Comparative Religion: Whither – and Why?' (1959: 31–58) and a small book written primarily for a popular audience, *The Faith of Other Men* (1963).

In this section, I outline Smith's key methodological premises, beginning with his discussion of faith as reified through history into the term 'religion', a term he insisted must be abandoned. I follow this by analysing Smith's interpretation of the faith–beliefs distinction and conclude by noting the implications of his dialogical approach to the study of religions, an approach which prompted Frank Whaling (1999: 246) to describe Smith's mature thinking as pioneering a new venture called 'a theology of religion'. I base my interpretations of Smith largely on his contribution to the Eliade and Kitagawa volume and *The Meaning and End of Religion* (1964), and additionally on two important books he wrote towards the end of his career, *Faith and Belief* (1979) and *Towards a World Theology* (1981).

The reification of faith into religion

The most significant concept for understanding religion, for Smith, is 'personal faith', which, he argued, cannot be defined, but which constitutes the 'locus' of religion (1964: 168). In *The Meaning and End of Religion*, Smith outlines in painstaking detail how the concept 'religion', as it is now portrayed in Western academic circles, has resulted from a process of reification, which has transformed faith from a living and dynamic factor in human existence into an objective system of beliefs, values and practices. In the nineteenth and twentieth centuries, when the comparative study of religions was born and grew in significance, Western scholarship began treating living faiths as if they were static, unchanging systems, the contents of which could be outlined under self-contained headings, such as Hinduism, Buddhism, Mohammedanism, Confucianism, Taoism and so on. This followed directly from applying the so-called laws of nature, as they had come to be understood since the advent of the Western Enlightenment in the seventeenth century, to human behaviour. Humans were regarded as objects and their institutions increasingly were analysed according to the same principles of objectification (1964: 50).

Understanding religions as total, self-contained systems of belief and practice thus is a relatively recent invention in human history, having resulted largely from Enlightenment thinking. Smith notes that in the Roman era, the Latin term *religio* conveyed an adjectival meaning rather than a nominative one with a substantive content (1964: 23). Hence, in ancient Rome, *religio* referred to sacred places, or devout people, that which was 'secondary to persons or things rather than ... things in themselves' (1964: 23). Early Christianity, likewise, constituted itself as a community of 'faith', which implied a certain attitude towards life and the transcendent, one marked by 'piety, reverence, devotion', with ramifications for 'every aspect of the believer's life, moral, social, intellectual, as well as liturgical' (1964: 27). Later, the Latin *religio* was used in the church as a term to designate ritual observances as well as the structural organization, the *ecclesia.*

This carried with it the sense of constructing boundaries between true and false, but it was not true and false belief so much that mattered as 'worshipping' God in a true way as opposed to worshipping false deities (1964: 29). The epitome of the early Christian understanding of religion is found in the attitude of St Augustine, for whom religion 'is a vivid and personal confrontation with the splendour and the love of God' (1964: 31). As Christian history progressed towards the Reformation, particularly under the influence of John Calvin, Smith asserts that the term 'Christian religion' was used with more frequency, but this can be misunderstood easily. Calvin's *Institutes* referred to 'instruction, instituting, setting up' and thus not to 'an overt institutional phenomenon nor an abstract system'. For Calvin, instruction in the Christian religion encouraged 'the sense of piety that prompts a man to worship' (1964: 37). It is only after the seventeenth century, when the process of reification took hold in earnest, that the term religion began to be applied to the 'Christian religion' and eventually referred to Christianity as one of the world's religions alongside other systems of belief and practice. Smith's review of Roman and Christian history thus was undertaken to demonstrate that religion in the West traditionally 'referred to something personal, inner, and transcendentally oriented, ... the nearest equivalent concept in modern English' being 'that of piety' (1964: 37).

Smith notes that today the concept religion is employed in popular, theological and academic contexts in four main ways. The first reflects the traditional usage, as in 'she is very religious', and refers to a quality of faith, piety or devotion. The second and third senses designate overt systems of beliefs, practices and values, either as an ideal (this is 'genuine' or real Christianity, Islam, Hinduism) or as historical or sociological phenomena (this is how Christianity, Islam, Hinduism are practised in reality). Finally, religion can be spoken of as a generic term in order to distinguish it from other aspects of life, such as the psychological, sociological, economic and political. Smith concludes that in each of these instances the term 'religion' should be 'dropped' since it is 'confusing, unnecessary and distorting' (1964: 48). Certainly, the experiences of faith, devotion and piety cannot be reified without distorting their meaning and significance. It is far more important to understand that people can 'be' religious, without ever employing the term 'religion' (1964: 22). The same can be said for different reasons about constructing religion as a generic term. After all, it is impossible to separate some objectified substance labelled 'religion' from the remainder of life. In the second and third senses, it is both confusing and distorting to distinguish an ideal from a real system of beliefs and practices, and to suggest that sociological and historical contexts describe reality as opposed to an ideal towards which adherents strive. Smith concludes:

> I have become convinced that the vitality of personal faith, on the one hand, and, on the other hand (quite separately), progress in understanding even at the academic level – of the traditions of other people throughout history and throughout the world, are both seriously blocked by our attempt to conceptualize what is involved in each case in terms of (a) religion. (Smith, 1964, 48–9)

Smith then reviews the category 'religion' as it applies outside the West and is led to the same conclusion. Everywhere, religion conceived of as an overt system of beliefs and practices is unknown. Only in the case of Islam is it possible to suggest that the self-designated term 'Islam' refers in the Qur'ān to a system of beliefs which every human being should embrace. It is portrayed in Arabic by the term *dīn*, which seems to refer in one sense to piety or devotion, but also carries with it the meaning of one religious system as opposed to other systems, that is, one's own which is true and all others false. Smith asserts: 'The Muslim world, then, is definitely and explicitly conscious of something that it calls, and is persuaded that it ought to call, a religion' (1964: 77). Smith devotes a great deal of space to analysing why this is so, and discovers that, although the meaning of *dīn* in seventh-century Arabia cannot be translated precisely into the modern meaning of religion (1964: 93), a process of reification was ongoing nonetheless in Arabia as in the West (1964: 78–80). In addition, there is the further complication that Islam regards itself as a religion of revelation in which God has spoken literally, even if the Word must be translated into a human language (1964: 95). This adds a sense of uniqueness and distinctiveness in Islam with respect to all other 'religions'.

Closer analysis, however, uncovers a more complicated situation. The word *islām* is used rather rarely in the Qur'ān, with the text preferring 'more dynamic and personal terms' (1964: 101). For example, the word 'God' is written 2,697 times, but *islām* is used just eight times. More importantly, for Smith, when the Qur'ān speaks of a relationship between God and human beings, 'the great term and concept is "faith"' (*īmān*) (1964: 101), 'which appears 45 times in the Qur'ān'. Variations of faith appear even more times, such as in the phrase 'man of faith' or 'faithful' (1964: 101–2). When the noun, *islām*, is compared with its version as a verb, *aslama*, which occurs 72 times in the Qur'ān, the deeper meaning of the nominative form is shown as referring to actions: submitting, surrendering, giving oneself in total commitment (1964: 102). Smith concludes from this terminological analysis that 'the Qur'ān is concerned, and presents God as being concerned, with something that persons do, and with the persons who do it, rather than with an abstract entity'(1964: 102). In this sense, the actual word *islām* should be regarded as a 'verbal noun', referring to the religion of those who surrender themselves to God. Faith (*īmān*) too is a 'verbal noun, since it includes the testimony to belief in Allah', a testimony that consists entirely of 'engagement' and involves 'the ability to see the transcendent, and to respond to it; to hear God's voice, and to act accordingly'(1964: 102). The so-called special case of Islam as a religion in the systematic sense thus evaporates. 'Vivid and dynamic – and personal: these are the qualities of the term *īslām* in the Qur'ān. What was proclaimed was a challenge, not a religion' (1964: 103).

Personal faith, cumulative tradition and the faith–beliefs distinction

At the core of what is meant by religion, we find everywhere the experience of faith, which under various historical and cultural influences has been objectified

198

progressively into systems of belief. Smith famously overcomes the problem of academic reification in *The Meaning and End of Religion* by suggesting that scholars need to understand, as I noted at the outset of this discussion, that the locus of what they call religion in all cases is the 'personal faith' of human beings, a faith directed towards the transcendent but expressed in outward and observable forms, such as myth, ritual, community, law, ethics, scripture and belief (1964: 141). These outward and observable forms display the dynamic character of faith operating through history as 'cumulative tradition' through which the devotees within the traditions collectively have expressed their faith in a transcendent reality. Smith's emphasis on the cumulative nature of the traditions is intended to avoid the very criticisms of reification he has so carefully identified and so roundly condemned as a distortion of inner faith. Since traditions build on the past, by introducing over time new expressions of faith, they testify to the living character and dynamic nature of faith. Scholars can never penetrate into the core of personal faith, since by definition it is personal (although not individualistic) (1981: 47, 62); hence, they are limited to observing and describing the outward manifestations of inner commitment. Nonetheless, by integrating the relationship between inner experience and outer expression into their analyses, scholars produce accurate and empathetic descriptions of the expressions of faith, which, at the same time, are replete with historical detail. Smith summarizes just this position in an oft-quoted section from *The Meaning and End of Religion*:

> Every religious person is the locus of an interaction between the transcendent, which is presumably the same for every man (though this is not integral to our analysis), and the cumulative tradition, which is different for every man (and this is integral). And every religious person is the active participant, whether little or big, in the dynamics of the tradition's development. (Smith, 1964: 168)

He adds emphatically: 'This, then, I make bold to suggest, is how what used to be called a religion actually works in human history' (1964: 168).

The important distinction between faith and belief, which I first described in my discussion of A. G. Hogg as a formative influence in the phenomenology of religion, is dealt with in detail in Smith's book, *Faith and Belief* (1979; reprint 1998), which he confesses took him 12 years to write (1998: acknowledgements). Smith stresses, as he did in *The Meaning and End of Religion*, that faith, although universal, is expressed through the many phenomena of religion, which include, but by no means are restricted to, beliefs. Faith, of course, cannot be defined in a quantifiable sense, since it reflects chiefly a human 'engagement' with what any tradition values most. Smith explains: '"Faith" then, I propose, shall signify that human quality that has been expressed in, has been elicited, nurtured, and shaped by, the religious traditions of the world' (1998: 6). He admits that this leaves faith 'unspecified', but he explains that we know where to look for it in the cumulative tradition and it is just this which makes academic enquiry possible. We see faith in:

the involvement of the Christian with God and with Christ and with the sacraments ...;
the involvement of the Hindu with caste and with the law of retributive justice and the
maya-quality of this mundane world and with the vision of a final liberation; the
involvement of the Buddhist with the image of the Buddha and with the moral law and
with an institutionalized monastic order and with the final dream of a further shore
beyond this sea of sorrow; the involvement of the primitive animist with the world
perceived in poetic, if bizarre, vitality and responsiveness. (Smith, 1998: 5–6).

Since faith is expressed differently in the traditions, it is important to determine
the precise relationship of belief to faith. Smith asserts that belief, as it is now
understood, refers to ideas, concepts and intellectual formulations of transcendence (1998: 12). Historically, however, it implied much more than a cognitive
idea; it referred to a testimony of faith, or at the very least to an act of believing
'with faith'. In this sense, belief gives shape and content to the deepest hopes and
aspirations of human beings, but the way it operates and the emphases it receives
vary greatly among religious traditions. For example, when Christians first
encountered people of other traditions, they wanted to know what they believed.
Smith says, 'Since they themselves believed something, they presumed that others
would too' (1998: 13). This reflects, according to Smith, a rather peculiar
Christian approach, since belief has played such an important part in Christian
faith resulting in systematic theologies and long debates over doctrinal
formulations. In other traditions, belief has not been a primary mode for
expressing faith. 'What theology is to the Christian Church, a ritual dance may
be to an African tribe; a central formulation of the human involvement with final
verity' (1998: 15).

Does this mean, at the highest level that, although the expressions of faith
differ, all faith is the same? Smith responds that this is not his thesis. He admits
that there are 'several types of faith' even within 'the arena of a single system of
belief' (1998: 11). Moreover, faith can change over time and vary quite
dramatically according to personal or historical circumstances. Nevertheless, a
case can be made for faith across all religious boundaries bearing some of the
same characteristics and as being much more similar than the outward
expressions that can be observed in the varying traditions. 'There is less
difference between the faith of Christians and that of Muslims and of Hindus
than there is among the formulae and symbols by which that faith is visibly
expressed' (1998: 11). Smith does not explain what causes faith to differ in type,
since in a fundamental sense, it is too personal to examine, but he implies that it
can be distinguished by its 'relation to its explicit and particular "object"' (1998:
7). The object of faith does not reveal faith itself, but indicates 'where it is
directed', for example, to God, to Dharma, to Christ and so on (1998: 7). It is at
this point that Smith almost falls into conceptualizing the object of faith, and
thus ascribing differences in type to differences in belief, much in the way A. G.
Hogg had done. Smith backs away from this view, however, because in the end
he regards the emphasis on beliefs as evidence of a secularizing process in
Western history that finally undermines faith by depersonalizing it, objectifying
it and making it susceptible to the rules of empirical investigation.

When belief comes to dominate faith, the sense of transcendence is lost. This is why a belief-orientated interpretation of religion reinforces the secularizing tendencies of the modern West. All judgements in this new world view are made on external matters, with little awareness being shown to the internal forces that gave rise to the external expressions. This leads Smith to launch into a litany of abuses he attributes to what he calls the 'new non-transcendence-oriented culture (the first such in humanity)' (1988: 143). It has 'drained' transcendence out of people's perceptions. Its terms are dictated by scientific methods so that experiences of transcendence can be reduced to empirical testing. Belief in this way has become 'the category of thought by which skeptics, reducing others' faith to manageability, translated that faith into mundane terms' (1988: 144). Religion as belief has been manipulated into 'an intellectual instrument for secularizing one's understanding of the human' (1988: 144). In a final scathing condemnation of modern Western secularism, Smith proclaims that 'believing' has become 'a category of thought calculated to denature the religious life' (1988: 144). It is clear from this series of emotive statements why Smith so vehemently insisted on a firm demarcation between faith and belief, and why the locus of religion must be seen as residing in a personal faith directed towards transcendence, and, further, why it must never be reduced to any of its manifestations in the cumulative tradition, particularly to its expression in belief.

Smith's move towards self-reflexivity

Smith's method for studying people's faith and their expressions of faith was informed by his underlying commitment to a religious world view. In his contribution to the Eliade and Kitagawa volume on methodologies, published in 1959, Smith anticipated his later argument about reification by observing that the study of religion in Western scholarship currently is undergoing a fundamental transformation from regarding religion impersonally as an 'it' to a more personal understanding of faith. Smith (1959: 34) explains that the first stage in personalizing the study of religions has already occurred, since scholars have begun describing personal faith in terms of what people, referred to as 'they', say, do or believe. This is even now advancing to a deeper level whereby scholars are becoming aware of their own involvement with those they are studying, so that 'we' are now talking about what 'they' say, do or believe. Smith then urges scholars to advance to the next phase in the personalization process by adopting a dialogical approach so that the 'they' is changed to a 'you' and the study becomes one of 'we' talking to 'you'. If this is accomplished, a scholar will finally understand that the study of human faith requires breaking down the old subject–object dichotomy so that the one doing the studying and the one being studied merge into a common enterprise consisting of ' "we all" ... talking with each other about "us" ' (1959: 34). The culmination of the dialogical approach results in the recognition that 'in comparative religion man is studying himself' (1959: 55).

In *Towards a World Theology*, published 22 years after his contribution to the

Eliade and Kitagawa volume, Smith enlarged on the dialogical method, calling it 'corporate critical self-consciousness' (1981: 59–60). By this he meant a form of self-reflexivity whereby the scholar adopts a 'critical, rational and inductive' self-conscious approach to the study of a community of persons, which he described as being comprised of at least two people, the one doing the studying and the one being studied (1981: 60). The community, what Smith called earlier the 'we' talking to 'us', becomes aware of 'any given particular human condition or action as a condition or action of itself as a community' (1981: 60). In other words, when scholars engage in a study of religion they include themselves as humans in their investigations as well as participants in the communities they are studying. This implies that the scholar experiences and understands the conditions or actions he or she is studying simultaneously, both subjectively as participant and objectively as observer. In this way, the subjective experience of the scholar, comprising a personal and existential involvement much like faith, is united with objective knowledge, which adopts an external, critical, analytical and scientific perspective (1981: 60). The results of scholarship are verified in this way both subjectively and objectively, experientially and empirically.

In his contribution to the Eliade and Kitagawa book, Smith (1959: 52–3) described the verification process, in line with his dialogical method, as two-fold and potentially three-fold. What a scholar says about any particular religion must be 'intelligible and acceptable' to those within the religion and, at the same time, 'intelligible and acceptable' to those outside the religion. Since scholars mostly work within the academic tradition, what they report must be intelligible and acceptable to the scholarly community, but, at the same time, to the religion about which they are speaking. A potential third level occurs when two religious traditions are involved. For example, if an academic is studying both Christianity and Islam, what is presented must be intelligible and acceptable to three communities: the Christian, the Muslim and the scholarly. Smith observes, 'This is not easy, but I am persuaded that both in principle and in practice it can be done' (1959: 53).

In *Towards a World Theology*, Smith reiterated this method and expanded on the three-fold verification process, referring to it as 'the verificationist principle' of 'humane knowledge' (1981: 97). The principle is applied in three stages. The first requires that an outside observer's statement be acceptable to the faith community being studied. 'No statement about Islamic faith is true that Muslims cannot accept. No personalist statement about Hindu religious life is legitimate in which Hindus cannot recognize themselves. No interpretation of Buddhist doctrine is valid unless Buddhists can respond, "Yes! That is what we hold"' (1981: 97). The second part of the principle applies to the outside observer, so that what is said about the faith communities 'must satisfy the non-participant, and satisfy all the most exacting requirements of rational inquiry and academic rigour' (1981: 97). Finally, the third aspect applies to people of other faiths, so that no statement about Muslims, for example, can be regarded as true that non-Muslims cannot accept. No account of Hinduism can be legitimate if the Hindu's neighbours cannot recognize the Hindu in the accounts. 'No statement about Buddhist doctrine is valid unless non-Buddhists can respond, "Yes – now we understand what those Buddhists hold" '(1981: 97).

Smith offers an example of how the verificationist principle works by describing an Indian temple at Madurai in south India (1981: 62–3). He says that if we are to appreciate the temple in its full significance, we must 'get inside the consciousness of those for whom it is a sacred space' and experience 'how it feels and what it means to be a worshipper within it' (1981: 66). At the same time, we must investigate the temple objectively, learning the facts about its history. In addition, we need to understand its impact on the surrounding community, comprehend its place in the larger city of Madurai, and even learn to know 'how it is perceived ... by the small iconoclastic Muslim group in the area, for whom temple worship is a sin' (1981: 66). In a secular state such as India, it will also be important to become aware of the perceptions of the temple held by Marxists, 'whose analysis of its economic role is impertinent in one sense but not in both' (1981: 66). Corporate critical self-consciousness thus, if applied faithfully and rigorously, provides true and verifiable knowledge of the temple as a human institution. Smith calls this an example of achieving 'humane knowledge', the aim of which is not pure objectivity, but 'disciplined corporate self-conscious-ness, critical, comprehensive, global' (1981: 78–9), a form of knowing that collapses once and for all the subject–object dichotomy that for so long had dominated Western approaches to the study of human faith (1981: 78–9).

W. C. Smith as a theologian and a phenomenologist of religion

From his discussion of religion as a reified form of faith, through his dismissal of belief as evidence of a secularizing tendency in Western scholarship, to his insistence on the scholar of religion becoming part of the subject matter of religious studies, Smith has brought into the open the relationship between theology and phenomenology. His overall emphasis on the importance of understanding the transcendent element in religion, and the way he regarded personal faith as directed towards the transcendent, suggests that he regarded varying religious traditions as 'windows' through which people both appre-hended the transcendent and towards which they directed their faith. As we have seen, he presumed the transcendent was the same for everyone, although he suggested this was not integral to his argument. Nevertheless, that he asserted what appears to be an ontological reality behind personal faith, suggests his approach was broadly theological, derived from his liberal Christian and ecumenical background. This view is confirmed as early as 1959 in Smith's contribution to the Eliade and Kitagawa volume, where he not only advocated a radical form of empathy consistent with the phenomenological assertion that nothing human is alien to another human, but implied that a scholar must possess personal faith in order to engage in the dialogical study of religion. This had the effect of moving religious studies away from a science of religion towards inter-faith dialogue. Smith made just this point when he asserted:

I have argued that one cannot study religion from above, only from alongside or from within – only as a member of some group. Today the group of which the student recognizes himself as a member is capable of becoming, even is in process of becoming, world-wide – and interfaith. (Smith, 1959: 55)

By 1981, when he published *Towards a World Theology*, this aim had become even more explicit and personal:

My aspiration is to participate Christianly in the total life of mankind – the intellectual life, and the religious, as well as (as we already all do) the economic and political. And I invite others to do so Jewishly, Islamically, Buddhistically, or whatever. (Smith, 1981: 129)

Smith, nonetheless, was a phenomenologist. This is demonstrated clearly by his stress on empathy for the religious person and his hostility towards the reductionistic tendencies of positivistic science. His efforts to promote under-standing, beginning with the scholarly community, and extending to religious communities in general, also followed broad phenomenological principles. Yet, it was in the way he analysed religion by giving it a structure, with personal faith at the core, directed towards a transcendent object and expressed dynamically in history through typological categories, that finally establishes him firmly within the phenomenological tradition. That he was also a liberal, ecumenical Christian theologian was not incidental to his phenomenology, a relationship we have now seen emerging consistently among the scholars that comprise this study, with the exception of J. Z. Smith. It is perhaps Cantwell Smith's unashamed and forthright embracing of a theological perspective, however, that distinguishes him from other phenomenologists, who either, as in the case of the Dutch scholars, visibly delineated phenomenology from theology, or as in the case of Eliade, employed theological assumptions without making these explicit. It is fitting, therefore, that Cantwell Smith comprises the last figure I consider among the North American phenomenologists since one of the most important subsequent debates in religious studies that he in part generated centres around the accusation that the category 'religion' is a Christian theological construct which, in line with Smith's own suggestion, should be dropped, but for very different reasons than Smith advocated.

Conclusion

The North American 'school' of phenomenology, perhaps much more so than the Dutch and British 'schools', has blurred the disciplinary boundaries between a phenomenological approach to the study of religions and other academic approaches. As a result, I have been forced to justify why I regard each figure included in this chapter as a phenomenologist as opposed to their preferred self-designation as respectively a sociologist of religion (Wach), a historian of religion (Eliade and J. Z. Smith) or a comparativist in the study of religions (W. C. Smith). I have argued that the persistent themes that have been associated with phenomenology – bracketing out prior assumptions, employing a fully empathetic approach, identifying typologies, interpreting the meaning of religious behaviour and insisting that religion comprises a category in its own right – consistently appear in various forms in the writings of the scholars I have considered in this chapter. In this sense, I have not been forced to 'fit' them into my own prior definition of a phenomenology of religion; they have written themselves into it. Certainly, the substantive contributions to theory and method in the study of religion each has made reflect broadly similar academic assumptions, distinguished in the Chicago school by a consistent hermeneutical method and by W. C. Smith's deconstruction of the term 'religion' in order to safeguard its irreducible core.

The issues raised by the North American school, as I have just suggested in my discussion of W. C. Smith, extend beyond terminology and strike at the root of one of the most critical problems now confronting the academic study of religions. On the one side, we have seen lying beneath the surface in the writings of the North American scholars a resistance to academic strategies that would relegate the study of religions either to a sub-category of the social sciences or as a branch of Christian theology. Both are seen as dangerous, although in the cases of Eliade and W. C. Smith, it is clearly the encroachments of a certain type of secularizing 'scientism' into the study of religions that poses the greatest danger. J. Z. Smith, on the other hand, by embracing a culturally contextualized approach, seen historically in terms of political choices, would appear to regard the theologizing of the study of religions as the more potent enemy. We have seen that Wach sought to combine sociological and phenomenological methods, but in so doing hardly escaped a religionist perspective, and in many ways, can be regarded as far removed from sociological reductionism as Eliade or W. C. Smith. The methodological problems so clearly identified by the scholars I have placed in the North American school, and which were echoed by those in the Dutch and British traditions of phenomenology, have anticipated and helped to define the subsequent debates in the study of religions to which I now turn in the final section of this book.

References

Allen, D. (1978), *Structure and Creativity in Religion: Hermeneutics in Mircea Eliade's Phenomenology and New Directions* (The Hague: Mouton).

Barbosa da Silva, A. (1982), *The Phenomenology of Religion as a Philosophical Problem* (Lund: CWK Gleerup).

Bettis, J. D. (ed.) (1969), *Phenomenology of Religion. Eight Modern Descriptions of the Essence of Religion* (London: SCM Press).

Cave, D. (1993), *Mircea Eliade's Vision for a New Humanism* (Oxford and New York: Oxford University Press).

Capps, W. H. (1995), *Religious Studies: The Making of a Discipline* (Minneapolis: Fortress Press).

Cox, J. L. (1991), *The Impact of Christian Missions on Indigenous Cultures: The 'Real People' and the Unreal Gospel* (Lewiston, New York: Edwin Mellen).

Cox, J. L. (1998), *Rational Ancestors. Scientific Rationality and African Indigenous Religions* (Cardiff: Cardiff Academic Press).

Eliade, M. (1954), *The Myth of the Eternal Return* (New York: Pantheon Books).

Eliade, M. (1958), *Birth and Rebirth: The Religious Meaning of Initiation in Human Culture* (New York: Harper and Row).

Eliade, M. (1959), 'Methodological remarks on the study of religious symbolism', in M. Eliade and J. M. Kitagawa (eds), *The History of Religions: Essays in Methodology* (Chicago and London: University of Chicago Press), pp. 86–107.

Eliade, M. (1966), *Rites and Symbols of Initiation. The Mysteries of Birth and Rebirth* (New York: Harper and Row).

Eliade, M. (1969), *The Quest: History and Meaning of Religion* (Chicago: University of Chicago Press).

Eliade, M. (1973), *Australian Religions: An Introduction* (Ithaca, New York: Cornell University Press).

Eliade, M. (1975) [1963], *Myth and Reality* (New York: Harper Torchbooks).

Eliade, M. (1981), *Autobiography: Volume I, 1907–1937, Journey East, Journey West* (San Francisco: Harper and Row).

Eliade, M. (1987) [1959], *The Sacred and the Profane. The Nature of Religion* (San Diego, New York and London: Harcourt).

Eliade, M. (1988), *Autobiography: Volume II, 1937–1960, Exile's Odyssey* (Chicago and London: University of Chicago Press).

Eliade, M. (1989) [1964], *Shamanism: Archaic Techniques of Ecstasy* (Harmondsworth: Arkana Penguin Books).

Eliade, M. (1996) [1958], *Patterns in Comparative Religion* (Lincoln and London: University of Nebraska Press).

Kitagawa, J. M. (1959), 'The history of religions in America', in M. Eliade and J. M. Kitagawa (eds), *The History of Religions: Essays in Methodology* (Chicago and London: University of Chicago Press), pp. 1–30.

McCutcheon, R. T. (2003), *The Discipline of Religion: Structure, Meaning, Rhetoric* (London and New York: Routledge).

Pals, D. L. (1996), *Seven Theories of Religion* (Oxford and New York: Oxford University Press).

Pettersson, O. and Åkerberg, H. (1981), *Interpreting Religious Phenomena. Studies with Reference to the Phenomenology of Religion* (Stockholm: Almqvist and Wiksell Internatonal).

Platvoet, J. G. (1988), *A Concise History of the Study of Religions* (Harare: University of Zimbabwe, Department of Religious Studies, Classics and Philosophy (Internal Publication)).

Rennie, B. S. (1996), *Reconstructing Eliade. Making Sense of Religion* (Albany: State University of New York Press).

Sharpe, E. J. (1986), *Comparative Religion: A History* (London: Duckworth, 2nd edn).

Smith, J. Z. (1978), *Map is Not Territory. Studies in the History of Religions* (Leiden: E. J. Brill).

Smith, J. Z. (1980), 'The bare facts of ritual', *History of Religions*, 20, 112–27.

Smith, J. Z. (1982), *Imagining Religion: From Babylon to Jonestown* (Chicago and London: University of Chicago Press).

Smith, J. Z. (1987), *To Take Place. Toward Theory in Ritual* (Chicago and London: The University of Chicago Press).

Smith, J. Z. (1990), *Drudgery Divine. On the Comparison of Early Christianities and the Religions of Late Antiquity* (Chicago and London: University of Chicago Press).

Smith, J. Z. (2000), 'Classification', in W. Braun and R. T. McCutcheon (eds), *Guide to the Study of Religion* (London and New York: Cassell), pp. 35–44.

Smith, W. C. (1943), *Modern Islam in India: A Social Analysis* (Lahore: Minerva Book Shop).

Smith, W. C. (1957), *Islam in Modern History* (Princeton: Princeton University Press).

Smith, W. C. (1959), 'Comparative religion: Whither – and why?', in M. Eliade and J. Kitagawa (eds), *The History of Religions: Essays in Methodology* (Chicago and London: University of Chicago Press), pp. 31–58.

Smith, W. C. (1963), *The Faith of Other Men* (New York: Harper and Row). (Reprinted under the title, *Patterns of Faith around the World*, 1998, Oxford and Boston: Oneworld Publications).

Smith, W. C. (1964), *The Meaning and End of Religion. A New Approach to the Religious Traditions of Mankind* (New York: Mentor Books).

Smith, W. C. (1981), *Towards a World Theology. Faith and the Comparative History of Religion* (Philadelphia: Westminster Press).

Smith, W. C. (1998), *Faith and Belief: The Difference between Them* (Oxford and Boston: Oneworld Publications. First published as *Faith and Belief*, 1979, Princeton: Princeton University Press).

Waardenburg, J. J. (1973), *Classical Approaches to the Study of Religion: Aims, Methods and Theories of Research. Volume I* (The Hague: Mouton).

Wach, J. (1951), *Types of Religious Experience: Christian and Non-Christian* (Chicago: University of Chicago Press).

Wach, J. (1958), *The Comparative Study of Religions* (edited with an introduction by J. M. Kitagawa) (New York: Columbia University Press).

Wach, J. (1962) [1944], *Sociology of Religion* (Chicago and London: University of Chicago Press).

Wach, J. (1967), 'Introduction: The meaning and task of the history of religions (Religionswissenschaft)', in J. M. Kitagawa (ed.), *The History of Religions: Essays on the Problem of Understanding* (Chicago and London: The University of Chicago Press), pp. 1–19.

Whaling, F. (1999), 'Theological approaches', in P. Connolly (ed.), *Approaches to the Study of Religion* (London and New York: Continuum), pp. 226–74.

Chapter 7

Phenomenology at the Crossroads: Subsequent Debates in the Academic Study of Religions

From the beginning of this book, I have sought to demonstrate how influences emanating from philosophy, theology and the social sciences have played formative roles in the thinking of key figures in the phenomenology of religion. I have noted that scholars disagree as to how deeply Husserl's analysis of consciousness directly influenced the phenomenology of religion, but I have argued that the epistemological problem outlined by Husserl, based on the subject–object dichotomy, has defined the parameters within which phenomenologists of religion have approached their subject. Following Husserl's use of the *epoché* to suspend judgements about the 'natural attitude', they attempted to bracket out potentially distorting presuppositions stemming both from confessional Christian theology and from positivistic science in order that, by using empathetic methods, they could enter into the experiences of believers to achieve understanding-in-depth (*Verstehen*). The essence of the phenomena 'appeared' to the perceiver in Husserl's phenomenology; the essence or core of religion manifested itself for phenomenologists of religion through particular historical and social data, which they then organized into categories or types. The theory of religion as essence and manifestation, so critical for a phenomenological interpretation of religion, thus confirms a crucial influence derived from Husserl.

I argued in Chapter 2 that the theological school of Albrecht Ritschl exercised an important influence over phenomenologies of religion, particularly through the Ritschlian judgement of value, whereby religions could be compared according to the potential quality of the experience they made possible for their devotees and the ethical motivations they induced within them. This led, I suggested, to the distinction between faith and belief in the thought of A. G. Hogg, which was very closely followed and employed in the writings of Wilfred Cantwell Smith. When placed alongside Schleiermacher's description of religion as a feeling of absolute dependence on the infinite, Hogg, in true Ritschlian fashion, suggested that the comparative value of religions could be determined solely by the ability of a religion's beliefs to ensure a lasting, intense and satisfying experience of faith. At its core, faith, for Hogg, was universal; the specific experiences of faith, including their intensity and duration, were determined by beliefs. Hogg's idea, although not quite identical to Otto's notion of the numinous as the essence of religion, shared with Otto's position a commitment to the universal experience of faith, differing only according to the various ways individuals apprehended the numinous. With religion now identified as a universal faith born out of experiences of a numinous ('wholly other') core, the study of religion could be described as unique, irreducible to any other academic discipline, and requiring its own methods of investigation. As we have seen, this idea ran consistently through the writings of the phenomenologists of religion, from the earlier efforts by Kristensen to delineate phenomenology from history, theology and the social sciences to Ninian Smart's key tool for the study of religions, which he called 'methodological agnosticism'. The most forceful exponents of an irreducible core of faith as the defining characteristic of religion, of course, were Eliade and Cantwell Smith, but, I have shown how this idea was adapted by others in the phenomenological tradition.

In Chapter 3, I stressed that, although phenomenologists frequently reacted strongly against what they regarded as the reductionistic tendencies of the social sciences, important positive influences could also be identified. Against the background of Hegel's dialectic philosophy, Troeltsch and Weber conceived patterns or types as developing within and reacting to social and historical contexts. This idea was taken over strongly by Bleeker, through his concept of *entelecheia*, by Wach in his organization of religious behaviours into social categories, and it can be seen operating beneath the surface in the way van der Leeuw employed ideal types to make sense of religious data and in Cantwell Smith's interpretation of the history of religions as 'cumulative tradition'. It also becomes evident in the thought of Jonathan Z. Smith, who identifies places of attention as becoming sacred largely around theories of political power, an idea that can be read into the way Weber analysed the influence of Calvinism on the development of capitalist economic systems. The social sciences influenced phenomenologists in the British school, at one level, by prompting a negative reaction to descriptions of Africans as 'fetishists' who lacked any belief in God, but also positively in the ways, for example, Edwin Smith and Geoffrey Parrinder sought to ground their conclusions on rigorous linguistic and ethnographic research. As a historian, Andrew Walls carefully analysed the colonial background to the descriptions of religions in Africa, and more broadly throughout the non-Western world, in order to create a typology that encompassed all 'primal' peoples as the base upon which the 'world religions' were constructed. In a way quite different from Parrinder and Walls, W. C. Smith's hostility towards the Western reification of faith into 'religion' foreshadowed the post-colonial argument that 'religion' was invented in the West as a means of asserting power over the largely non-Christian, non-Western world. Finally, the emphasis on myth, symbol and the universal human unconscious in the writings of C. G. Jung made it possible for Eliade to posit that the archaic human being, the so-called primitive, is not deficient in religious understanding, but defines the prototypical religious person who can be contrasted with contemporary secularized humanity.

By the 1970s, serious methodological questions were being asked about the assumptions that had motivated many of the key thinkers in the phenomenology of religion I have outlined in this book. At the study conference to which I referred earlier held in 1973 in Turku, Finland on methodologies in the science of religion, sponsored by the International Association for the History of Religions, Lauri Honko (1979: xvii) observed: 'At the turn of the 1970s it was becoming clear that Western science, not least in the humanities and social sciences, was undergoing a profound process of self-examination, which seemed to be leading to the breakdown of certain older paradigms and even, possibly, to the emergence of new ones'. Armin Geertz and Russell McCutcheon (2000: 18) contend that the Turku Conference was an attempt to save the science of religion 'from its theoretical naiveté and encourage interdisciplinary cooperation'. Certainly, they were referring in large measure to the phenomenology of religion, which by the 1970s increasingly was being criticized for combining theological assumptions with a simplistic understanding of Husserl's phenomenology in support of its claim that religion constitutes a unique subject matter, irreducible to any other dimension within human experience. In the remainder of this chapter, following the order I set out in Chapters 1–3 of this book on the formative influences in the phenomenology of religion, I trace how these criticisms have developed into and have spawned three subsequent debates that currently are engaging scholars: 1) the debate over phenomenology as a viable philosophical method on which to base the study of religions; 2) the religion–theology debate; and 3) the debate over the socially engaged scholar of religion. At the conclusion of this chapter, I offer my own responses to these debates based on my conviction that the phenomenological tradition, interpreted in a broad and dynamic way, continues to offer important insights on some of the most critical issues now confronting students of religion.

The debate over the continued philosophical viability of the phenomenology of religion: The critique of Gavin Flood

New perspectives on the philosophical standpoints that originally motivated key thinkers in the phenomenology of religion have cast serious doubts over its continued usefulness as a methodological tool in the study of religions. The philosophical problem at the root of the phenomenology of religion, as I traced it in my discussion of Husserl in Chapter 1, has resulted from the classical formulation of the subject–object dichotomy. When translated into religious studies, this has been formulated in terms of the personal faith of the scholar of religion and how that faith influences the scholar's understanding and interpretation of religious data. We have seen that the sympathetic approach running consistently through the writings of key phenomenologists has been based on their own personal religious convictions. This was stated explicitly by Kristensen, who believed that students of religion could achieve understanding only if they had some experience of religion and that, by studying religion, their personal religious commitment would be strengthened. This same theme has

211

recurred repeatedly when the core of religion, depicted variously as 'personal faith' by W. C. Smith, 'the sacred' by Mircea Eliade or 'power' by Gerardus van der Leeuw, is described as being expressed through its observable manifestations. The manifestations can be described, classified and compared, but never understood apart from scholars possessing some personal sense of the essence of religion itself. In classical phenomenology, only by cultivating a feeling for religion, could the subjective observer enter into the object of study and thereby overcome the otherwise insurmountable distance between the scholar of religion and religious adherents. Of course, phenomenologists insisted that their interpretations of religions should not replicate what believers, or even their theologians, said about their traditions, but they maintained nonetheless that scholarly accounts should never be offensive to believing communities.

In other contexts, this has been called the 'insider–outsider' problem in the study of religion, a topic, as we have seen, that has been dealt with comprehensively in a recent collection of articles edited by Russell McCutcheon (1999), in which it becomes clear once again how theology has exercised a lasting influence over religious studies. The insider–outsider problem, however, extends beyond theology to fundamental philosophical issues, as has been demonstrated by Gavin Flood in his book *Beyond Phenomenology: Rethinking the Study of Religion* (1999). For Flood, the problem begins with Husserl, from whom the phenomenology of religion derived its fundamental analytic principles: ' "bracketing" (*epoché*), the "eidetic reduction", and "empathy" (*Einfühlung*)' (1999: 92). Flood argues that these concepts originally played a positive role in liberating 'the study of religions from theological dogmatism', but now they unduly limit 'the range of methodological possibilities within the study of religions' (1999: 93).

The first problem phenomenologists inherited from Husserl, according to Flood, is that by bracketing out 'the natural attitude', Husserl separated 'meaning' from 'existence' (1999: 97). When the natural attitude is in play, existence is taken for granted, but when it is put in brackets, 'the ego "abstains" from the world's affirmation or denial' (1999: 100). This procedure is undertaken in order that the meaning of the phenomena apprehended in perception can be understood directly by the observer, who has become freed by the *epoché* from considering the question of the existence of the phenomena. The central problem entailed in such a separation for Flood is that it privileges 'cognition over affect' and 'speech over writing'. Flood explains: 'This denial of the distinction between being and meaning is to privilege the status of language and the sign: all meaning is constructed within language and its referents cannot be separated from it' (1999: 101). Within the phenomenology of religion, the separation between existence and meaning contradicts the 'idea of phenomenological neutrality' because it 'is implicitly a denial of the being expressed in language' (1999: 102). Describing facts while remaining 'agnostic' about the 'truth content' or 'existence' of that to which the facts refer, in Flood's view, 'creates an immediate tension if the claim of phenomenology is that an accurate understanding of religious statements can be made by its method' (1999: 102). In other words, to withhold judgement about the truth claims of a religion follows precisely the Husserlian division between existence and meaning, but in the process it restricts understanding on a believer's own terms, since believers insist on asserting the very truth claims on which phenomenologists refuse to comment. 'There is ... a limit to phenomenology, and a barrier beyond which it cannot venture into the realm of insider discourse, because of its separation of meaning from existence' (1999: 102).

A second problem the phenomenology of religion inherited from Husserl, according to Flood, results from Husserl's emphasis on the 'transcendental ego' or 'unified subject'. For Husserl, the unified subject is at the same time particular and universal, operating under individual constraints but possessing a kind of universal rationality that combines passive receptions of the data of the world with an active ordering of them into coherent patterns. As I noted in Chapter 1, the transcendental ego, although particular and individual, asserts a common understanding of the world with others, or obtains intersubjectivity, through empathy. In the phenomenology of religion, this same process operates when the subjective observer, in this case the scholar of religion, is able to penetrate into the inner meaning of religious facts. This also follows Husserl's pattern whereby the scholar of religion (like the transcendental ego) observes passively the actual data of religion, but constructs them (again like the transcendental ego) actively into meaningful patterns, in this case into religious typologies. What is bracketed in this process also follows the Husserlian analysis because, according to Flood, it 'entails the idea of a universal, rational subject who does the bracketing and who is also the subject of religious experience' (1999: 107). The phenomenologist of religion overcomes the subject–object dichotomy by interrupting the 'isolated subjectivity' of the transcendental ego and by 'focusing upon the experience of the flow of consciousness or the self's intentional structure' (1999: 107). For Flood, this is seen clearly in the writings of Eliade, whose 'experience of the sacred' constitutes individual religious experience in particular situations, but at the same time delineates the structure of the religious consciousness generally (1999: 107).

The application of a Husserlian analysis of consciousness to religious studies has resulted in the overriding emphasis among phenomenologists on subjective states, conveyed in terms of numinous experience, faith or inner enlightenment. This is clear, as Flood has noted, in the case of Eliade, but it can be traced also to the work of Joachim Wach and before him to Otto and Schleiermacher where religion is defined 'as the subject's affective apprehension – or "creature feeling" – of the "wholly Other"'. For Flood, this turns the study of religion into a study of the structure of the religious 'consciousness' at the expense of what he calls 'intersubjective performance', by which he means studying the structure of religious acts, such as rituals or narratives (1999: 108). He argues: 'It is ritual structure and performed narratives which are primary in the transmission of traditions through the generations and not any individual experience or state of consciousness' (1999: 108). For example, he says, in the possession dances performed in Kerala, India, what is important for a scholar of religion to understand is 'intersubjectively agreed narrative performance', not the inner state of consciousness of the possessed dancers. The phenomenology of religion misses this because it is wed to the idea it imported from Husserl that 'assumes the universality of the rational subject ... who can, through objectification, have access to a truth external to any particular historical and cultural standpoint' (1999: 108).

Flood's own resolution to the problem of the phenomenology of religion inherited from Husserl has already been implied in his idea of 'intersubjective performance', which later in his book he relates to the work of Mikhail Bakhtin, the Russian philosopher of language and literature, particularly to Bakhtin's concepts of 'dialogism', 'utterance' and 'heteroglossia' (1999: 150). Flood explains that for Bakhtin the self is 'constructed purely in relationship' and it is relationship, rather than Husserl's 'transcendental ego', which determines the constraints of understanding (1999: 154). All understanding is 'embedded' in the dialogical encounter, at the core of which is utterance, defined by Flood as a 'speech act occurring in a language within a specific social, cultural and historical situation' (1999: 156). Dialogism is the discipline of understanding the interconnection between utterances; 'indeed, there can be no utterance without a relation to others' (1999: 156). Flood identifies this closely to the concept of 'intertextuality' in which all texts, like a mosaic, reflect other texts and transform them. The ideas of diversity of language and intertextuality stand beneath the concept of 'heteroglossia' or the 'unending diversity and fragmentation' of discourse (1999: 157). Understanding occurs through a gradual unfolding of meaning in which discourse is 'illumined in the activity of dialogue'. In the end, Flood calls for a 'dialogical religious studies' based on language, the contextual nature of research and the 'dialogic self'. These ensure that religious studies will be conducted in an ethic of sensitivity to power relations because it removes the researcher's 'epistemic privilege' granted in traditional phenomenologies of religion (1999: 168). Moreover, the 'dialogic self' overcomes the old subject–object dichotomy of classical phenomenology by constructing understanding through interaction between situated participants. Flood concludes: 'Dialogism understood in these terms, as drawing on Bakhtinian dialogism and linguistic anthropology, provides a coherent and forward-looking research programme in religious studies' (1999: 185).

Flood's major contribution to the current debates in religious studies thus challenges the viability of phenomenology as a philosophical tradition on which research in religion can be founded. The methods introduced by phenomenologists, which sought to limit the effect of potentially distorting biases, described by Kristensen and Parrinder as the application of evolutionary theories to religions and cultures, and by Eliade and Smart as the reductionistic tendencies within the social sciences, have been shown by Flood to be based on a philosophical theory that introduces even deeper, but more subtle, biases into the way knowledge is obtained and organized. By positing a universal human experience at the core of all religion that is intuited by the 'isolated subject', the phenomenologist ignores, or at least minimizes, the importance of localized social, historical and cultural contexts. In addition, the 'epistemic privilege' phenomenology grants to the researcher remains hidden, since it veils the power relations between the researcher and the communities that are being researched. By performing the phenomenological bracketing to eliminate every type of prejudice, the scholar of religion paradoxically remains in control of knowledge and thereby dictates the rules for interpreting religious phenomena. This makes phenomenology, at the very least, vulnerable to the charge that it actually promulgates a method for maintaining power over the objects of academic study, notwithstanding the virtual unanimity among phenomenologists that their personal religious experience provides them with privileged access to the mind of the religious practitioner. This latter claim to religious insight, unavailable to other scholars, strongly implies a theological agenda beneath the phenomenology of religion and thus leads to the second subsequent debate I am considering in this chapter, the increasingly tense relationship between the academic study of religions and theology.

The religion–theology debate: Setting the context

During the 1993 Conference of the British Association for the Study of Religions, following a paper presented by one of the participants outlining different functions for religious studies and theology, the Dutch scholar of Buddhism, Ria Kloppenborg, turned to me and pronounced: 'We solved that problem in The Netherlands more than thirty years ago'. Undoubtedly, she was referring to the close attention given to this issue, and the apparent resolution provided, by Dutch scholars from Kristensen to Bleeker. Despite Kloppenborg's objection, the connection between theology and religious studies has never been severed in academic circles, although there have been numerous efforts to dispose of it once and for all. One of the earliest institutional attempts to separate the study of religions from theology occurred at the Marburg Congress of the International Association for the History of Religions in 1960, where a declaration composed by the Israeli scholar, Zwi Werblowsky, was approved by the delegates. Point two of the five-part declaration (Geertz and McCutcheon, 2000: 15) asserted that 'the awareness of the numinous or the experience of transcendence (where these happen to exist in religions) are – whatever else they may be – undoubtedly empirical facts of human existence and history, to be studied like all human facts, by the appropriate methods'. The empirical study of human responses to the transcendent meant that 'the discussion of the absolute value of religion is excluded by definition, although it may have its legitimate place in other, *completely independent* disciplines, such as e.g., theology and philosophy of religion' (emphasis mine). The Marburg Declaration was approved by the Congress, with even Eliade signing it, but it did not pass without objection. Geertz and McCutcheon note that Bleeker, as General Secretary of the IAHR, 'replied critically to Werblowsky's indirect critique of his own standpoint, and repeated that historians of religion must take it seriously when believers claim to have been in contact with the transcendent' (2000: 16).

That the problem of relating the academic study of religion to theology persists is demonstrated by the spate of recent publications dealing with this subject. For example, Jan Platvoet (1998) has written on the history of this relationship in Dutch universities; Steven Sutcliffe and I have done the same in a forthcoming article about the teaching of religion in traditional Scottish Divinity faculties; and Tim Jensen and Armin Geertz are working on a similar project in the case of Denmark.[1] On more theoretical levels, Donald Wiebe (1999), Ivan Strenski (1993), Ursula King (1984), Russell McCutcheon (1997, 2003), Richard King (1999) and Robert Segal (1983), to name just a few, have made important recent contributions to this debate. Moreover, in much broader terms, universities and academic institutions have reinforced the need for clarity by situating the study of religions within departments of theology, a fact that is underscored by the British umbrella organization, The Association of University Departments of Theology and Religious Studies (AUDTRS).

In the United States, during the mid-1980s, the American Academy of Religion conducted a review entitled, 'Religion and Theological Studies in American Education'. It produced a 'pilot study' based on questionnaires sent to lecturers in various types of public, private and church-related academic institutions in the United States. The results were compiled by Ray Hart, Professor of Religion and Theology at Boston University, and published in the *Journal of the American Academy of Religion* in 1991 (Hart, 1991: 715–827). Although the study examined a wide range of issues, of particular interest to the present discussion is the section headed: 'What is the relation between the *study* of religion and theology and the *practice* of religion?' (emphasis his). Hart reports that this question 'evoked voluminous comments' among respondents and that it produced a great deal of passion (1991: 778). Some respondents, however, were simply impatient with the question, answering tersely: 'Dated issue' or 'Get on with the *real* work!' or 'What a question!' (emphasis his) (1991: 778). Despite these complaints, Hart's survey disclosed that this issue was a real one for many academics. He reports that three views were rather evenly represented among those questioned:

1. The study of religion and the practice of religion are not internally related. The academic institution is the proper place for the study of religion; religious institutions (called cult centres in the report) define the proper venue for religious practice (1991: 780–1).
2. The relationship is 'completely open'; each may affect the other. If one remains true to the data of religion, the study and the practice of religion must be in persistent dialogue (1991: 781).
3. The study of religion 'presupposes practice'; one cannot do the former without having experience of the latter (1991: 781).

Religious studies in American universities and theological seminaries, of course, has developed against a quite different historical background and with somewhat more diverse organizational structures than in Britain and other European countries, but the issues relating to the aims and purposes of the study of religions are very much the same, and I would suggest still very much alive.

Richard King, formerly Professor of Religion in the University of Derby and now at Vanderbilt University in the United States, writing from a British perspective, has observed that the teaching of religion in universities as 'non-confessional', secular and pluralistic has clear political implications for academic institutions in which theology, understood as the study (with some presumed allegiance) of religions in the Jewish and Christian traditions, still predominates (if not in numbers, in terms of influence) over religious studies. King (1999: 53) calls theology the 'older sibling' of religious studies, thereby drawing attention to the historical development of the study of religions in Europe within departments of theology and which Armin Geertz says has created the struggle within European academic circles 'to bring religion out of the shadows of ... theology' (1999: 509). Just as in the study conducted by the American Academy of Religion, King identifies three positions outlining the relationship between theology and religious studies, what he calls paradigms or models (1999: 53). These are:

1. The separatist paradigm. Religious studies and theology are entirely separate, because religious studies adopts a non-confessional, secular stance that is foreign to theology. This emphasizes the value of the outsider's position who may be willing to take stands that are unpopular with believers or even antithetical to the accepted truths of a particular faith community.
2. Religious studies as 'broad church' model. Theology is regarded as merely one among many methodologies in the study of religions. Others are drawn from the human and social sciences, including psychology, sociology, anthropology and philosophy. Theology, simply, is added to the list.
3. Theology as 'broad church' model. On this model, religious studies falls within theology. The study of religions is undertaken, even in its secular forms, for the purposes of informing and developing theological reflection. Or from another angle, religious studies is seen as just another form of modern theology, but in a secularist, post-Christian guise.

Certain similarities can be found between King's models and the different positions outlined in the American study. King (1999: 54–5) notes, for example, that an advocate of the separatist model would argue that there is no place for the practice of religion in religious studies, as would be allowed in paradigm 2 where the theologian (as a practitioner) is included among those who define appropriate methodologies, and in model 3 where religious studies becomes a servant of practice. These correspond, in some measure, to Hart's summary of differing views: no relationship; a dialogical relationship; and a fundamental and necessary connection between study and practice. I draw attention to the analyses provided by Hart and King to underscore that the problem of separating religion from theology persists both theoretically and practically among scholars of religion and in academic institutions. Geertz and McCutcheon attribute this persistence to assumptions inherent within the phenomenology of religion. They argue that 'because many of its earlier practitioners were educated in theology ... the phenomenology of religion became a useful discipline that allowed students of religion to pursue their theological interests' (2000: 17). This same point has been made even more forcefully by the Canadian scholar of religions, Donald Wiebe, who accuses the academic study of religions of a 'failure of nerve' (1999: 139) when called upon to resist pressures exerted on it by theology.

Donald Wiebe's anti-theological critique of van der Leeuw, Eliade and Smart

Wiebe's main targets within the phenomenology of religion are Gerardus van der Leeuw, Mircea Eliade and Ninian Smart, each of whom he accuses of theologizing the academic study of religions. From his analysis of van der Leeuw's *Religion in Essence and Manifestation*, Wiebe (1999: 186–7) concludes that van der Leeuw, ostensibly in support of academic objectives, insisted that every scholar must come from a 'cultural orientation to life', very much akin to a personal position of faith. According to Wiebe, van der Leeuw maintained that because the scholar is situated in a particular context, the scientific enterprise cannot be separated from the scholar's 'own religio-cultural quest'. To argue otherwise for van der Leeuw is naïve and deceptive, since 'it prevents the investigator's biases from being recognized and critically (scientifically) clarified'. This argument, which Wiebe dubs 'circular', undermines the academic objective it claims to support precisely because it 'ignores the critical differences between religion and the academic, scientific study of religion'. This means that van der Leeuw, who had sought to move the Dutch tradition begun by Chantepie de la Saussaye and C. P. Tiele beyond theology, actually accomplished just the opposite by directing it back into theology. 'With van der Leeuw ... the study of religion not only does not advance beyond the stage already reached by that discipline in Holland, but rather it returns to an earlier theological approach – one that amounts to a subversion of the scientific study of religion' (1999: 186–7).

Wiebe is likewise scathing of Eliade's hermeneutical method, calling it, not an interpretation leading to an understanding of religious traditions, but an attempt 'to recover the abandoned transcendental values and meanings once provided to their devotees by those traditions' (1999: 60). This view is supported by Eliade's insistence that the primitive and archaic forms of religion are paradigmatic for religious life in general because they disclose 'fundamental existential situations that are directly relevant to modern man' (Wiebe, 1999: 60; Eliade, 1969: ii–iii). Wiebe observes that Eliade's interest in primitive and archaic traditions does not derive from a scientific approach to the study of religions, since this 'would entail a reductionistic distortion *of* the truth of religion and therefore, it seems, of the truth *about* religion' (emphasis his) (1999: 60). In a separate article, Wiebe (1999: 197–204) discusses the first edition of Eliade's new *Encyclopedia of Religion* (1987), in which Eliade, as General Editor, repeated in the preface his oft-declared assertion, that religion constitutes a subject in its own right, *sui generis*, and that students of religion should resist all efforts to reduce it to other aspects of human life or to treat it as an epiphenomenon of other disciplines. Wiebe argues that Eliade's anti-reductionist stance veils 'a hidden theological agenda' (1999: 198), in support of which he cites Eliade's vision that by interpreting religion 'religiously', scholars contribute to the 'salvation' of 'modern man'. The 'knowledge of religion' becomes thereby 'religious knowledge', confirming in Wiebe's view that Eliade's hermeneutical approach 'is indistinguishable from the religio-theological' (1999: 198).

As we have noted earlier, both van der Leeuw and Eliade maintained that the religious person acknowledges a transcendent force or numinous reality as the source for human religious experience. Wiebe argues that not only do van der Leeuw and Eliade assert that this is true for believers, but they go beyond this by affirming that the transcendent constitutes an ontological reality, which in some sense the scholar must experience personally if genuine understanding of religion is to be achieved and communicated. In this way, van der Leeuw and Eliade are susceptible to charges that their phenomenologies of religion amount to theologies of religion. Ninian Smart is much less open to this charge, since his 'methodological agnosticism' would seem to distance him from a personal commitment to a religious world view, in a way that neither van der Leeuw nor Eliade appeared to accept. According to Wiebe (1999: 53–67), this is not the case. Smart's ideas, when analysed carefully, expose the same theological assumptions beneath the phenomenology of religion that he discovered in the writings of van der Leeuw and Eliade. This is particularly evident in Smart's refusal to follow Peter Berger's position of 'methodological atheism' on the grounds that it potentially offends believers. Although in many ways the differences between methodological atheism and Smart's brand of agnosticism are slight, the key objection lodged by Smart centres on the insider's point of view, something he attempts to overcome by his variations on van der Leeuw's interpretation of the phenomenological *epoché*.

Wiebe (1999: 53) contends that Smart's underlying theological motive is expressed in his concluding analysis to *The Science of Religion and the Sociology of Knowledge* (1973), where Smart insists that 'learning about religion' and 'feeling the living power of religion' not only can go together but 'must go together if the study of religion is to enter boldly into its new era of promise' (Wiebe, 1999: 53; Smart, 1973a: 160). Rather than leading the study of religions into a new era, Wiebe insists that by confusing study with experience, Smart's advice 'is more likely to entrench the traditional, religio-theological study of religion from which the academic, scientific study of religion first emerged' (1999: 53). Wiebe goes on to probe what he considers Smart's ambiguous description of religious studies as including questions of truth, which extend beyond scientific interests, despite Smart's contention that religious studies is quite distinct from theology. Wiebe argues that by raising the issue of truth in religion, Smart has once again posed a problem for academic studies in religion by re-establishing 'earlier ties between the academic study of religion and piety' (1999: 54). Wiebe asserts that Smart's return to a theological perspective is confirmed by the way he deals with *epoché* or bracketing in *The Phenomenon of Religion* (1973), which Wiebe describes as the book in which Smart most 'consciously addresses the issue of methodology in the study of religion' (1999: 54).

We have seen in Chapter 5 that Smart analysed in depth the idea of bracketing as it was derived from van der Leeuw, which Wiebe interprets as Smart's way of going beyond a neutral position in the study of religions. Wiebe says that Smart's understanding of a 'flat neutralism' corresponds to the external descriptions of religion, which do not reveal the inner sense of what it is to be religious (1999: 56). This could be one interpretation of how the *epoché* should be employed, where all forms of religious belief, feeling, sentiment or experience are put in brackets, left aside and not considered in a purely empirical study. What Smart called 'bracketing of expressions' or 'expressive bracketing' (1973b: 32), by contrast, allows the scholar to include in the suspended judgements the feelings expressed by religious adherents, without endorsing those feelings. Wiebe says it is at this point that the ambiguity in Smart's position emerges. If the scholar of religion aims not simply to gain knowledge of religions but in some way to testify to religious values and sentiments, then something more than a science of religion is implied, if not theology, at least metaphysics (1999: 57).

When Smart refers to the 'Focus' of faith, as the chief interest of the believer, Wiebe notes that he has raised the proper question: 'Is the Focus ... part of the phenomenon which the observers witness?' (1999: 57). To answer this question affirmatively would cause the scientist to violate a theoretical boundary, since to include the Focus itself as a part of the empirical data of religion is to make theological or metaphysical objects a part of the study. Smart's response to this, as we have seen, is simply to refuse to ask the question about the reality of the Focus, on the grounds that the scholar of religion is not interested in affirming or denying its existence. By using the method of expressive bracketing, the observer is able to allow the believer's own point of view to form part of the phenomena without ever asking if that point of view is true or false, right or wrong. This would seem to be perfectly acceptable to an empirical method, but Wiebe says it veils religionist sentiments. Why else, he asks, would Smart reject Berger's notion of methodological atheism as requiring 'a priori reductionistic explanations of religion' (1999: 57)? Methodological atheism seems to rule out of hand any possibility that the believer may be apprehending a real or true Focus. By not asking the question, the phenomenologist leaves open the possibility that the Focus of faith for the adherent may actually exist. And it is here, Wiebe asserts, that Smart makes the telling admission: 'If it does exist, then the phenomenological description does actually describe a manifestation of the Focus, or in other words it is not just a description of human events' (Smart, 1973b: 68; Wiebe, 1999: 58). Wiebe contends that in this statement, Smart has contravened the very phenomenological principle he has just affirmed, since 'the phenomenologist is one who ought not to pose the question in the first place' (1999: 58).

Wiebe argues that quite obviously Smart is trying to avoid the naturalistic presuppositions entailed in empirical science, which would rule out, before any investigations were begun, the supernatural explanations entailed in religious interpretations of the world. By adopting an agnostic perspective within the notion of bracketing, Smart has distinguished religious studies as a phenomenological undertaking from a purely positivistic science. Wiebe understands this, but pushes Smart further on the question by asking if the Focus can be included among the phenomena of religion. As we have just seen, it would appear that refusing to ask whether the Focus is real or not would imply that only the believer's view would be included among the phenomena of religion, not the object of a believer's faith. This would mean that the scholar of religion is limited to describing and interpreting human events only, and has no concern with anything outside human responses to what believers and the theologians within the believers' traditions postulate to be transcendentally real. Wiebe says that it is just at this point that Smart's argument becomes inconsistent, since 'surprisingly' he moves away from just such a conclusion, by allowing a theological explanation to complement a phenomenological one. By remaining agnostic about the object of faith, 'the scientist need not deny the cognitive value of such complementary theological accounts' (1999: 65). Methodological agnosticism, as opposed to methodological atheism, thus not only leaves open the possibility that the object of a believer's faith is real, it 'predisposes the student of religion to assume the "truth" of religion (that is, to assume that the Focus of religion exists)' (1999: 65). Phenomenological bracketing, particularly when interpreted by Smart as 'expressive' bracketing, assumes that the believer is always right. Explanations which oppose the adherent's perspective are forbidden, thereby, in Wiebe's words, lending 'more coherence to the phenomenon'. Wiebe admits that the descriptions allowed by Smart 'are far more complex than those usually provided by the kind of empiricist students he criticizes', but in the end, the way he employs 'the principle of bracketing' ensures that his 'is a kind of crypto-theological enterprise' (1999: 65).

Theological assumptions in the thought of Ninian Smart and W. C. Smith: The controversial appraisal by Timothy Fitzgerald

Wiebe is not alone in suggesting that the phenomenology of religion, rather than providing a scientific methodology for religious studies, actually restricts it within boundaries dictated by theological assumptions. Ivan Strenski (1993: 2), for example, has argued that Eliade, as a non-reductionist, has 'tried to smuggle into his methodological approaches to the study of religion his own religious faith in the transcendental existence of the Sacred'. Robert Segal (1983: 108), in a way similar to Wiebe, has criticized the emphasis within phenomenology on the believer's point of view, arguing that the scholar of religion can never appreciate the *reality* of religion for believers without endorsing the very assumptions that motivate the devotee's faith. Similar criticisms have been developed and expanded more recently by Timothy Fitzgerald of the University of Stirling in his controversial book, *The Ideology of Religious Studies* (2000), but Fitzgerald carries his argument to even more radical conclusions than previous scholars by contending that the term 'religion' as an analytical category should be abandoned as an essentialist, ideological concept created by theologians for theological purposes. If I can push his argument to the limit, for Fitzgerald, the study of religion, by definition, is theology. If we are studying something other than theology, what we actually mean when we use the word 'religion', is the study of cultures and societies.

Fitzgerald's detailed analysis of this issue supports my contention that the religion–theology debate generated by key figures in the phenomenology of religion remains alive today, and continues to play a crucial role in methodological discussions among scholars of religion. Fitzgerald dismisses as academically fatal the impatience of those who say, like the American respondent to the AAR questionnaire, 'Let's get on with the real work', because if we ignore the inherent problems in defining what we mean by religion (theology, culture or something else), 'we are rendered impotent by our ignorance of the semantic and ideological bias of our own tools of analysis' (2000: 53). In order to help us understand the seriousness of the problem, Fitzgerald devotes a chapter in his book to Ninian Smart and the phenomenology of religion. He does so, partly because he admits that Smart is associated with the movement to make the study of religion non-reductive to other disciplines in the social sciences, but at the same time to retain its non-theological approach. Fitzgerald also acknowledges the importance Smart held in the academy, calling him 'probably the most persuasive British writer to formulate some kind of theoretical model for studying religion' (2000: 55).

Fitzgerald argues that Smart's thinking has been influenced strongly by his forebears in the phenomenology of religion, particularly van der Leeuw and Otto. Although he by no means followed his predecessors uncritically, according to Fitzgerald, Smart maintained with them 'an essentialist, reified concept of religion and religions' based on the idea that religion is 'a distinctive and analytically separable kind of thing in the world that can be identified and distinguished from non-religious institutions throughout the vast range of human cultures' (2000: 55). Like van der Leeuw, Smart analysed religion in terms of its essence and its social and cultural manifestations. This becomes particularly clear when Smart applies his dimensional categories to religions, whereby the core of religion is expressed in various observable forms, such as myths, rituals, art and legal institutions. Fitzgerald claims that 'the imagery is of a primary substance, an essence, taking on some of the secondary properties of the institutional media through which it manifests itself' (2000: 56). This essentialist notion of religion is also implied in Smart's distinction between religions and world views or between what Fitzgerald calls 'religions proper (Islam, or Buddhism, for instance) and religion-like ideologies such as nationalism, Marxism, Maoism, and Freudianism' (2000: 56). Maoism, as a secular ideology in Smart's thinking, for example, can be analysed according to the seven-dimensional model, just as can Buddhism, but only Buddhism can be regarded as encompassing a genuinely religious world view because 'it is centred on the transcendent', which defines its core or irreducible essence (2000: 58).

As we have seen in our discussion of the British school in Chapter 5, Smart attempted to overcome the problem of essentialism by applying to religion Wittgenstein's idea of organizing specific uses of language within categories he called 'family resemblance' (Smart, 1986: 46). According to Fitzgerald (2000: 57), Smart used this in support of his argument that 'meaning ... is to be found not in some essential characteristic such as belief in God, or in a unique kind of personal experience, but in the actual usage of the word in real contexts' (1986: 57). If we push this concept very far, however, we see that it is either meaningless or retains an essentialist notion. For example, Religions 'A' (those originating in India) may share certain characteristics in common with Religions 'B' (classical Chinese religions), which share certain characteristics with Religions 'C' (shamanistic religions of northern Asia), which share some characteristics with Religions 'D' (African religions). Religions A (Indian) and D (African) look so different that they do not seem to fall under the same category. Smart thought that by tracing the connections in this way, even the most dissimilar religions could still be classified under the category 'religion', without insisting that they share the same things in common (see Cox, 1996: 13–15). Fitzgerald disagrees. He asserts that such an approach makes it impossible to say what is included or excluded under the term 'religion', unless the scholar retains an essentialist notion. What connects Religions A with Religions D cannot be located in the historical, cultural or social contexts out of which they emerged, since these are unrecognizable and probably unidentifiable, but they are united by a common core, the essence of religion: its transcendental referent (2000: 58).

Finally, Fitzgerald asserts that Smart's use of the phenomenological *epoché* implies that special methods are needed by the scholar of religion that are not available to other scholars. Although Smart urges the student of religion to remain agnostic about the ontological status of the transcendental focus of religion by bracketing out its reality or lack of reality, this approach unwittingly supports the notion that religion constitutes a category of its own in need of unique methods to carry out its investigation. Since researchers in religion cannot know if the transcendental referent exists or not, the empirical tools of investigation employed in the social sciences are not available to them. This means that religion cannot be understood without special theoretical and methodological approaches that at the very least draw from the philosophy of religion, but in most cases require theological tools of analysis. Hence, if scholars want to distinguish something which is religious as opposed to the non-religious, they look for such things as 'a heaven, an ancestral other world, a supreme being, or generally to unseen mystical powers' (2000: 62). For Fitzgerald, this is indistinguishable from a liberal, ecumenical Christian theology which historically was responsible for constructing the category we know in academic institutions in the West as 'world religions' (2000: 63).

Fitzgerald also devotes a section of his book to the thought of Wilfred Cantwell Smith, whom he credits with advancing discussions around the religion–theology debate by insisting that the category 'religion' should be dropped. According to Fitzgerald, Smith quite rightly has shown that the category 'not only distorts western understanding of the rest of the world but also distorts western self-understanding' (2000: 43). But this is as far as Fitzgerald is willing to go, since, in the first instance, he accuses Smith of obscuring his argument in 'verbiage', a failing that causes him to use a host of terms as imprecisely as scholars have employed the term 'religion' uncritically. Examples cited by Fitzgerald, which he extracted from Smith's Presidential address to the American Academy of Religion in 1983, include the phrases: 'the complex world history of humankind's religious and spiritual life', 'the worldviews of religious complexes', 'traditional religious ideologies', 'the long and wide history of human religiousness' (Smith, 1983: 3–18; Fitzgerald, 2000: 45). Fitzgerald observes that Smith is not being 'ironic' when he uses such language; 'they are used as though they all mean something important' (2000: 45).

Smith's lack of terminological precision, of course, is not his major failing for Fitzgerald. He, like Smart, has smuggled into his analysis of religion a transcendental element. Fitzgerald observes: 'Smith is defining our subject of study in terms of fundamental values that are represented as transcendental' (2000: 47). These values in turn are interpreted theologically, since Fitzgerald argues that in Smith's writings the term 'transcendent' 'refers to an absolute reality, a monotheistic Being who creates the world and invests human life and history with purpose' (2000: 47). In this way, Smith has imposed his own Christian assumptions on to a field he has accused of being undermined by the increasing power of Western scientific thinking. Fitzgerald counters that it is not only the secularizing tendencies in the West that have reified human faith into a category called religion; the process has been extremely useful to 'a new class of preachers, missionaries, social workers, ecumenicists, and theology of religion academics' (2000: 47). Into this last category, Fitzgerald inserts phenomenologists of religion, who take as their 'tenet of faith ... that there are many religions in the world that are all equally (more or less) responses to the transcendent God' (2000: 47). In the end, Smith falls prey to a fundamental contradiction of thought. On the one hand, he offers a biting critique of the category 'religion' while 'as a modern liberal ecumenical' theologian, he is caught up in promoting the very category he has rejected. Fitzgerald concludes: 'He has himself proclaimed his own faith, which he uses as the framework of basic presuppositions in terms of which he wishes to explain the data of other ideologies' (2000: 47).

The theological problem Fitzgerald identifies in the writings of Ninian Smart and Wilfred Cantwell Smith can be applied to each of the key phenomenologists of religion we have identified, with the likely exception of Jonathan Z. Smith, whom Fitzgerald does not discuss. The phenomenological error, for Fitzgerald, does not result from the careful descriptions and the resulting classifications phenomenologists have employed to interpret a variety of social and cultural activities, but from their insisting that for something to qualify as religion, it must refer to a transcendental entity. This makes the category 'religion' indistinguishable from theology, since the study of religion is maintained as a distinct category, *sui generis,* in a classification of its own, requiring its own methodologies and its own department within universities on the basis of one criterion only: its numinous, sacred, transcendental core. If scholars want to avoid the problems they have inherited from phenomenology and study religion non-theologically, Fitzgerald insists that they must focus on the real object of study: the social, 'understood as the values of a particular group and their institutionalization in a specific context, including the way power is organized and legitimated' (2000: 71). Fitzgerald concludes that the social is not some dimension of religion, some aspect that can be studied as if it were an 'optional extra', but 'the actual locus of a nontheological interpretation' (2000: 71).

The biting critiques of the phenomenology of religion provided by Wiebe and Fitzgerald define the terms on which the current debate is proceeding about the methods employed and the subject matter considered by scholars of religion. For Wiebe, a non-theological approach means a full commitment to an empirical, rational study of human activities and their related institutions. It must proceed in an objective, open, but neutral way using fully the methods available to any social scientist. For Fitzgerald, the problem goes beyond keeping theological commitment out of the study of religion and extends to the category itself, which he contends is so loaded with theological content that it must be abandoned in favour of other socially rooted and culturally specific categories. Drained of its theological content, religion, for Fitzgerald, loses its unique character and is indistinguishable from social contexts. The religion–theology debate thus extends beyond terminological issues and penetrates to the very core of what it means to be a student of religion. Both Wiebe and Fitzgerald have challenged powerfully one of the principal and overriding assumptions held by scholars in the phenomenological tradition, in Fitzgerald's words, that religion is located 'in some distinctive *sui generis* set of characteristics that can be analytically separated from ritual and symbolism in general' (2000: 249–50).

The debate over the socially engaged scholar of religion

I am calling the third critical debate provoked by the phenomenology of religion the controversy created by the contexts in which the study of religions occurs, which often demand scholars to make personal decisions as to whether or not they should become involved *as scholars* in the social, political and economic issues that affect directly the religious communities they study. Article Five of the Marburg Declaration, approved at the Congress of the International Association for the History of Religions in 1960, affirmed that 'students of religion' may wish to 'join with others in order to contribute their share towards the promotion of certain ideals – national, international, political, social, spiritual and otherwise'. It warned, however, that 'this is a matter of individual ideology and commitment, and must under no circumstances be allowed to influence or colour the character of the IAHR' (cited by Geertz and McCutcheon, 2000: 15–16). As we saw in Chapter 4, this position was supported at the Turku Conference in 1973 by C. J. Bleeker (1979: 174), who noted that some people 'think that the history of religions and the phenomenology of religion should mainly serve to foster world-peace and social harmony, by creating mutual understanding among the adherents of the different religions'. However noble such ideals may be, he reminded the conference, the history and phenomenology of religion 'are a purely scholarly affair'. These issues have persisted since Marburg and Turku as underlying concerns for students of religion, but they have re-emerged recently, partly under the influence of the attacks on the World Trade Center in New York on September 11 2001, into an increasingly contentious debate over the legitimate place in the academy of the socially engaged scholar of religion.

This debate differs markedly from the one promoted by Timothy Fitzgerald, since his aim remained largely an analytical one focusing on the category 'religion'. The debate also diverges from the argument over social scientific reductionism, whereby phenomenologists fought against defining the uniquely religious core of religion in terms of social, economic, political, cultural or other tools of social scientific analysis. It is closely related, however, to Flood's dialogic resolution to the subject–object dichotomy since contexts often dictate, for example, that scholars study religious communities in situations of conflict, or engage with groups whose religious ideologies oppose international development aims. As a participant in a dialogic process, the scholar in such situations is implicated by definition. One way of clarifying the debate over the socially engaged scholar of religion is to ask if the study of religion, which is always undertaken by human beings, can proceed if the human being conducting the research refuses to become involved in the sometimes life-threatening situations affecting their human 'objects' of research. A clear case in point relates to witchcraft in Africa, which has been studied, interpreted and explained according to many theories, but generally has not involved the scholar as scholar, in efforts to ameliorate the oftentimes dehumanizing effects of witchcraft beliefs, usually on women.

Warnings against scholars of religion confusing their academic tasks with social, moral or political goals, as expressed in Marburg and Turku, are in some senses contradicted by the underlying humanistic concerns of the phenomenology of religion, which were implied early on in the attempts by Tiele and Chantepie de la Saussaye to separate out confessional theology from a comparative study of religions, but which became explicit in Mircea Eliade's 'new humanism', as outlined in the The Quest (1969), and in Wilfred Cantwell Smith's 'world theology'. In our review of key phenomenologists of religion, we have already seen in Eliade's writings a conviction that the study of religion produces at the very least, in Jungian fashion, a sense of meaning in life that is conducive to positive mental health, which, due to the absence of sacred orientation in contemporary Western life, in part explains the anguish of 'modern man'. David Cave (1993: 27) argues that Eliade's hermeneutics contained a 'spiritual dimension' through which religious facts were understood religiously, 'on their own plane of reference'. This means that the phenomenology of religion advocated by Eliade was, in his own terms, a 'new humanism' (1969), defined by Cave as 'a spiritual, humanistic orientation toward totality capable of modifying the quality of human existence itself' (1993: 27). Cantwell Smith, in a similar vein, concluded his discussion in *Towards a World Theology* (1981) with a passionate appeal for 'the intellectual questions of the meaning of religious diversity', which are now taking place in contexts of 'modern secularity', to penetrate within the very walls of the religious institutions that are affected by such questions and not be left to outsiders (1981: 194). Comparable goals were expressed in missionary language by Edwin W. Smith and Geoffrey Parrinder, who, by identifying a universal belief in God in Africa, were not simply articulating a missionary theology, but on one level at least, were disputing on humanistic grounds prior descriptions of Africans as primitive, animistic and intellectually and genetically deficient.

The sometimes opposing strands within the phenomenology of religion towards the social involvement of the scholar have contributed directly to the current debates surrounding this issue. On the one hand, the underlying humanistic tendency running consistently through the writings of the key phenomenologists suggests that, despite their emphases on neutral descriptions, the bracketing of prior assumptions and the position of methodological agnosticism, they directed the academic study of religions towards an engagement with the issues most affecting religious devotees. On the other hand, by asserting a universal sacred core of religion, they spiritualized religious experience in a way that made the study of religions irrelevant to social contexts. These issues have been addressed from different perspectives in two important books, both of which shed light on the role of phenomenology in the debate over the socially engaged scholar of religion. The first is Russell McCutcheon's treatment of the matter in his book, *Critics Not Caretakers* (2001) and the second is by David Chidester, who has described the study of religion from the perspective of apartheid South Africa in his volume, *Savage Systems* (1996). I will review the positions taken by McCutcheon and Chidester before describing how the debate they have articulated emerged fully into the open at the recently concluded Congress of the International Association for the History of Religions held in 2005 in Tokyo.

McCutcheon on the public role of the scholar

McCutcheon considers the issue of social engagement by the student of religion under the theme, 'The Scholar of Religion as Public Intellectual' (2001: 125). He argues that the first error committed by scholars of religion is to have insisted on the autonomy of their own discipline, which he calls an 'indefensible assumption' based on a 'metaphysical reduction' (2001: 129). This has had the effect of isolating religious studies institutionally from other disciplines in universities, with the exception of theology. The methods of Eliade in particular, as a prime example of a phenomenologist of religion, may have contributed to individual self-enlightenment, but they have not provided scholars of religion with the tools for developing a critical commentary on the societies in which they live and work. This failing can be traced precisely to Eliade's thesis that 'studying religion provides deep, essential, absolute, or otherworldly insights into the very nature of things' (2001: 129). In McCutcheon's view, Eliade's theory of sacred intrusions into profane space and time is nothing other than an attempt to 'historicize all ahistorical claims' (2001: 129). By defining the focus of religion as the unknowable, non-historical, atemporal, transcendent sacred, Eliade and other phenomenologists have surrendered their right to make comments about political and social issues. This follows not just from the theologizing of the study of religion, but from the assumption, clearly stated in Eliade, that what is religious is good, healthy, positive and redemptive (2001: 129). If McCutcheon is right, the scholar holding such a view lacks the critical capacity to comment on the social reality of religion as a contributing factor to a variety of human actions, both positive and negative, including the relationship between religion and violence.

A second problem attributable to phenomenology has resulted from its insistence on describing religious data and offering interpretations only in ways that believers themselves can affirm. McCutcheon likens this method to 'nuanced descriptions' or 'reflexive autobiography', which reduces the role of the scholar to one of 'a reporter repeating the insider's unsubstantiated claims' (2001: 130). By refusing to offend adherents, such scholars have cut themselves off from any debate that entails explanations or critical analysis, not just on matters of meaning, but on issues of public concern. On questions such as 'charting the course of a public school curriculum, a welfare agency, or even a policy for war', phenomenologists, as 'translators' of the believer's point of view, 'have no voice and little, if anything, to add' (2001: 131).

The theory that the core of religion is 'personal faith', 'the numinous', 'the sacred' or 'the transcendent', leads to a highly individualized interpretation of religion, that for McCutcheon lies at the root of the problematic distinction between religion and the secular (2001: 131). The religion–secular dichotomy is based on the assumption that religion belongs to the private sphere, which, because it has its foundation in an otherworldly authority, can be affected only in a limited way by worldly concerns. At the same time, it has little to say to those same worldly concerns. McCutcheon puts it this way: ' "Religion" is construed as an independent variable occupying the untainted realm of pure and private moral insight that is opposed to, and the salvation of, the messy public worlds of politics and economics' (2001: 131–2). In a similar way, scholars who define the study of religion as *sui generis*, isolate themselves from sociologists, anthropologists, economists, political scientists and all others like them who do not possess the methods required to investigate such a unique subject matter. Nonetheless, the boundaries affect both sides: what outsiders cannot penetrate, the insider cannot abandon. In this way, the scholar of religion lives in privilege and isolation, studying what no other scholar considers, but at the same time remains uniquely unqualified to comment on matters where there are 'no gods, no myths, no rituals, and no hierophanies' (2001: 132). It is clear from this line of argument that McCutcheon sees only institutional and intellectual decline for the academic study of religions if the same principles of social non-involvement are perpetuated. He concludes that the challenge to scholars of religion today is whether they will accept a public role 'or through our efforts to spiritualize and dehistoricize the people and practice we study ... we will continue to obscure both our data and our social role' (2001: 141).

But what precisely defines the public role of the scholar of religion for McCutcheon? Does it translate into a practical engagement with specific social and political issues? It would seem not. He reserves the public role of the intellectual to that of a critic who lays bare the 'mechanisms of power and control' (2001: 141). This applies in the first instance to the way religions themselves are studied (exemplified by his own challenges to the phenomenology of religion), and hence refers to a critical self-examination of theory and method in the study of religions. McCutcheon says, 'our role is unfailingly to probe beneath the rhetorical window dressings that authorize conceptual and social constructions of our own making' (2001: 141). This responsibility applies more generally beyond analysing institutions that study religion to a critique of social 'mechanisms of power', which the scholar of religion uncovers, brings to light and where appropriate challenges' (2001: 140). In the public sphere, the scholar of religion questions 'self-evidencies', promotes 'intellectual freedom' and by cooperating 'in a cross-disciplinary way' with other scholars who employ methods drawn from their own fields of study, identifies 'those homogenizing, ideological strategies so necessary for the manufacture and management of human communities' (2001: 142).

In the end, McCutcheon defines the appropriate engagement of the scholar of religion with institutions in society as 'cultural criticism' (2001: 142). To return to my example of witchcraft in Africa, this implies that the scholar would not (as a scholar) participate in activities aimed at eradicating the practice, but would raise critical questions about it, uncover the colonial motives that created laws to suppress it, and expose how patriarchal institutions continue to benefit from it. To my mind, this differs little from Bleeker's comments at Turku or from Article 5 of the Marburg Declaration, in each case where the study of religions is described as a scholarly enterprise, not as a partisan one. In this sense, it would seem that McCutcheon's argument is levelled not so much against those who maintain academic neutrality in the face of social or political movements – it would appear he agrees with this – but, like Fitzgerald, he directs his harshest criticism against the surreptitious theological agenda he believes has been smuggled into the study of religions by advocates of the *sui generis* theory. This is seen clearly in McCutcheon's concluding comments, where he prescribes for scholars of religion 'the role of critic, rather than Eliade's role of savior, for our work is carried out within the material contestations of history rather than in the mists of primordial time' (2001: 142).

David Chidester: Studying religion in the light of colonial history

If McCutcheon's discussion tends towards a highly theoretical analysis of the public role of the scholar, David Chidester (1996) roots his argument squarely in the context of South Africa and its long history of apartheid. Chidester is not directly confronting the problem of the socially engaged scholar of religion, but he demonstrates how social, historical and political contexts in the study of religions affect scholarly interpretations and dictate the categories through which those interpretations are filtered. He cites with approval Jonathan Z. Smith's (1990: 34) observation that the history of the study of religions 'has been by no means an innocent endeavour'. Chidester's opening line in his book makes this point absolutely clear: 'This book is … a critical analysis of the emergence of the conceptual categories of *religion* and *religions* on colonial frontiers' (emphasis his) (1996: 1). The analyses which follow underscore repeatedly the fact that the study of religions looks very different when seen from the perspective of the marginalized than it does from within academic structures of power operating in Western universities. The main point I draw from his line of reasoning is this: if scholars of religion have produced their analytic categories historically by siding with powerful political and social interests, the argument cannot be maintained today that the scholar, for reasons of academic neutrality or methodological agnosticism, should remain aloof from engaging with the pressing social problems that affect the lives of those who constitute the subject matter of religious studies.

Chidester supports this contention by showing in a detailed historical account of missionary and colonial activities in South Africa that the findings of 'comparative religionists' became powerful tools of colonialism to establish and exercise 'local control' by constructing a 'discourse about others that reinforced colonial containment' (1996: 2). He shows how attitudes towards indigenous people changed from declaring initially that they had no religion, through ones with their roots in ancient religions, usually degenerated from ancient Judaism, to a religion that could be compared with Western religion. In each case, he argues, these interpretations served colonial interests. For example, sixteenth and seventeenth-century travel literature describing the Africans as having no religion resulted not just from ignorance, but it served to justify 'an intervention in local frontier conflicts over land, trade, labor and political autonomy' (1996: 14). By the eighteenth century, when trade routes had been established with Portuguese, Dutch and English traders, the usefulness of referring to African religion as 'fetishism' is seen since it reinforced the idea that Africans 'overvalued trifling objects … but they undervalued trade goods' (1996: 15). Chidester points out that by the nineteenth century, when European colonization was gaining momentum, the denial that Africans had religion perpetrated the notion of the 'lazy savage', one who lacked initiative and industry (1996: 15). Finally, when religious 'systems' were acknowledged in Africa, they were asserted in order to establish boundaries, to fix the parameters between the religion of the white and that of the savage. This reification of religion created frontiers, whereby the other could be contained, but at the same time, where negotiations across frontiers could occur (1996: 21). This means, as I read Chidester, that although a study of the history of religions in South Africa needs to be traced in terms of category formation, religious typologies, morphologies and structures of religion, these always need to be understood in the context of 'the apartheid system of ethnic separation' (1996: 22).

This same principle applies to a study of the history of religions generally, particularly when scholars outline the emergence of the world religions paradigm. Chidester notes that nineteenth-century missionary reports, often based on accounts by converts, contributed to theories about the newly emerging idea of world religions, through which technologically simple and kinship-based societies were depicted as being on a lower rung in an evolutionary scheme leading upwards towards civilized societies and their religions (1996: 244–6). If we follow Chidester in this argument, we would need to interpret our earlier discussion of the African roots to British phenomenology in these same colonial contexts. For example, when Edwin W. Smith described Africans as children in their religious apprehensions, he was taking a line that Chidester traces to seventeenth and eighteenth-century reports which portray Africans as 'permanent children' (1996: 56). In this sense, Smith's analogy was an improvement, since he envisaged Africans as 'growing up' in Christ, but as Chidester notes, this is very much like being 'credited with a religion that discredited them' (1996: 70).

On Chidester's reasoning, the charge that missionaries undermined indigenous religious perspectives can be lodged also against the contention, maintained both by Smith and Parrinder, that Africans everywhere believe in a Supreme Being. Conceived in Christian terms, the idea of the universal African God enabled missionaries to portray Africans, largely for Western audiences and colonial objectives, as naturally accepting the Christian message and thus as being made into good citizens by embracing the moral values the missionaries preached. Chidester illustrates this in his discussion of the term *uNkulunkulu* (1996: 132–6), which Western missionaries and academics attributed as the name of the Zulu Supreme Being. Chidester shows that the assignment of a single, monotheistic idea of God to *uNkulunkulu* emerged from the colonial encounter with African cultures as a strategy to 'civilize' the native population, but he demonstrates that in reality the term referred to disparate beliefs with numerous regional variations, often rendering *uNkulunkulu* as an original ancestor of all humanity, or more likely as an original ancestor of a particular people (1996: 135). This, of course, underscores Chidester's central argument that we cannot understand the creation of the category 'religion', African or otherwise, outside colonial contexts where power relations were assembled and reinforced by ideological constructions of the 'other'. On this line of reasoning, religion as a Western, Christian invention should be understood not as a phenomenological error in category formation, but as resulting from real-life historical and colonially inspired categories of power: high and low, civilized and primitive, central and peripheral.

The debate at the Nineteenth Congress of the IAHR in Tokyo

The issues posed by McCutcheon and Chidester became the subject of heated debates at the Nineteenth Congress of the International Association for the History of Religions meeting in Tokyo in 2005, where over 1,700 delegates from around the world gathered to consider the theme 'Religion: Conflict and Peace'. The opening symposium of the Congress featured papers on 'Religion and Dialogue among Civilizations'. Although the opening symposium did not form part of the official academic agenda of the Congress, since it had been organized locally by the Japanese Committee and was meant to involve the larger community beyond the delegates themselves, it was described prominently in the programme and entailed the first exposure most delegates had to the considerations of the Congress. The aim of the opening session was described by the organizers as considering the 'wars and conflicts in many parts of the world' and how 'there have been attempts among the religious communities of the world to undertake a dialogue between religions' (Congress Secretariat, 2004: 8). The opening symposium could be portrayed as precisely the application of religion to issues of conflict and peace rather than restricting its discussions to an academically neutral study of how religion has or is contributing to societal disorder. That this was a critical issue was raised at several of the panels on theory and method in the study of religions and it emerged at the meeting of the IAHR International Committee, where some delegates voiced serious objections to the approach adopted during the opening symposium.

One of the academic panels at the Congress was organized by Donald Wiebe, who presented a paper entitled, 'Disentangling the Role of the Scholar-Scientist from that of the Public Intellectual in the Modern Academic Study of Religion' (2005: 308–9). During his presentation, in an apparent reference to McCutcheon, Wiebe noted that an argument widely advanced in recent times maintains 'that an essential aspect of the study of religion involves engagement in public analysis and critique of its economic, political, and cultural environment' (2005: 308). Wiebe's position at the Congress was unequivocal in its opposition to this line of thinking. In the printed abstract of his paper, he wrote: 'The case for the student of religion as public intellectual is unsound' (2005: 309). During the panel discussion, he maintained that the role of the scholar of religion entails critical analysis of issues and theories relating to the study of religions, but this critical stance does not extend to involvement in matters of political and social concern. Of course, he admitted, the scholar of religion as a private citizen has every right, and as a good citizen an obligation, to participate in such matters. The scholar does this, however, not as a scholar, since that role is quite closely prescribed by academic rules, but as any other concerned citizen living and participating in a social democracy would do. The counter-argument to this, expressed by some delegates in response to Wiebe's paper, emphasized Chidester's position that the context for studying religion very much determines the methodology the scholar employs.

At another panel, some delegates focused on religious violence in Nigeria and challenged scholars, who are living in the midst of conflict between Muslims and Christians in the area, to become involved by promoting peaceful reconciliation between the opposing factions. An example of just this position is found in a paper sent to the Congress by Umar Danfulani of the Department of Religion in the University of Jos, Nigeria. In his paper, Danfulani (2005: 36) analyses 'the Jos crisis', which he describes as events occurring on 7 September 2001 that resulted in widespread violence between Muslims and Christians, the ramifications of which are still felt to this day in the region. The sectarian conflicts in Nigeria, of course, were overshadowed by world attention directed to New York on September 11, but Danfulani argues passionately that academics need to turn their attention to other places which may expose even more clearly the impact of religious conflict on local communities than did the attacks on the World Trade Center. Danfulani (2005: 36) concludes in a way that sharply contrasts with Wiebe's position by calling for 'a strategy that will lead both Muslims and Christians living in Jos to say "Never Again" to the gloomy events that started on the 7th of September, 2001'.

The issues that surfaced in Tokyo emphasize the controversy surrounding the increasing significance that is being attributed by scholars of religion, many of them working in non-Western institutions, to social contexts as determining the subject matter for religious studies. When placed alongside the argument that religious adherents and scholars of religion participate dialogically in the research environment, it becomes clear how the debate born out of contesting claims, which traditionally have dogged the phenomenology of religion, between academic neutrality and 'insider' knowledge have come to the fore. This traditional problem is being addressed increasingly today not as a debate over scholarly objectivity, but as a dichotomy between personal and academic choices regarding the scholar's own involvement in the sometimes overwhelming tragedies that affect human communities as a result of religious behaviour.

Response to the debates: A philosophical reconfiguring of the phenomenology of religion

The debate about the continued philosophical viability of phenomenology as a method in the academic study of religions disputes whether or not questions about knowledge should be couched in terms of subjects and objects, or whether they should be conceived narratively and dialogically, in a way anticipated much earlier by Wilfred Cantwell Smith. In one sense, Gavin Flood's contribution to this debate follows directly from the emphasis in the phenomenology of religion on incorporating the perspectives of believers into any interpretation the scholar offers about the believers. Flood attempts to go beyond this characteristic strand of phenomenology by insisting that empathy actually perpetuates the distinction between subjects and objects. I would argue in response that so long as the rules by which academic research is conducted are applied, the best that can be achieved by the scholar is a kind of radical empathy, rooted in self-reflexivity, but which acknowledges the fundamental distinction between the 'self' (the researcher) and the 'other' (the objects of research). This fact does not preclude dialogue since the scholar proceeds necessarily according to an explicit commitment to scientific rationality, which operates alongside the religious commitments of the communities the scholar studies.

In a previous publication (Cox, 2003a: 25–37), I drew attention to the fact that the study of 'alterity', which is usually contrasted with the concept of 'identity', has become an important cross-disciplinary topic, particularly throughout the social sciences. In my analysis, I drew on the description of the relationship between alterity and identity provided by Ernst van Alphen (1991: 1–16), a scholar of literary theory, who problematizes the subject–object dichotomy in a way that directly relates to Flood's critique of the phenomenology of religion. In traditional anthropological and ethnographic literature, van Alphen suggests, scholars have tended either to describe 'the other' in wholly exotic terms or in ways that denigrate the other. In either case, how the scholar interprets the other reflects far more about the identity of the scholar who offers the interpretation than it does about the ones who constitute the objects of the scholar's interpretation. This implies that interpretations do not rest on neutral descriptions of the activities of those being studied, but on what the scholar reads into the descriptions derived from his or her own identity (cultural, personal, religious, gendered and so on). A second point follows from this. The identity of the other is affected directly by scholarly interpretations, both in terms of a community's self-perceptions and in the way identities are imputed by outsiders on the basis of the interpretations the scholar provides. In this sense, scholarly descriptions and interpretations always entail power. Drawing on psychoanalytic theory, van Alphen then adds a third dimension by demonstrat-ing that the scholar's descriptions and interpretations of the other also affect the identity of the scholar, since the self and the other (identity and alterity) are not distinct. If what I say about the other reveals at least as much about me as it does about the one I am studying, the descriptions and interpretations I offer of the other directly influence my own self-identity.

What does van Alphen's analysis of alterity and identity imply for Flood's critique of the phenomenology of religion from a philosophical perspective? Van Alphen suggests the answer: 'The only way to know the other is by letting the other speak about me, by giving the other the position of "I"' (1991: 15). The reason for this is that when I speak about the other, I am immediately involved in demarcating my own self-image. By extension, when the other speaks about me, the other also is engaged in a process of self-definition. For methodologies aimed at examining cultures and identities, this implies that research proceeds legitimately only when contracts of understanding are formed between the researcher and the researched communities. The contract entails dialogue in the fullest sense of the term, where the researcher outlines the purpose and the process of the research to what traditionally would have been regarded as the passive 'objects' of the research. The assumption beneath the contract is that when a researcher poses questions or engages in field observations, the identity of the researcher is revealed. On a philosophical plane, this means that when researchers speak about the objects of research (alterity), they transport the other into themselves and thereby make their own identities transparent. The counterpart to this occurs when a researcher invites (not compels) members of the researched community to speak about their perceptions of the researcher and the research project. In this way, a dialogical research plan is created, since the constructed 'other' issues from both the researcher and the researched community, lending credence to W. C. Smith's idea that 'humane knowledge' can be achieved only from communal self-consciousness. This kind of knowledge cannot convey knowledge of the other in some objectified and reified understanding of knowing, but only as a result of the interplay between differing identities and alterities, each of which, scholarly and religious, is understood as socially and culturally embedded.

When this method is applied directly to the phenomenology of religion, it clearly reconfigures the method away from thinking strictly in terms of that which is 'inside' and that which is 'outside', but it does not ignore this distinction. By bringing to awareness the fact that one's self-identity is revealed by the questions posed and the ensuing interpretations offered, a scholar of religion is applying the technique of *epoché*, since by bringing scholarly self-identity to consciousness, researchers are less prone to distort others in their descriptions of them. Recognizing the power the scholar holds over the other adds a further layer of a radical empathy, since not only is the other affected by the scholar's findings, but, as we have just seen, the identity of the scholar is likewise affected. By inviting the community into the research project, the results of the research are altered in a way that gives new meaning to the phenomenological 'understanding-in-depth' (*Verstehen*). These reformulations of classical phenomenology still wrestle with the original problem as to how best to incorporate believers' perspectives into scholarly interpretations, but they add a dialogical dimension that credits devotees with a stake in the hermeneutical task.

For many believers, the engagement with research will be in service to deeply felt commitments and oftentimes to non-negotiable beliefs. Scholars of religion also acknowledge that they maintain a non-negotiable ideological commitment to scientific research. These methods differ from naïvely assuming that the scholar can bracket out all prior assumptions, but they assert, in the phenomenological tradition, that researchers should attempt to bring to their own consciousness their personal and academic assumptions, while making these assumptions transparent to the communities they are studying. This latter point not only acknowledges that academic neutrality is impossible, but quite emphatically rejects it as an ideal for academic research. I see no fundamental contradiction between these positions and that of the attitudinal perspectives advocated traditionally by phenomenologists of religion. I follow in this way a loosely applied use of the *epoché*, by adopting a self-reflexive stance, and commit myself to involving the community in any interpretation I provide, so that, again loosely applied, the phenomena can be allowed to speak for themselves. In this way, the task of interpretation results from a combination of scholarly selfvity and empathy, but it is saved from Flood's epistemic privileging of the 'er by accepting fully a dialogical approach to the meaning of knowledge on a mutually constructed interchange between alterity and identity.

Response to the debates: Recapturing religion from theologians and advocates of cultural studies

As has become evident throughout this book, phenomenologists of religion consistently defined their field of study as distinct from other academic disciplines, usually by delineating history, philosophy and phenomenology from theology on the one hand, and from the social sciences on the other. Armin Geertz (1999: 509) argues that today this is not just a pleasant discussion about how to divide academic tasks, but 'a battle ... that employs all the discursive, organizational, economic, and political strategies that human beings normally employ in any conflict situation'. If Geertz is right, we are engaged in an ideological warfare between those who advocate a unique place for religious studies in academic institutions, and those who see it either as a branch of theology or as a kind of study that should be transferred into departments of cultural studies or ritual studies or even be shifted into centres of African, Asian or gender studies. As we have seen, the debate hinges on attitudes scholars adopt towards the idea, articulated forcefully by Fitzgerald, that religion as a classification of its own is essentialist, decontextualized and depends on a transcendental referent.

In a contribution I made to a volume devoted to defining religion, edited by Jan Platvoet and Arie Molendijk (Cox, 1999: 267–84), I argue that we need a preliminary definition of what we mean by religion in order to study what people do who engage in it. In other words, we start with something assumed in order that the parameters of our study can be defined pragmatically. The preliminary definition I proposed includes the key phrase, 'non-falsifiable postulated alternate reality'. I argue that identifiable communities around the world perform certain activities, believe certain things, invest authority in certain personalities, hallow certain texts, tell various stories and legitimate morality by reference to a non-falsifiable and purely postulated alternate reality (or as has been pointed out to me recently, realities). It is possible to study the activities in which communities specifically engage in response to their belief in postulated non-falsifiable realities. These alternate realities are postulated by those in the community, not by me.

The problem of how to interpret what communities do and say in cultural contexts other than the scholar's own is not unique to the student of religion, but the religious studies scholar is interested specifically in how communities organize themselves in relation to their postulated non-falsifiable alternate realities. If a psychologist wants to do the same thing, he or she might apply psychological methodologies to the study of a community, or more likely to individuals in the community, who postulate certain things and organize some of their activities around non-falsifiable alternate realities. The sociologist, economist, political scientist, geographer and anthropologist each use the tools of her or his discipline to offer insights on the community's beliefs, institutions and activities relevant to the postulated non-falsifiable alternate realities. Psychologists, sociologists, political scientists, geographers and anthropologists do much more than study the way their discipline understands how communities relate to what they postulate to be alternate realities, but when they do so, we say they are applying their discipline to religion (the psychology of religion, the sociology of religion, the anthropology of religion, and so on). I am not a psychologist, sociologist, anthropologist, geographer or economist, but a scholar, I say, of religions. By this, I am referring to my primary task of identifying and describing human behaviours that help me interpret how communities, and individuals in communities, relate to what they postulate to be alternate realities. In the interpretative role, I need to develop structures that help me convey what I mean and how I make sense of the data I am studying. By a structure, I am referring to patterns or models, even typologies, for understanding aspects of what communities do in relation to their postulated non-falsifiable alternate realities.

For example, I have studied and written on death rituals in Zimbabwe, a fact that is not missed by Fitzgerald (Cox, 1995: 339–55; Fitzgerald, 2000: 84–7). There are many ways of studying death rituals drawing on, for example, anthropological literature (classically, Victor Turner's interpretation of the ritual process) or political analyses (examining the power and gender roles of various members of the community in the funeral rituals). In my research, I was interested to identify the nature of and the beliefs about postulated non-falsifiable realities. I concluded that the fundamental focus for such rituals was always the ancestor, who, by being brought back home to the family through rituals, was perceived by the community as continuing to play a fundamental role in safeguarding community well-being, as a deceased person, but as a living force nonetheless. I related how these understandings of the ancestor affected interpretations within the community of illness, misfortune and social relationships. In all of this, I needed knowledge of the social structure of Shona society. I required an understanding of the political and economic background. I even needed some awareness of Shona psychology. My central focus, however, was on none of these, but on identifying and interpreting a structure of inter-relationships between the community and its postulated non-falsifiable alternate realities, and how the community institutionalized its activities around those non-falsifiable realities, what I had defined in a preliminary and pragmatic way as religion.

In my view, this is not theology; neither is it essentialist or decontextualized. Yet, it is not strictly anthropology, sociology, psychology, economics or politics, although it does not ignore these. Scholars of religion do not study alternate realities themselves, since these cannot be studied, but they analyse what is postulated about that which is non-falsifiable and in what ways this affects the communities which do the postulating. As a scholar of religion, my aim is to describe, fairly represent and interpret what identifiable communities say, do and believe as a direct result of postulating alternate realities. My descriptions are accountable to the data; my representations need to include the communities I am representing and my interpretations need to be grounded in my own acknowledged and transparent personal constructs. If this means that I am pinning the study of religion on a transcendental referent, I admit it (although I think the term 'transcendental' is the wrong word to use), but I am doing so because that is precisely what the members of the communities themselves do.

To my mind, this is scientific, because it is a study of a human activity, which is entirely testable, and indeed capable of falsification. All scientific disciplines do the same thing. Sociologists study the social structures that comprise human behaviour. Psychologists examine the personality and adjustment of individuals in societies. Economists look at patterns of markets and forces that influence investments. Political scientists examine structures of power that legitimate decision-making. We could argue that each discipline is artificial because it extracts one aspect out of complex and inter-related factors, none of which can be understood fully without the other. Nevertheless, we do so for reasons of analysis and interpretation, knowing fully that we need to integrate what has been extracted artificially back into the whole to achieve understanding. Religion is no different. To focus on how communities relate to postulated non-falsifiable realities is artificial and extracted, but we do so, like all academic disciplines, for analytic and interpretative reasons.

Religious studies, on this analysis, includes theology as a part of its arena of study, much like King's separatist paradigm. Theologians, at least on one definition, are practitioners. They study, analyse and interpret, generally from within one tradition, the meaning of what that tradition maintains is a revelatory act. Scholars of religion regard theologies as ways that some communities reflect on their own non-falsifiable alternate realities. In other words, theologies, like ritual, morality, myth, scripture, community, law and art, form part of the data on which religious studies carries out its work. This is not to assert a position of superiority, but merely to define quite different and distinctive roles for religious studies and theology. It is for this reason that I resist efforts to push religious studies into theology by definition or to situate it in something else, perhaps also vaguely defined, as cultural studies. I remain unconvinced by the arguments posited by Wiebe and Fitzgerald and counter that what we do in the phenomenological study of religions, as analyses and interpretations of communities that institutionalize behaviours around their postulated non-falsifiable alternative realities, is closely aligned to what is done in the social sciences, and yet not the same as any one other social science. Religious studies, in my view, exists alongside theology in university departments as an accident of history, and not, as the new wave anti-religionists would contend, as part of a deeper ideological commitment to a transcendental referent. It is possible, on this line of thinking, to remain committed to the classical phenomenological position that the study of religions can be non-theological and yet non-reductive to any of the social sciences.

Response to the debates: A sociological perspective that engages and disengages the scholar of religion

In modern, secular societies, religion is expressed widely in intensely individualistic and non-institutional ways (see Bruce, 2002: 82–5). As a result, the term 'spirituality' increasingly is being substituted for the word 'religion' by practitioners, particularly in the West. This history has been outlined in detail by Philip Sheldrake (1995: 41), who refers to the renewed interest in spirituality as resulting from a 'paradigm shift in the general approach to theology towards a greater reflection on human experience as an authentic source of divine revelation'. If scholars accept the term 'spirituality' as an appropriate alternative to religion, in my view, this would confirm Wiebe's charge that we are now experiencing a 'loss of nerve' and are engaged in a process of re-theologizing religious studies. This is why in several recent publications, I have sought to apply to my discussions of religion in contemporary society a distinction proposed by the French sociologist, Danièle Hervieu-Léger between religion and 'sacredness', or what I think translates into the same thing, the distinction between religion and spirituality (see Cox, 2004: 259–64; Cox, 2003: 69–87).

In her important book, *Religion as a Chain of Memory* (2000), Hervieu-Léger contends that the meaning conveyed to individuals by religion is connected inextricably with institutions that transmit authority. This leads her to identify religion as existing whenever 'the authority of tradition' has been invoked 'in support of the act of believing' (2000: 76). For her, changes in the context of believing define fundamentally the transformations that are occurring in contemporary society. She deliberately refers to 'believing' as opposed to 'beliefs' because she wants to underscore the dynamic process that is occurring when people affirm their faith. Her definition of believing corresponds in some ways to my own preliminary definition of religion by stressing that believing refers to 'the body of convictions – both individual and collective – which are not susceptible to verification' (2000: 72). Acts of believing play a major role in social processes in the contemporary world because they construct meanings for individuals and for society. This implies that religion cannot be equated with the 'sacred', in an Eliadian sense, because religion denotes institutions of society, which bind adherents together through a shared allegiance to an overwhelming, authoritative tradition. Many people in contemporary society, of course, testify to profound 'religious' experiences, even though these are not connected to an authoritative tradition. Hervieu-Léger calls such testimonies evidence of experiences of sacredness, not religion. I would classify these under the term 'spirituality', understood not ontologically, but as emotional responses to particular socio-cultural contexts, including, as Hervieu-Léger notes, a wide variety of activities including sporting events, political rallies, rock concerts and even gatherings dominated by speaking in tongues (glossolalia) (2000: 58–9).

Although Hervieu-Léger's distinction between religion and sacredness leads to her primary concerns with the process of secularization and the question of the persistence of religion in the modern world, I find her analysis helpful for clarifying the debate over the socially engaged scholar of religion. This is precisely because secularization entails processes which, under the influence of increasing globalizing forces, create an overarching homogeneity of power, resulting, in Hervieu-Léger's words, 'from the eclipse of the idiosyncrasies rooted in the collective memory of differentiated concrete groups' (2000: 128). The homogeneity of modern society differs markedly from what Jan Platvoet (2002: 83) describes as the nineteenth-century Dutch (and by extension, European) 'monochromatic Christian scene' because secularization aligns all social life, in Hervieu-Léger's words, with 'the sphere of production' (2000: 128). This, accompanied by mass and almost instant communication, has produced a collective memory that is far different from the traditional religious memory; secularized memory is 'surface memory, dull memory, whose normative, creative capacity' has been dissolved (2000: 128). This process is embodied today in the dominating and total power of world capitalism, which coerces all differentiated societies with their various collective memories into one mould, defined and exploited by market forces.

The very factors which create a homogeneous economic system lead at the same time to a radical fragmentation of what Hervieu-Léger calls the 'individual and group memory' (2000: 129). In modern societies, individuals belong to numerous, specialized and atomized groups. Although each operates within a general framework dictated by market forces, none is able to sustain a singular commitment to an all-consuming authority. In other words, individuals in modern societies are unable to coalesce around a unified memory. Because modern institutions are fragmented in space and time, individuals relate to them in a piecemeal fashion. No real central, organizing tradition can be sustained. As soon as a bit of memory is established, it is nearly always immediately destroyed. For Hervieu-Léger, 'the collective memory of modern societies is composed of bits and pieces' (2000: 129). The secular, as opposed to religion, therefore, is best defined as the homogenization in the contemporary global society of a surface memory, promulgated under the guise of the world economic system, which in turn is accompanied by a radical fragmentation of all collective memories.

If Hervieu-Léger's sociological analysis is correct, it means that the debate over the socially engaged scholar of religion has been couched in far too simplistic terms, since it has been regarded largely as an extension of the insider–outsider problem, asking if the scholar, as one who in the phenomenological tradition sides with the believer's point of view, should encourage actions that directly affect the quality of life experienced by believing communities. Or, in another light, it asks if the scholar should promote policies that attempt to change religious practices that adversely affect members of the religious community, and possibly the wider society, including such far-ranging activities as suicide bombings or female circumcision. If we follow Hervieu-Léger, the scholar of religion neither advocates on behalf of the religious community, nor attempts to alter what might be considered harmful practices within or without that community. Rather, the role of the scholar is limited to describing social processes which have resulted from institutions of power, both religious and secular, and to identifying what effects these processes are having on religion and experiences of sacredness (spirituality) in contemporary contexts. In this way, a scholar may expose the power structures that are undermining religion through a process of radical individualization, and at the same time may discover that a different kind of authority based on powerful political and economic interests has been substituted in its place. By interpreting religion as the authoritative transmission of tradition in social contexts, scholars of religion are provided with a key method of analysis through which they can offer an informed public comment and submit an incisive social critique.

This means, at a fundamental level, that I agree with McCutcheon when he describes the scholar of religion as a 'critic, not a caretaker'. As one whose particular skills entail analyses of religious contexts, the scholar bears a responsibility to apply these to vital issues affecting contemporary society. This means that the role of the scholar as public critic never occurs in a way that is disengaged from social contexts, as Chidester's history of religion in South Africa has demonstrated. Yet, the difference between scholarly action and social action remains unambiguous in the writings of both McCutcheon and Chidester. It would be difficult to read Chidester's survey of the constructed term 'religion' in colonial South Africa while remaining indifferent to the oppressive forces that used religion eventually as one of the building-blocks to establish the apartheid state. Chidester's analysis, in other words, influences public attitudes and perceptions, which in turn inform and motivate action in society. Yet, the scholarly critique remains one step removed from advocating actual political programmes for change. This is where, I think, Danfulani is wrong. His role as a scholar of religion is to uncover the way authoritative traditions are being utilized to foment violence between Christians and Muslims in Nigeria, but his academic project does not entail setting up projects of mediation or inter-community centres of dialogue.

Recently, I contributed a paper to a conference on the religion–secular dichotomy in colonial contexts in which I outlined how the capitalist reinterpretation of land in Alaska under the terms of the Alaska Native Claims Settlement Act has placed land into the hands of indigenous shareholders in Native corporations.[2] Although I did not directly criticize the Act, I attempted to show how its passage completed a long process of assimilation of indigenous peoples in the region and how it was conceived by those who framed it as a means of finally making Alaskan indigenous peoples fully 'Americanized'. I noted that the indigenous transmission of authority, which resided traditionally in a loosely organized social structure that employed various specialists, including shamans, and was constructed in order to secure the means of subsistence, had become subject to the radically homogenizing power of the capitalist system, which paradoxically had created an equally radical heterogeneous system in the form of individual shares in corporations. My analysis followed very closely Hervieu-Léger's distinction between religion and sacredness and how religion is embedded in social processes. My aim clearly was to act as a critic of a system which in many ways has had devastating effects on the indigenous peoples of Alaska, but I made no specific proposals, as had been done, for example in the 1980s, by Justice Thomas Berger on behalf of the Inuit Circumpolar Conference (1985). In my paper, I portrayed the scholarly role as one of informing policies for change, but not as extending to devising detailed proposals for social and/or legal action.

My position represents a very different approach from that taken by Scott Appleby and James Busuttil in their contributions to a recent book on religious fundamentalism and social change, entitled *The Freedom to Do God's Will* (ter Haar and Busuttil, 2003). Appleby argues that 'certain principles are true apart from their level of inculturation in any given society' (2003: 205), which leads Busuttil in a concluding chapter to propose 'policy responses' to fundamentalism, including, among others, encouraging a multi-religious framework to discuss common problems (2003: 233), exposing young fundamentalists to positive experiences (2003: 234) and strengthening democratic resources within fundamentalist communities (2003: 235). In her introductory chapter to the book, Gerrie ter Haar, who is Professor of Human Rights in the Institute of Social Studies in The Hague and who was the International Programme Chairperson for the 2005 IAHR Congress in Tokyo, notes that Appleby has set his discussion of fundamentalism in terms of 'the universality of moral values' and guides a 'middle course' between a 'thoughtless universalism' and cultural relativism, 'in order to enhance participation of cultural and religious groups in the formulation of shared moral values' (2003: 22). Ter Haar then commends Busuttil's policy proposals on the grounds that 'the aim of this project is not to theorize for academic purposes only, but to show the relevance of academic analysis for those who are called upon to formulate policies in response to the existence and growth of fundamentalist movements' (2003: 22–3). Ter Haar's position sounds very close to one I am advocating, and indeed it even appears consistent with McCutcheon's view. It is clear, however, that ter Haar, Appleby and Busuttil go further than either McCutcheon or I are prepared to go, by transforming the role of the scholar from one that critically analyses religion in social contexts into one that participates actively in projects aimed at promoting social change. In my view, the sociological theory advanced by Hervieu-Léger, when accompanied by McCutcheon's description of the academic as public intellectual and Chidester's emphasis on understanding the context for the study of religions, engages the scholar *as scholar* appropriately by avoiding the error of confusing political and social action with critical analysis.

Conclusion

In this chapter, I have identified three debates in the study of religions, which I contend have been created largely by the formative philosophical and theological, and to a lesser extent, the social scientific influences over the phenomenology of religion and which have been articulated in various ways by many of the key figures I have identified in the Dutch, British and North American schools of phenomenology. I have attempted to demonstrate that the

subsequent debates engendered by the phenomenology of religion l
centre of religious studies today, since how they are resolved will c
future directions in the academic study of religions. Just as I have li:
overall study in this book to phenomenologists of religion, I have restricted the
subsequent debates to those that have been fostered within that tradition.
Many other important discussions are going on in religious studies beyond the
problems I have identified in this chapter, one of the most significant of which
has been initiated by feminist writers on religion, such as Ursula King (1995),
and which Gavin Flood (1999: 226–30) considers in his concluding comments
in *Beyond Phenomenology* under the theme, 'Gender and Religious Studies'. It
is clear that many of the ways religion has been approached reflect what
Malory Nye (2000: 9) calls 'gendered knowledge', a perspective that is made
evident by the overwhelmingly male domination in the history of the
phenomenology of religion I have described in this book. This history
prompted Brian Bocking (2004: 108) to observe recently that 'the phenomenology of religion [at Lancaster under Smart]... did not anticipate feminist
theory; it didn't try to understand gender'. Other issues that relate to the
personal perspective and involvement of scholars of religion, as Nye notes,
include questions of sexuality and the body, how we represent others and
power and agency (2000: 9–10).

Many of these issues nonetheless have appeared in the debates I have identified
in this chapter, such as the call for self-reflexivity in Flood's narrative account of
religious scholarship, power relations in religious construction as exposed in
Chidester's post-colonial critique, and the need to address the public role of the
scholar on wide-ranging social problems. Certainly, if philosophical clarity over
the subject–object dichotomy is demanded (as asserted by Flood), if the concern
over the relationship between theology and religious studies still defines an
underlying tension for scholars of religion (as argued by Wiebe and Fitzgerald),
and if the role of the scholar of religion in relation to the social and historical
contexts that define religious communities requires careful consideration (as
attested in different ways by McCutcheon, Chidester, Danfulani and ter Haar),
understanding how these debates have emerged from the phenomenology of
religion is critical for engaging in them thoughtfully, analytically and in an
informed manner, as I have sought to do in my own responses to each. If I am
correct in my concluding analysis in this chapter, the phenomenology of religion
continues to play a crucial role as a central method in the study of religions
today, either as the subject of intense criticism or as the springboard for new
interpretations and advances in religious studies. In my view, this confirms that I
was correct to limit my focus in this book to analysing the thinking of key
phenomenologists of religion, not simply because they were important
historically, but because the competing resolutions to the debates they fostered
will determine if, and in what form, the study of religions will persist as a credible
academic discipline, and if, in the process, it can be construed as germane to
other social and cultural contexts.

References

Alphen, E. van (1991), 'The other within', in R. Corbey and J. Th. Leerssen (eds), *Alterity, Identity, Image: Selves and Others in Society and Scholarship* (Amsterdam and Atlanta: Rodopi), pp. 1–16.

Appleby, R. S. (2003), 'Religions, human rights and social change', in G. ter Haar and J. J. Busuttil (eds), *The Freedom to Do God's Will. Religious Fundamentalism and Social Change* (London and New York: Routledge), pp. 197–229.

Berger, T. R. (1985), *Village Journey: The Report of the Alaska Native Review Commission* (New York: Hill and Wang).

Bleeker, C. J. (1979), 'Evaluation of previous methods: Commentary', in L. Honko (ed.), *Science of Religion: Studies in Methodology. Proceedings of the Study Conference of the International Association for the History of Religions, held in Turku, Finland, August 27–31, 1973* (The Hague, Paris and New York: Mouton), pp. 173–7.

Bocking, B. (2004), 'The study of religions: The new queen of the sciences?', in S. Sutcliffe (ed.), *Religion: Empirical Studies* (Aldershot and Burlington: Ashgate), pp. 107–19.

Bruce, S. (2002), *God is Dead. Secularization in the West* (Oxford: Blackwell).

Busuttil, J. J. (2003), 'Policy responses to religious fundamentalism', in G. ter Haar and J. J. Busuttil (eds), *The Freedom to Do God's Will. Religious Fundamentalism and Social Change* (London and New York: Routledge), pp. 230–7.

Cave, D. (1993), *Mircea Eliade's Vision for a New Humanism* (New York and Oxford: Oxford University Press).

Chidester, D. (1996), *Savage Systems. Colonialism and Comparative Religion in Southern Africa* (Charlottesville and London: University Press of Virginia).

Congress Secretariat for the 19th World Congress of the IAHR (2004), *Call for Papers* (Tokyo: Department of Religious Studies, University of Tokyo).

Cox, J. L. (1995), 'Ancestors, the sacred and God: Reflections on the meaning of the sacred in Zimbabwean death rituals', *Religion*, 25, 339–55.

Cox, J. L. (1996), *Expressing the Sacred: An Introduction to the Phenomenology of Religion* (Harare: University of Zimbabwe Publications, 2nd edn).

Cox, J. L. (1999), 'Intuiting religion: A case for preliminary definitions', in J. G. Platvoet and A. L. Molendijk (eds), *The Pragmatics of Defining Religion: Contexts, Concepts and Contests* (Leiden: Brill), pp. 267–84.

Cox, J. L. (2003a), 'African identities as the projection of Western alterity', in J. L. Cox and G. ter Haar (eds), *Uniquely African? African Christian Identity from Cultural and Historical Perspectives* (Trenton, New Jersey: Africa World Press), pp. 25–37.

Cox, J. L. (2003b), 'Contemporary Shamanism in global contexts: "Religious" appeals to an archaic tradition?', *Studies in World Christianity: The Edinburgh Review of Theology and Religion*, 9 (1), 69–87.

Cox, J. L. (2004), 'Afterword – Separating religion from the "sacred":

Methodological agnosticism and the future of religious studies', in S. J. Sutcliffe (ed.), *Religion: Empirical Studies* (Aldershot: Ashgate), pp. 259–64.

Danfulani, U. H. D. (2005), 'Abstract. The cobra is running wild: Narrating the events and evaluating causes of the Jos crisis since September 7th 2001', in Congress Secretariat of the 19th World Congress of IAHR, *The Book of Abstracts* (Tokyo: Department of Religious Studies, University of Tokyo), p. 37.

Eliade, M. (1969), *The Quest: History and Meaning in Religion* (Chicago: University of Chicago Press).

Eliade, M. (Editor in Chief) (1987), *Encyclopedia of Religion* (16 volumes) (New York: Macmillan).

Fitzgerald, T. (2000), *The Ideology of Religious Studies* (New York and Oxford: Oxford University Press).

Flood, G. (1999), *Beyond Phenomenology: Rethinking the Study of Religion* (London and New York: Cassell).

Geertz, A. W. (1999), 'Review of Russell T. McCutcheon *Manufacturing Religion*', *The Journal of Religion*, 79 (3), 508–9.

Geertz, A. W. and McCutcheon, R. T. (2000), 'The role of method and theory in the IAHR', in A. W. Geertz and R. T. McCutcheon (eds), *Perspectives on Method and Theory in the Study of Religion* (Leiden: Brill), pp. 3–37.

Haar, G. ter (2003), 'Religious fundamentalism and social change: A comparative inquiry', in G. ter Haar and J. J. Busuttil (eds), *The Freedom to Do God's Will. Religious Fundamentalism and Social Change* (London and New York: Routledge), pp. 1–24.

Hart, R. L. (1991), 'Religious and theological studies in American higher education: A pilot study', *Journal of the American Academy of Religion*, 59 (4), 715–827.

Hervieu-Léger, D. (2000), *Religion as a Chain of Memory* (translated by S. Lee) (Oxford: Polity Press).

Honko, L. (1979), 'Introduction', in L. Honko (ed.), *Science of Religion: Studies in Methodology. Proceedings of the Study Conference of the International Association for the History of Religions, held in Turku, Finland, August 27–31, 1973* (The Hague, Paris and New York: Mouton Publishers), pp. xv–xxix.

King, R. (1999), *Orientalism and Religion. Postcolonial Theory, India and 'The Mystic East'* (London: Routledge).

King, U. (1984), 'Historical and phenomenological approaches to the study of religion. Some major developments and issues under debate since 1950', in F. Whaling (ed.), *Contemporary Approaches to the Study of Religion. Volume I: The Humanities* (Berlin, New York and Amsterdam: Mouton Publishers), pp. 29–164.

King, U. (1995), 'A question of identity: Women scholars and the study of religion', in U. King (ed.), *Religion and Gender* (Oxford: Blackwell), pp. 219–43.

McCutcheon, R. T. (1997), *Manufacturing Religion. The Discourse on Sui Generis*

Religion and the Politics of Nostalgia (New York and Oxford: Oxford University Press).

McCutcheon, R. T. (ed.) (1999), *The Insider/Outsider Problem in the Study of Religion: A Reader* (London and New York: Cassell).

McCutcheon, R. T. (2001), *Critics Not Caretakers. Redescribing the Public Study of Religion* (Albany: State University of New York Press).

McCutcheon, R. T. (2003), *The Discipline of Religion. Structure, Meaning, Rhetoric* (London and New York: Routledge).

Nye, M. (2000), 'Editorial: Culture and religion', *Culture and Religion*, 1 (1), 5–12.

Platvoet, J. G. (1998), 'Close harmonies: The science of religion in Dutch *Duplex Ordo* theology, 1860–1960', *Numen*, 45, 115–61.

Platvoet, J. G. (2002), 'Pillars, pluralism and secularisation: A social history of Dutch sciences of religions', in G. Wiegers (ed.), *Modern Societies and the Science of Religions* (Leiden: Brill), pp. 83–148.

Segal, R. (1983), 'In defense of reductionism', *Journal of the American Academy of Religion*, 51 (1), 97–124.

Sheldrake, P. (1995), *Spirituality and History: Questions of Interpretation and Method* (London: SPCK, 2nd edn).

Smart, N. (1973a), *The Science of Religion and the Sociology of Knowledge* (Princeton: Princeton University Press).

Smart, N. (1973b), *The Phenomenon of Religion* (New York: The Seabury Press).

Smart, N. (1986), *Concept and Empathy. Essays in the Study of Religion* (edited by D. Wiebe) (London: Macmillan).

Smith, J. Z. (1990), *Drudgery Divine. On the Comparison of Early Christianities and the Religions of Late Antiquity* (Chicago: University of Chicago Press).

Smith, W. C. (1981), *Towards a World Theology* (Philadelphia: Westminster Press).

Smith, W. C. (1983), 'The modern West in the history of religion', *Journal of the American Academy of Religion*, 52 (1), 3–18.

Strenski, I. (1993), *Religion in Relation: Method, Application and Moral Location* (London: Macmillan).

Wiebe, D. (1999), *The Politics of Religious Studies: The Continuing Conflict with Theology in the Academy* (London: Macmillan).

Wiebe, D. (2005), 'Abstract: Disentangling the role of the scholar-scientist from that of the public intellectual in the modern academic study of religion', in Congress Secretariat of the 19th World Congress of IAHR, *The Book of Abstracts* (Tokyo: Department of Religious Studies, University of Tokyo), pp. 308–9.

Notes

[1] James L. Cox and Steven J. Sutcliffe, 'Religious Studies in Scotland: A Persistent Tension with Divinity'; Tim Jensen, 'From History of Religions to the Study of Religions: Trends and Tendencies in Denmark', a paper

presented at the 50th Anniversary Conference of the British Association for the Study of Religions, 15 September 2004. The Cox–Sutcliffe article will form part of a forthcoming special issue of *Religion* on the study of religions in Europe, expected in 2006.

[2] James L. Cox, 'Secularising the Land: The Impact of the Alaska Native Claims Settlement Act on Indigenous Understandings of the Land'. Paper delivered at a Conference on 'The Religion–Secular Dichotomy in Colonial Contexts', University of Stirling, 4–6 July 2003.

Bibliography

Adams, J. L. (1972), 'Introduction', in E. Troeltsch, *The Absoluteness of Christianity and the History of Religions* (London: SCM Press Ltd), pp. 7–20.

Adams, J. L. (1976), *On Being Human Religiously. Selected Essays in Religion and Society* (edited and introduced by M. L. Stackhouse) (Boston: Beacon Press).

Allen, D. (1978), *Structure and Creativity in Religion: Hermeneutics in Mircea Eliade's Phenomenology and New Directions* (The Hague: Mouton).

Alphen, E. van (1991), 'The other within', in R. Corbey and J. Th. Leerssen (eds), *Alterity, Identity, Image: Selves and Others in Society and Scholarship* (Amsterdam and Atlanta: Rodopi), pp. 1–16.

Appleby, R. S. (2003), 'Religions, human rights and social change', in G. ter Haar and J. J. Busuttil (eds), *The Freedom to Do God's Will. Religious Fundamentalism and Social Change* (London and New York: Routledge), pp. 197–229.

Barbosa da Silva, A. (1982), *The Phenomenology of Religion as a Philosophical Problem* (Lund: CWK Gleerup).

Bell, D. (1990), *Husserl* (London and New York: Routledge).

Benét, W. R. (1977), *The Reader's Encyclopedia* (London: Book Club Associates, 2nd edn).

Berger, P. L. (1969), *The Sacred Canopy. Elements of a Sociological Theory of Religion* (Garden City, New York: Doubleday Anchor Books).

Berger, T. R. (1985), *Village Journey: The Report of the Alaska Native Review Commission* (New York: Hill and Wang).

Berkeley, G. (1998) [1710], *A Treatise Concerning the Principles of Human Knowledge* (Oxford: Oxford University Press).

Bettis, J. D. (ed.) (1969), *Phenomenology of Religion. Eight Modern Descriptions of the Essence of Religion* (London: SCM Press).

Biezais, H. (1979), 'Typology of religion and the phenomenological method', in L. Honko (ed.), *Science of Religion: Studies in Methodology. Proceedings of the Study Conference of the International Association for the History of Religions, held in Turku, Finland, August 27–31, 1973* (The Hague, Paris and New York: Mouton Publishers), pp. 143–61.

Bleeker, C. J. (1963), *The Sacred Bridge: Researches into the Nature and the Structure of Religion* (Leiden: E. J. Brill).

Bleeker, C. J. (1972), 'The contribution of the phenomenology of religion to the study of the history of religions', in U. Bianchi, C. J. Bleeker and A. Bausani (eds), *Problems and Methods of the History of Religions* (Leiden: E. J. Brill), pp. 35–54.

Bleeker, C. J. (1975), *The Rainbow. A Collection of Studies in the Science of Religion* (Leiden: E. J. Brill).

Bleeker, C. J. (1979), 'Evaluation of previous methods: Commentary', in L. Honko (ed.), *Science of Religion: Studies in Methodology. Proceedings of the Study Conference of the International Association for the History of Religions, held in Turku, Finland, August 27–31, 1973* (The Hague, Paris and New York: Mouton Publishers), pp. 173–7.

Bocking, B. (2004), 'The study of religions: The new queen of the sciences?', in S. Sutcliffe (ed.), *Religion: Empirical Studies* (Aldershot and Burlington: Ashgate), pp. 107–19.

Bosch, L. P. van den (1999), 'Friedrich Max Müller: His contribution to the science of religion', in E. R. Sand and J. P. Sorensen (eds), *Comparative Studies in History of Religions* (Copenhagen: Museum Tusculanum Press, University of Copenhagen), pp. 11–39.

Braun, W. and McCutcheon, R. T. (eds) (2000), *Guide to the Study of Religion* (London and New York: Cassell).

Bruce, S. (2002), *God is Dead. Secularization in the West* (Oxford: Blackwell).

Busse, R. P. (1995), 'Religious cognition in light of current questions', in D. Jodock (ed.), *Ritschl in Retrospect: History, Community, and Science* (Minneapolis: Fortress Press), pp. 166–85.

Busuttil, J. J. (2003), 'Policy responses to religious fundamentalism', in G. ter Haar and J. J. Busuttil (eds), *The Freedom to Do God's Will. Religious Fundamentalism and Social Change* (London and New York: Routledge), pp. 230–7.

Capps, W. H. (1995), *Religious Studies: The Making of a Discipline* (Minneapolis: Fortress Press).

Carman, J. B. (2000), 'Modern understanding of ancient insight: Distinctive contributions of W. B. Kristensen's phenomenology of religion', in S. Hjelde (ed.), *Man, Meaning, and Mystery: Hundred Years of History of Religions in Norway. The Heritage of W. Brede Kristensen* (Leiden: E. J. Brill), pp. 157–72.

Cave, D. (1993), *Mircea Eliade's Vision for a New Humanism* (New York and Oxford: Oxford University Press).

Chantepie de la Saussaye, P. D. (1891), *Manual of the Science of Religion* (London and New York: Longmans, Green, and Co., 1st edn).

Chidester, D. (1996), *Savage Systems. Colonialism and Comparative Religion in Southern Africa* (Charlottesville and London: University Press of Virginia).

Chryssides, G. (2001), 'Unrecognised charisma? A study of four charismatic leaders'. Paper presented at the International Conference, 'The Spiritual Supermarket: Religious Pluralism in the Twenty-first Century', organized by INFORM and CESNUR. www.cesnur.org/london2001/chryssides.htm.

Clayton, J. P. (ed.), *Ernst Troeltsch and the Future of Theology* (Cambridge: Cambridge University Press).

Colpe, C. (1979), 'Symbol theory and copy theory as basic epistemological and conceptual alternatives in Religious Studies', in L. Honko (ed.), *Science of Religion: Studies in Metholodology. Proceedings of the Study Conference of the International Association for the History of Religions, held in Turku, Finland, August 27-31, 1973* (The Hague, Paris and New York: Mouton Publishers), pp. 161–73.

Congress Secretariat for the 19th World Congress of the IAHR (2004), *Call for Papers* (Tokyo: Department of Religious Studies, University of Tokyo).

Cox, J. L. (1979), 'Faith and faiths: The significance of A. G. Hogg's missionary thought for a theology of dialogue', *Scottish Journal of Theology*, 32, 241–56.

Cox, J. L. (1991), *The Impact of Christian Missions on Indigenous Cultures: The 'Real People' and the Unreal Gospel* (Lewiston, New York: Edwin Mellen).

Cox, J. L. (1993), *Changing Beliefs and an Enduring Faith* (Gweru, Zimbabwe: Mambo Press).

Cox, J. L. (1995), 'Ancestors, the sacred and God: Reflections on the meaning of the sacred in Zimbabwean death rituals', *Religion*, 25, 339–55.

Cox, J. L. (1996), *Expressing the Sacred: An Introduction to the Phenomenology of Religion* (Harare: University of Zimbabwe Publications, 2nd edn).

Cox, J. L. (1998a), 'Religious typologies and the postmodern critique', *Method and Theory in the Study of Religion*, 10, 244–62.

Cox, J. L. (1998b), *Rational Ancestors. Scientific Rationality and African Indigenous Religions* (Cardiff: Cardiff Academic Press).

Cox, J. L. (1999), 'Intuiting religion: A case for preliminary definitions', in J. G. Platvoet and A. L. Molendijk (eds), *The Pragmatics of Defining Religion: Contexts, Concepts and Contests* (Leiden: Brill), pp. 267–84.

Cox, J. L. (2003a), 'African identities as the projection of Western alterity', in J. L. Cox and G. ter Haar (eds), *Uniquely African? African Christian Identity from Cultural and Historical Perspectives* (Trenton, New Jersey: Africa World Press), pp. 25–37.

Cox, J. L. (2003b), 'Contemporary Shamanism in global contexts: "Religious" appeals to an archaic tradition?', *Studies in World Christianity: The Edinburgh Review of Theology and Religion*, 9 (1), 69–87.

Cox, J. L. (2004a), 'From Africa to Africa: The significance of approaches to the study of African religions at Aberdeen and Edinburgh Universities from 1970 to 1998', in F. Ludwig and A. Adogame (eds), *European Traditions in the Study of Religion in Africa* (Wiesbaden: Harrassowitz Verlag), pp. 255–64.

Cox, J. L. (2004b), 'Afterword. Separating religion from the "sacred": Methodological agnosticism and the future of religious studies', in S. J. Sutcliffe (ed.), *Religion: Empirical Studies* (Aldershot: Ashgate), pp. 259–64.

Cunningham, A. (2001), 'Obituary: Ninian Smart', *Religion*, 31 (4), 325–6.

Danfulani, U. H. D. (2005), 'Abstract. The cobra is running wild: Narrating the events and evaluating causes of the Jos crisis since September 7th 2001', in Congress Secretariat of the 19th World Congress of IAHR, *The Book of Abstracts* (Tokyo: Department of Religious Studies, University of Tokyo), p. 37.

Daniels, J. (1995), 'How new is neo-phenomenology? A comparison of the methodologies of Gerardus van der Leeuw and Jacques Waardenburg', *Method and Theory in the Study of Religion*, 7, 43–55.

Darwin, C. (1964) [1859], *On the Origin of the Species* (with an introduction by Ernst Mayr) (Cambridge, Massachusetts: Harvard University Press).

Descartes, R. (1912) [1637], *A Discourse on Method* (translated by J. Veitch and with an introduction by A. D. Lindsay) (London and New York: Everyman's Library).

Descartes, R. (1931) [1911], *The Philosophical Works of Descartes, Vol. I and II* (rendered into English by E. S. Haldane and G. R. T. Ross) (Cambridge: Cambridge University Press).

Descartes, R. (1989), *Meditations I and II*, in J. E. White, *Introduction to Philosophy* (St Paul, Minnesota: West Publishing Company).

Drescher, H.-G. (1992), *Ernst Troeltsch: His Life and Work* (London: SCM Press).

Eliade, M. (1959), 'Methodological remarks on the study of religious symbolism', in M. Eliade and J. M. Kitagawa (eds), *The History of Religions: Essays in Methodology* (Chicago and London: University of Chicago Press), pp. 86–107.

Eliade, M. (1966), *Rites and Symbols of Initiation. The Mysteries of Birth and Rebirth* (New York: Harper and Row).

Eliade, M. (1969), *The Quest: History and Meaning of Religion* (Chicago: University of Chicago Press).

Eliade, M. (1975), *Myth and Reality* (New York: Harper Torchbooks).

Eliade, M. (1981), *Autobiography: Volume I, 1907–1937, Journey East, Journey West* (San Francisco: Harper and Row).

Eliade, M. (1987) [1959], *The Sacred and the Profane. The Nature of Religion* (San Diego, New York and London: Harcourt).

Eliade, M. (Editor in Chief) (1987), *Encyclopedia of Religion* (16 volumes) (New York: Macmillan).

Eliade, M. (1988), *Autobiography: Volume II, 1937–1960, Exile's Odyssey* (Chicago and London: University of Chicago Press).

Eliade, M. (1989) [1964], *Shamanism: Archaic Techniques of Ecstasy* (Harmondsworth: Arkana Penguin Books).

Eliade, M. (1996) [1958], *Patterns in Comparative Religion* (Lincoln and London: University of Nebraska Press).

Farr, R. M. (1996), *The Roots of Modern Social Psychology: 1872–1954* (Oxford: Blackwell).

Feuerbach, L. (1893), *The Essence of Christianity* (translated from the second German edition by Marian Evans) (London: Kegan Paul, Trench, Trübner, and Co.).

Fitzgerald, T. (2000), *The Ideology of Religious Studies* (New York and Oxford: Oxford University Press).

Flood, G. (1999), *Beyond Phenomenology: Rethinking the Study of Religion* (London and New York: Cassell).

Forward, M. (1998), *A Bag of Needments. Geoffrey Parrinder and the Study of Religion* (Bern: Peter Lang).

Freud, S. (1938) [1919], *Totem and Taboo* (Harmondsworth: Penguin Books).

Freud, S. (1949) [1940], *An Outline of Psycho-Analysis* (translated by J. Strachey) (London: Hogarth Press).

Geertz, A. W. (1999), 'Review of Russell T. McCutcheon *Manufacturing Religion*', *The Journal of Religion*, 79 (3), 508–9.

Geertz, A. W. and McCutcheon, R. T. (2000), 'The role of method and theory in the IAHR', in A. W. Geertz and R. T. McCutcheon (eds), *Perspectives on Method and Theory in the Study of Religion* (Leiden: Brill), pp. 3–37.

Haar, G. ter (2003), 'Religious fundamentalism and social change: A comparative inquiry', in G. ter Haar and J. J. Busuttil (eds), *The Freedom to Do God's Will. Religious Fundamentalism and Social Change* (London and New York: Routledge), pp. 1–24.

Hamilton, M. (2001), *The Sociology of Religion. Theoretical and Comparative Perspectives* (London and New York: Routledge, 2nd edn).

Hart, R. L. (1991), 'Religious and theological studies in American higher education: A pilot study', *Journal of the American Academy of Religion*, 59 (4), 715–827.

Hastings, A. (2001), 'Review Article: Geoffrey Parrinder', *Journal of Religion in Africa*, XXXI (3), 354–9.

Hefner, P. (1972), 'Albrecht Ritschl: An introduction', in A. Ritschl, *Three Essays* (translated by P. Hefner) (Philadelphia: Fortress Press), pp. 3–50.

Hegel, G. W. F. (1910) [1807], *The Phenomenology of Mind* (translated, with introduction and notes by J. B. Baillie) (London: S. Sonnenschein).

Hegel, G. W. F. (1969) [1812–16], *Science of Logic* (translated by A. V. Miller) (London: George Allen and Unwin Ltd).

Hegel, G. W. F. (1977) [1807], *Phenomenology of Spirit* (translated by A. V. Miller) (Oxford: Clarendon Press).

Heidegger, M. (1962), *Being and Time* (translated by J. Macquarrie and E. Robinson) (Oxford: Basil Blackwell).

Herrmann, W. (1895), *The Communion of the Christian with God: A Discussion in Agreement with the View of Luther* (translated by J. S. Stanyon) (London: Williams and Norgate, 3rd edn).

Herrmann, W. (1904), *Faith and Morals* (translated by D. Matheson and R. W. Stewart) (London: Williams and Norgate).

Hervieu-Léger, D. (2000), *Religion as a Chain of Memory* (translated by S. Lee) (Oxford: Polity Press).

Hjelde, S. (2000), 'Introduction', in S. Hjelde (ed.), *Man, Meaning, and Mystery: Hundred Years of History of Religions in Norway. The Heritage of W. Brede Kristensen* (Leiden: E. J. Brill), pp. xii–xxii.

Hofstee, W. (2000), 'Phenomenology of religion versus anthropology of religion? The "Groningen School" 1920–1990', in S. Hjelde (ed.), *Man, Meaning, and Mystery: Hundred Years of History of Religions in Norway. The Heritage of W. Brede Kristensen* (Leiden: E. J. Brill), pp. 173–90.

Hogg, A. G. (1903), 'Agnosticism and faith', *Madras Christian College Magazine*, New Series 3 (2), 75–84.

Hogg, A. G. (1904), 'The Christian interpretation of mediation', *Madras Christian College Magazine*, New Series 3 (7), 357–69.

Hogg, A. G. (1909), *Karma and Redemption. An Essay Toward the Interpretation of Hinduism and the Re-Statement of Christianity* (London, Madras and Colombo: Christian Literature Society).

Hogg, A. G. (1911), *Christ's Message of the Kingdom. A Course of Daily Study for Private Students and for Bible Classes* (Edinburgh: T. and T. Clark).

Hogg, A. G. (1917), 'The God that must needs be Christ Jesus', *International Review of Missions*, vols 21–4, 62–73, 221–32, 383–94, 521–33.

Hogg. A. G. (1922), *Redemption from this World or the Supernatural in Christianity* (Edinburgh: T. and T. Clark).

Hogg, A. G. (1947), *The Christian Message to the Hindu* (London: SCM Press).

Honko, L. (1979), 'Introduction', in L. Honko (ed.), *Science of Religion: Studies in Methodology. Proceedings of the Study Conference of the International Association for the History of Religions, held in Turku, Finland, August 27–31, 1973* (The Hague, Paris and New York: Mouton Publishers), pp. xv–xxix.

Husserl, E. (1931), *Ideas. General Introduction to Pure Phenomenology* (translated by W. R. B. Gibson) (London: George Allen and Unwin Ltd).

Husserl, E. (1965), *Phenomenology and the Crisis of Philosophy* (translated with an introduction by Q. Lauer) (New York: Harper and Row).

Husserl, E. (1970) [1913–1922], *Logical Investigations. Volumes I and II* (translated by J. N. Findlay) (London: Routledge and Kegan Paul).

Husserl, E. (1977), *Cartesian Meditations. An Introduction to Phenomenology* (translated by D. Cairns) (The Hague: Martinus Nijhoff).

Idowu, E. B. (1962), *Olódùmarè. God in Yoruba Belief* (London: Longmans).

Idowu, E. B. (1995), *Olódùmarè. God in Yoruba Belief* (revised and expanded edition) (Old Bethpage, New York: Original Publications).

Ingarden, R. (1975), *On the Motives which Led Husserl to Transcendental Idealism* (translated by A. Hannibalsson) (The Hague: Martinus Nijhoff).

James, G. A. (1995), *Interpreting Religion. The Phenomenological Approaches of Pierre Daniel Chantepie de la Saussaye, W. Brede Kristensen, and Gerardus van der Leeuw* (Washington, DC: The Catholic University of America Press).

Jaspers, K. (1953), *The Origin and Goal of History* (translated by Michael Bullock) (London: Routledge and Kegan Paul).

Jodock, D. (ed.) (1995), *Ritschl in Retrospect: History, Community, and Science* (Minneapolis: Fortress Press).

Jung, C. G. (1915) [1911], *The Psychology of the Unconscious* (translation with introduction by B. M. Heindel) (London: Kegan Paul).

Jung, C. G. (1923) [1921], *Psychological Types or the Psychology of Individuation* (translated by H. G. Baynes) (London: Routledge and Kegan Paul).

Jung, C. G. (1936) [1907], *The Psychology of Dementia Praecox* (translated by A. A. Brill) (New York: Nervous and Mental Disease Publishing Company).

Jung, C. G. (1960), *The Structure and Dynamics of the Psyche* (*Collected Works*, Volume 8, translated by R. F. C. Hull) (London: Routledge and Kegan Paul).

Jung, C. G. (1963), *Mysterium Coniunctionis. An Inquiry into the Separation and*

Synthesis of Psychic Opposites in Alchemy (*Collected Works*, Volume 14, translated by R. F. C. Hull) (London: Routledge and Kegan Paul).

Jung, C. G. (1968), *Analytical Psychology: Its Theory and Practice. The Tavistock Lectures* (London: Routledge and Kegan Paul).

Jung, C. G. (1972), *Four Archetypes: Mother, Rebirth, Spirit, Trickster* (London: Routledge and Kegan Paul).

Jung, C. G. (2002) [1958], *The Undiscovered Self* (London: Routledge).

Kant, I. (1909) [1785], *Fundamental Principles of the Metaphysic of Morals*, in T. K. Abbott (translator), *Kant's Critique of Practical Reason and other Works on the Theory of Ethics* (London, New York and Bombay: Longmans, Green, and Co., 6th edn), pp. 1–84.

Kant, I. (1909) [1788], *Critical Examination of Practical Reason*, in T. K. Abbott (translator), *Kant's Critique of Practical Reason and other Works on the Theory of Ethics* (London, New York and Bombay: Longmans, Green, and Co., 6th edn), pp. 87–262.

Kant, I. (1929) [1787], *Critique of Pure Reason* (translated by N. K. Smith) (London: Macmillan and Company, 2nd edn).

Kant, I. (1977) [1783], *Prolegomena to Any Future Metaphysics That Will be Able to Come Forward as Science* (the Paul Carus translation extensively revised by J. W. Ellington) (Indianapolis: Hackett Publishing Company).

Kenny, A. (ed.) (1994), *The Oxford Illustrated History of Western Philosophy* (Oxford: Oxford University Press).

King, R. (1999), *Orientalism and Religion. Postcolonial Theory, India and 'The Mystic East'* (London: Routledge).

King, U. (1982), 'Current state of the study of religions in British universities. Report on the response of the July questionnaire', *British Association for the History of Religions Bulletin*, 38 (December), 1–5.

King, U. (1984), 'Historical and phenomenological approaches to the study of religion. Some major developments and issues under debate since 1950', in F. Whaling (ed.), *Contemporary Approaches to the Study of Religion. Volume I: The Humanities* (Berlin, New York and Amsterdam: Mouton Publishers), pp. 29–164.

King, U. (1995), 'A question of identity: Women scholars and the study of religion', in U. King (ed.), *Religion and Gender* (Oxford: Blackwell), pp. 219–43.

Kitagawa, J. M. (1959), 'The history of religions in America', in M. Eliade and J. M. Kitagawa (eds), *The History of Religions: Essays in Methodology* (Chicago and London: University of Chicago Press), pp. 1–30.

Kolakowski, L. (1975), *Husserl and the Search for Certitude* (New Haven and London: Yale University Press).

Kraemer, H. (1938), *The Christian Message in a Non-Christian World* (London: Edinburgh House Press).

Kristensen, W. B. (1954), *Religionshistorisk Studium* (Oslo: Olaf Norlis Forlag).

Kristensen, W. B. (1960), *The Meaning of Religion* (translated by J. Carman) (The Hague: Martinus Nijhoff).

Lauer, Q. (1965), 'Introduction', in E. Husserl, *Phenomenology and the Crisis of Science* (translated with an introduction by Q. Lauer) (New York: Harper and Row), pp. 1–68.

Leeuw, G. van der (1963), *Religion in Essence and Manifestation. Volumes I and II* (translated by J. E. Turner with Appendices incorporating the additions of the second German edition by H. H. Penner) (New York and Evanston: Harper and Row Publishers). (Original publication in one volume, 1938, London: George Allen and Unwin Ltd).

Locke, J. (1924) [1690], *An Essay Concerning Human Understanding* (abridged and edited by A. S. Pringle-Pattison) (Oxford: Clarendon Press).

McCutcheon, R. T. (1997), *Manufacturing Religion. The Discourse on Sui Generis Religion and the Politics of Nostalgia* (New York and Oxford: Oxford University Press).

McCutcheon, R. T. (ed.) (1999), *The Insider/Outsider Problem in the Study of Religion: A Reader* (London and New York: Cassell).

McCutcheon, R. T. (2001), *Critics Not Caretakers. Redescribing the Public Study of Religion* (Albany: State University of New York Press).

McCutcheon, R. T. (2003), *The Discipline of Religion. Structure, Meaning, Rhetoric* (London and New York: Routledge).

Mackintosh, H. R. (1937), *Types of Modern Theology: Schleiermacher to Barth* (London: Nisbet and Co. Ltd).

Mackintosh, H. R. and Macaulay, A. B. (eds) (1900), 'Editors' Preface', in A. Ritschl, *The Christian Doctrine of Justification and Reconciliation* (Edinburgh: T. and T. Clark), pp. v–vi.

Madras Christian College (1924), 'College Notes', *Madras Christian College Magazine*, Quarterly Series 4 (1), 70.

Mbiti, J. S. (1969), *African Religions and Philosophy* (London: Heinemann Educational Books Ltd).

Mbiti, J. S. (1970), *Concepts of God in Africa* (London: SPCK).

Meyer, E. (1894), *Geschichte des Alterthums* (Stuttgard: J. G. Cotta).

Meyer, E. (1902), *Zur Theorie und Methodik der Geschitchte* (Halle a.S.: M. Niemeyer).

Molendijk, A. L. (2000), 'At the cross-roads: Early Dutch science of religion in international perspective', in S. Hjelde (ed.), *Man, Meaning, and Mystery: Hundred Years of History of Religion in Norway. The Heritage of W. Brede Kristensen* (Leiden: E. J. Brill), pp. 19–56.

Moran, D. (2000), *Introduction to Phenomenology* (London and New York: Routledge).

Morgan, R. (2001), 'Religious studies in Britain: Lancaster in the sixties', *Religion*, 31 (4), 349–52.

Morris, B. (1987), *Anthropological Studies of Religion: An Introductory Text* (Cambridge: Cambridge University Press).

Müller, F. M. (1893) [1873], *Introduction to the Science of Religion* (London: Longmans, new edn).

Müller, F. M. (1891), *Physical Religion* (London: Longmans, Green, and Co.).

Müller, F. M. (1897), *Contributions to the Science of Mythology. Volumes I and II* (London: Longmans, Green, and Co).

Müller, F. M. (1898), *Theosophy or Psychological Religion* (London: Longmans, Green, and Co).

Naguib, S.-A. (2000), 'Lieblein, Kristensen and Schencke and the quest for Egyptian monotheism', in S. Hjelde (ed.), *Man, Meaning, and Mystery: Hundred Years of History of Religion in Norway. The Heritage of W. Brede Kristensen* (Leiden: E. J. Brill), pp. 101–13.

Nye, M. (2000), 'Editorial: Culture and religion', *Culture and Religion*, 1 (1), 5–12.

Otto, R. (1926), *The Idea of the Holy: An Inquiry into the Non-Rational Factor in the Idea of the Divine and Its Relation to the Rational* (translated by J. W. Harvey, 4th impression, revised with additions) (London: Humphrey Milford and Oxford University Press).

Palmer, M. (1997), *Freud and Jung on Religion* (London and New York: Routledge).

Pals, D. L. (1996), *Seven Theories of Religion* (New York and Oxford: Oxford University Press).

Parrinder, G. (1949), *West African Religion: Illustrated from the Beliefs and Practices of the Yoruba, Ewe, Akan and Kindred Peoples* (London: Epworth Press).

Parrinder, G. (1962), *Comparative Religion* (London: George Allen and Unwin Ltd).

Parrinder, G. (1964), *The World's Living Religions* (London: Pan Books Ltd).

Parrinder, G. (1969), *Religion in Africa* (London: Pall Mall Press and Harmondsworth: Penguin) (Reprinted, 1976, *Africa's Three Religions*, London: Sheldon Press).

Parrinder, G. (1974), *African Traditional Religion* (London: Sheldon Press, 3rd edn).

Parrinder, G. (1980), *Sexual Morality in the World's Religions* (Oxford: Oneworld).

Pettersson, O. and Akerberg, H. (1981), *Interpreting Religious Phenomena. Studies with Reference to the Phenomenology of Religion* (Stockholm: Almqvist and Wiksell Internatonal).

Platvoet, J. G. (1988), *A Concise History of the Study of Religions* (Harare: University of Zimbabwe, Department of Religious Studies, Classics and Philosophy, Internal Publication).

Platvoet, J. G. (1996), 'From object to subject: A history of the study of religions of Africa', in J. Platvoet, J. Cox and J. Olupona (eds), *The Study of Religions in Africa: Past, Present and Prospects* (Cambridge: Roots and Branches), pp. 105–38.

Platvoet, J. G. (1998), 'Close harmonies: The science of religion in Dutch *duplex ordo* theology, 1860–1960', *Numen*, 45, 115–61.

Platvoet, J. G. (2002), 'Pillars, pluralism and secularisation: A social history of Dutch sciences of religions', in G. Wiegers (ed.), *Modern Societies and the*

Science of Religions. Studies in Honour of Lammert Leertouwer (Leiden: E. J. Brill), pp. 82–148.

Popkin, R. H. and Stroll, A. (1986), *Philosophy. Made Simple* (London: Heinemann, 2nd edn).

Preus, J. S. (1987), *Explaining Religion. Criticism and Theory from Bodin to Freud* (New Haven and London: Yale University Press).

Rennie, B. S. (1996), *Reconstructing Eliade. Making Sense of Religion* (Albany: State University of New York Press).

Richmond, J. (1978), *Ritschl: A Reappraisal* (London and Glasgow: Collins).

Rinsum, H. J. van (2003), '"Knowing the African": Edwin W. Smith and the invention of African Traditional Religion', in J. L. Cox and G. ter Haar (eds), *Uniquely African? African Christian Identity from Cultural and Historical Perspectives* (Trenton, New Jersey: Africa World Press), pp. 39–66.

Ritschl, A. (1900), *The Christian Doctrine of Justification and Reconciliation* (translated and edited by H. R. Mackintosh and A. B. Macaulay) (Edinburgh: T. and T. Clark).

Ritschl, A. (1972), *Albrecht Ritschl. Three Essays* (translated with an introduction by P. Hefner) (Philadelphia: Fortress Press).

Schleiermacher, F. (1922) [1821], *The Christian Faith in Outline* (translated from the German by D. M. Baillie) (Edinburgh: W. F. Henderson).

Schleiermacher, F. (1988) [1799], *On Religion. Speeches to Its Cultured Despisers* (Introduction, translation and notes by R. Crouter) (Cambridge: Cambridge University Press).

Schutz, A. (1972) [1932], *The Phenomenology of the Social World* (translated by G. Walsh and F. Lehnert) (London: Heinemann Educational Books).

Scruton, R. (1982), *Kant* (Oxford and New York: Oxford University Press).

Scruton, R. (1984), *A Short History of Modern Philosophy: From Descartes to Wittgenstein* (London and New York: Ark Paperbacks).

Scruton, R. (2000), *On Husserl* (Belmont, California: Wadsworth).

Segal, R. (1983), 'In defense of reductionism', *Journal of the American Academy of Religion*, 51 (1), 97–124.

Sharot, S. (2001), *A Comparative Sociology of World Religions. Virtuosos, Priests, and Popular Religion* (New York and London: New York University Press).

Sharpe, E. J. (1965), *Not to Destroy but to Fulfil. The Contribution of J. N. Farquhar to Protestant Missionary Thought in India before 1914* (Uppsala: Gleerup).

Sharpe, E. J. (1971), *The Theology of A. G. Hogg* (Madras: Christian Literature Society).

Sharpe, E. J. (1986), *Comparative Religion: A History* (London: Duckworth, 2nd edn).

Sharpe, E. J. (1994), 'A. G. Hogg', in G. H. Anderson (ed.), *Mission Legacies: Biographical Studies of Leaders of the Modern Missionary Movement* (Maryknoll, New York: Orbis), pp. 330–8.

Sharpe, E. J. (1999), *Alfred George Hogg, 1875–1954: An Intellectual Biography* (Chennai, India: Christian Literature Society).

Sheldrake, P. (1995), *Spirituality and History: Questions of Interpretation and Method* (London: SPCK, 2nd edn).

Shils, E. A. (1949), 'Foreword', in M. Weber, *The Methodology of the Social Sciences* (translated and edited by E. A. Shils and H. A. Finch) (New York: The Free Press), pp. iii.

Smart, N. (1973a), *The Science of Religion and the Sociology of Knowledge* (Princeton: Princeton University Press).

Smart, N. (1973b), *The Phenomenon of Religion* (New York: The Seabury Press).

Smart, N. (1977), *The Religious Experience of Mankind* (Glasgow: Collins Fount Paperbacks). (First published, 1969, New York: Charles Scribner's Sons).

Smart, N. (1986), *Concept and Empathy. Essays in the Study of Religion* (edited by D. Wiebe) (London: Macmillan).

Smart, N. (1992), *The World's Religions. Old Traditions and Modern Transformations* (Cambridge: Cambridge University Press).

Smart, N. (1997), *Dimensions of the Sacred. An Anatomy of the World's Beliefs* (London: Fontana Press).

Smart, N. and Pye, M. (1981), 'Press Release: Religion under attack on campuses', *Bulletin of the British Association for the History of Religions*, 35 (October), 2.

Spencer, H. (1880), *First Principles* (London and Edinburgh: Williams and Norgate, 4th edn).

Smith, E. W. (1936), *African Beliefs and Christian Faith: An Introduction to Theology for African Students, Evangelists and Pastors* (London: The United Society for Christian Literature).

Smith, E. W. (ed.) (1950), *African Ideas of God: A Symposium* (London: Edinburgh House Press).

Smith, J. Z. (1978), *Map is Not Territory. Studies in the History of Religions* (Leiden: E. J. Brill).

Smith, J. Z. (1980), 'The bare facts of ritual', *History of Religions*, 20, 112–27.

Smith, J. Z. (1982), *Imagining Religion: From Babylon to Jonestown* (Chicago and London: University of Chicago Press).

Smith, J. Z. (1987), *To Take Place. Toward Theory in Ritual* (Chicago and London: The University of Chicago Press).

Smith, J. Z. (1990), *Drudgery Divine. On the Comparison of Early Christianities and the Religions of Late Antiquity* (Chicago and London: University of Chicago Press).

Smith, J. Z. (2000), 'Classification', in W. Braun and R. T. McCutcheon (eds), *Guide to the Study of Religion* (London and New York: Cassell), pp. 35–44.

Smith, W. C. (1943), *Modern Islam in India: A Social Analysis* (Lahore: Minerva Book Shop).

Smith, W. C. (1957), *Islam in Modern History* (Princeton: Princeton University Press).

Smith, W. C. (1959), 'Comparative religion: Whither – and why?', in M. Eliade and J. Kitagawa (eds), *The History of Religions: Essays in Methodology* (Chicago and London: University of Chicago Press), pp. 31–58.

Smith, W. C. (1963), *The Faith of Other Men* (New York: Harper and Row). (Reprinted under the title, *Patterns of Faith around the World*, 1998, Oxford and Boston: Oneworld Publications).

Smith, W. C. (1964), *The Meaning and End of Religion. A New Approach to the Religious Traditions of Mankind* (New York: Mentor Books). (First published 1962, New York: Macmillan.)

Smith, W. C. (1981), *Towards a World Theology. Faith and the Comparative History of Religion* (Philadelphia: Westminster Press).

Smith, W. C. (1983), 'The modern West in the history of religion', *Journal of the American Academy of Religion*, 52 (1), 3–18.

Smith, W. C. (1998), *Faith and Belief: The Difference between Them* (Oxford and Boston: Oneworld Publications). (First published as *Faith and Belief*, 1979, Princeton: Princeton University Press).

Stewart, R. W. (1904a), 'Biographical note', in W. Herrmann, *Faith and Morals* (translated by D. Matheson and R. W. Stewart) (London: Williams and Norgate), pp. v–xii.

Stewart, R. W. (1904b), 'Introduction by the Translator', in W. Herrmann, *Faith and Morals* (translated by D. Matheson and R. W. Stewart) (London: Williams and Norgate), pp. 3–5.

Stoecker, H. (2004), ' "The gods are dying": Diedrich Westermann (1875–1956) and some aspects of his studies of African religions', in F. Ludwig and A. Adogame (eds), *European Traditions in the Study of Religion in Africa* (Wiesbaden: Harrassowitz Verlag), pp. 169–74.

Storr, A. (1997), *Feet of Clay: A Study of Gurus* (London: HarperCollins).

Strenski, I. (1993), *Religion in Relation: Method, Application and Moral Location* (London: Macmillan).

Taylor, J. B. (ed.) (1976), *Primal World Views: Christian Dialogue with Traditional Thought Forms* (Ibadan, Nigeria: Daystar Press).

Tempels, P. (1959), *Bantu Philosophy* (translated by C. King) (Paris: Presence Africaine).

Thrower, J. (1999), *Religion: The Classical Theories* (Edinburgh: Edinburgh University Press).

Tiele, C. P. (1973) [1896 and 1898], 'Extracts from *Elements of the Science of Religion*', in J. Waardenburg, *Classical Approaches to the Study of Religion: Aims, Methods and Theories of Research. I. Introduction and Anthology* (The Hague and Paris: Mouton), pp. 96–104.

Troeltsch, E. (1972) [1902], *The Absoluteness of Christianity and the History of Religions* (London: SCM Press).

Troeltsch, E. (1991), *Religion in History: Essays* (translated by J. L. Adams and W. F. Bense, with an introduction by James Luther Adams) (Edinburgh: T. and T. Clark).

Voelkel, R. T. (1972), 'Introduction', in W. Herrmann, *The Communion of the Christian with God: Described on the Basis of Luther's Statements* (London: SCM Press, 2nd edn), pp. iv–lxviii.

Waardenburg, J. (1973), *Classical Approaches to the Study of Religion: Aims,*

Methods and Theories of Research. I. Introduction and Anthology (The Hague and Paris: Mouton Publishers).

Waardenburg, J. (1978), *Reflections on the Study of Religion* (The Hague, Paris and New York: Mouton Publishers).

Waardenburg, J. (2000), 'Progress in research on meanings in religion (1898–1998)', in S. Hjelde (ed.), *Man, Meaning, and Mystery: Hundred Years of History of Religion in Norway. The Heritage of W. Brede Kristensen* (Leiden: E. J. Brill), pp. 255–85.

Wach, J. (1951), *Types of Religious Experience: Christian and Non-Christian* (Chicago: University of Chicago Press).

Wach, J. (1958), *The Comparative Study of Religions* (edited with an introduction by J. M. Kitagawa) (New York: Columbia University Press).

Wach, J. (1962) [1944], *Sociology of Religion* (Chicago and London: University of Chicago Press).

Wach, J. (1967), 'Introduction: The meaning and task of the history of religions (Religionswissenschaft)', in J. M. Kitagawa (ed.), *The History of Religions: Essays on the Problem of Understanding* (Chicago and London: The University of Chicago Press), pp. 1–19.

Walls, A. F. (1980), 'A bag of needments for the road: Geoffrey Parrinder and the study of religion in Britain', *Religion*, 10 (Autumn), 141–50.

Walls, A. F. (1982), 'The Gospel as the prisoner and liberator of culture', *Missionalia*, 10 (3), 93–105.

Walls, A. F. (1983), 'Centre for the Study of Christianity in the Non-Western World, University of Aberdeen, Scotland', *British Association for the History of Religions Bulletin*, 39 (April), 10–11.

Walls, A. F. (1987), 'Primal religious traditions in today's world', in F. Whaling (ed.), *Religion in Today's World* (Edinburgh: T. and T. Clark), pp. 250–78.

Walls, A. F. (1990), 'The translation principle in Christian history', in P. C. Stine (ed.), *Bible Translation and the Spread of the Church: The Last 200 Years* (Leiden: E. J. Brill), pp. 24–39.

Walls, A. F. (2000), 'Eusebius tries again: Reconceiving the study of Christian history', *International Bulletin of Missionary Research*, 24 (July), 105–11.

Walls, A. F. (2004), 'Geoffrey Parrinder (1910) and the study of religion in West Africa', in F. Ludwig and A. Adogame (eds), *European Traditions in the Study of Religion in Africa* (Wiesbaden: Harrassowitz Verlag), pp. 207–15.

Ward, K. (2002), 'The study of religions', in E. Nicholson (ed.), *A Century of Theological and Religious Studies in Britain* (Oxford: Oxford University Press), pp. 271–94.

Weber, M. (1949), *The Methodology of the Social Sciences* (translated and edited by E. A. Shils and H. A. Finch) (New York: Free Press).

Weber, M. (1952), *Ancient Judaism* (translated and edited by H. H. Gerth and D. Martindale) (Glencoe, Illinois: Free Press).

Weber, M. (1958), *The Religion of India: The Sociology of Hinduism and Buddhism* (translated and edited by H. H. Gerth) (Glencoe, Illinois: Free Press).

Weber, M. (1968), *The Religion of China: Confucianism and Taoism* (translated and edited by H. H. Gerth) (New York: Free Press).

Weber, M. (1992) [1930], *The Protestant Ethic and the Spirit of Capitalism* (translated by T. Parsons, with an introduction by A. Giddens) (London and New York: Routledge).

Westermann, D. (1937), *Africa and Christianity* (London: Oxford University Press).

Whaling, F. (1999), 'Theological approaches', in P. Connolly (ed.), *Approaches to the Study of Religion* (London and New York: Continuum), pp. 226–74.

Wiebe, D. (1999), *The Politics of Religious Studies: The Continuing Conflict with Theology in the Academy* (London: Macmillan).

Wiebe, D. (2001), 'Ninian Smart: A Tribute', *Religion*, 31, 379–83.

Wiebe, D. (2005), 'Abstract: Disentangling the role of the scholar-scientist from that of the public intellectual in the modern academic study of religion', in Congress Secretariat of the 19th World Congress of IAHR, *The Book of Abstracts* (Tokyo: Department of Religious Studies, University of Tokyo), pp. 308–9.

Young, W. J. (2002), *The Quiet Wise Spirit. Edwin W. Smith 1876–1957 and Africa* (Peterborough: Epworth Press).

Index

Printed in the United Kingdom by
Lightning Source UK Ltd., Milton Keynes
141187UK00001B/54/P

9 780826 452894